SYSTEMS ANALYSIS FOR MANAGERIAL DECISIONS

A COMPUTER APPROACH

SYSTEMS ANALYSIS FOR MANAGERIAL DECISIONS

A COMPUTER APPROACH

P. Ramalingam

ASSOCIATE PROFESSOR OF INDUSTRIAL ENGINEERING
California State Polytechnic University, Pomona.

JOHN WILEY & SONS
New York / Santa Barbara / London / Sydney / Toronto

Copyright © 1976 by John Wiley & Sons, Inc.

All rights reserved. Published simultaneously in Canada.

No part of this book may be reproduced by any means, nor transmitted, nor translated into a machine language without the written permission of the publisher.

Library of Congress Cataloging in Publication Data:

Ramalingam, P   1941–
    Systems analysis for managerial decisions.

    Includes bibliographies and index.
    1. System analysis.   2. Industrial management—
Mathematical models.   I. Title.
HD20.5.R35   658.4′032   76–10534
ISBN 0-471-70710-4

Printed in the United States of America

10 9 8 7 6 5 4 3 2 1

To my parents

The complexities of modern organizations and their dynamic environment have created new challenges to decision making by management. Systems analysis—a valuable aid to making decisions in complex situations—has made a significant impact in arriving at better solutions for a great many vital problems during the last three decades. The development of automated digital computers with tremendous capabilities in computational speed and information storage has made it practical to solve numerous managerial problems with several variables. This book is an introduction to computer-aided systems analysis. Systems analysis deals with the study of a system and is often intended to determine the effective ways of planning and allocating resources to achieve desired goals. Systems analysis has been utilized by industry, government, and nonprofit organizations in making better decisions for numerous complex situations.

There are several text books available on systems analysis that present highly generalized viewpoints of systems or develop mathematical models in a rigorous and abstract manner. There are also many texts providing an introduction to automated computer systems and computer languages. However, none have lucidly combined mathematical concepts and models and the development of computer programs. This book is intended to satisfy that need. It has three purposes. First, it is written to illustrate many widely used mathematical models that utilize the computer and that are employed in a variety of functional areas. Second, it develops step by step transformation procedure of structuring computer programs for these mathematical models. Third, it demonstrates the use of computers in systems analysis studies. The decision-making situations can be classified into certainty, risk, or uncertainty, based on the kind of information possessed by a decision maker. The concepts and methods of systems analysis are presented in this book by organizing the material around the decision-making situations encountered by management. Several mathematical methods that are widely used in analyzing decision-making situations under certainty, risk, and uncertainty are presented. The concepts and methods of analysis are discussed using concrete examples. The analytical techniques are clearly illustrated by step-by-step instructions and numerical examples. Computer models are developed for the analytical techniques, and their applications are demonstrated by real-life problems. The problems at the end of each chapter provide a means for reviewing the subject matter and indicate additional applicable areas of the techniques. The problems are structured to challenge the reader's comprehension of the mathematical models and computer logic.

This book is intended especially as a textbook for an introductory course or a sequence of courses on systems analysis for juniors, seniors, or graduate students in business, economics, and engineering curricula. It is also suitable as a supplementary text for courses in operations research, management science, and quantitative methods for business decisions. Situational approach is utilized in presenting management problems and solutions procedures. Emphasis is on analytical methods for making managerial decisions under certainty, risk, and uncertainty situations. Although the focus is on mathematical models, mathematical requirements are at an elementary level. A knowledge of high school algebra is sufficient mathematical preparation. The reader need not have previously studied matrix algebra and probability theory because these subject areas are introduced and developed to the extent that are needed. The computer models are formulated in FORTRAN IV language. Although some exposure to FORTRAN IV would be helpful, Chapter 2 presents the features of FORTRAN IV using simple illustrations for readers lacking formal training in the language. For those readers experienced in using FORTRAN IV, this chapter is useful as a reference while writing programs.

I thank my good friend and mentor Professor Michael S. Inoue at Oregon State University, who considerably influenced my own thinking about systems analysis and allied subject areas. I am grateful to several of my students who have struggled through earlier drafts of this book and made many valuable suggestions. In particular, I thank my students Judd Tattershall, Sui-ming Chow, and David Thompson who have eagerly assisted in the refinement of this material. I appreciate also the encouragement of Dean Beaumont Davison during the difficult period of writing. I am deeply grateful to Professors Don Phillips and David Levy, both of Purdue University, for their helpful suggestions and corrections. I thank Eugene Patti for his valuable assistance in editing the manuscript. I also appreciate the help of Karen Winston, Colleen Johnson, Debbie Coleman, and Debbie Sanchez for preparing portions of the preliminary edition. Last but not least, I am indebted to my wife, Leah, for her patience and encouragement.

Pomona, California   Panchatcharam Ramalingam
August 1976

CONTENTS

4 Matrix Models   122

5 Allocation of Resources — LP   168

6 Allocation of Resources — Assignment Method   218

15 Simulation 503

16 Decision Making Under Uncertainty 533

17 Systems Analysis — An Overview 571

APPENDIXES

systems analysis - an introduction

THE MODERN ORGANIZATION is operating in an ever-changing environment. The changes result from many factors that include technological improvements, new materials, new products, growing markets, and government regulations. The managerial functions of planning, organizing, coordinating, and controlling the operations in the modern organization for achieving stability and growth are difficult and challenging to the managers. The organization can be a business, industry, government (federal, state, or local), or nonprofit organization (such as school, hospital, and library). Systems analysis can be a valuable aid to the managers in making better decisions. The development of computers has enabled the systems analysts to study large-scale systems and establish better solutions for them. We shall discuss in this chapter the nature of systems and models.

WHAT IS A SYSTEM?

"System" is currently a fashionable word. It means different things to different people. We are familiar with the terms solar system, irrigation system, transportation system, and electric power system. In a business situation, terms such as marketing system, management information system, manufacturing system, and inventory system are frequently used. For our purpose the general definition suggested by Hall (1962) is adequate.

A system is a set of objects with relationships between the objects and between their attributes.

The objects may be physical or abstract elements of the system. Some objects in a factory are the raw materials, labor force, buildings, equipments, expenditures, and company goals. Attributes are the characteristics of the objects. Attributes for a milling machine in the factory include its manufacturer's name, size, color, cutting time, speed, and purchase price. Relationships are the bonds that tie objects and attributes together.

The boundary of a system denotes a physical or conceptual limitation for the system. The objects, attributes, and relationships of the system are contained within the closed boundary. Everything outside the boundary is called the environment of the system. In analyzing a system, it is usually assumed that the objects and attributes of the environment are "fixed" or "given." The system cannot operate or control the environment. However, a change in the attributes of the environment can affect the system. If the performance of the system is changed, it might alter the attributes of the objects in the environment.

Figure 1.1 illustrates a system. Energy, materials, and information that flow across the boundary comprise inputs and outputs of the system. Inputs are the flows that pass from the environment into the system. Outputs are the flows that pass from the system to the environment. For a production system, the inputs include raw materials, electric power, capital, work force, and customer orders. The outputs comprise finished products, obsolete equipment, scrap, and industrial waste.

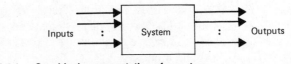

FIGURE 1.1   Graphical representation of a system.

SYSTEMS CLASSIFICATIONS

Systems can be classified into different categories for the convenience of description and analysis. We shall discuss the classifications that are useful for analyzing systems in business and industry.

Physical and abstract systems
Open and closed systems
Natural and manmade systems

Permanent and temporary systems
Adaptive and nonadaptive systems
Subsystems and supersystems.

Physical and Abstract Systems   A physical system consists of elements that are real objects such as people, machines, materials, and energy. In contrast, an abstract system deals with concepts such as organizational theory, philosophic systems, and the ideas under investigation. The elements, attributes, and relationships in an abstract system are developed by utilizing symbols, definitions, explanations, and assumptions.

Open and Closed Systems   In an open system energy, materials, or information are exchanged with the environment. Some examples of open systems are the school system, steel mill, and national economy. In other words, systems with inputs and outputs are called open systems. A closed system does not interact with the environment. It is completely self-contained. There is no flow of energy, materials, or information across the boundaries of the closed system.

Natural and Manmade Systems   A natural system, such as the solar system, climate, and biological structure, occurs in nature. A manmade system is developed by man presumably for some beneficial use. Computers, marketing systems, and economic theory are examples of such systems.

Permanent and Temporary Systems   A permanent system for example, the solar system, operates for a long period of time. However, systems that function for a relatively longer duration of time than that of human operations can also be considered permanent systems. For practical purposes the federal government can be assumed to be a permanent system. Opposed to the permanent system is the temporary system. A temporary system is planned to function for a limited period of time. A committee formed to elect a person in the national election operates until election day. The committee is in fact a temporary system.

Adaptive and Nonadaptive Systems   An adaptive system reacts to environmental changes. The nature of reactions in the adaptive system is conducive to the continued operation of the system. The behavior of adaptation for manmade systems can be completely explained, unlike that of natural systems. The environmental changes affect a nonadaptive system passively, and the system does not react to adapt itself to the modified environment.

Subsystems and Supersystems   Each system is always embedded in a larger system. For example, the manufacturing system of a factory is contained in the company. In turn, the company is nested in the corporation. Assume that the corporation owns and operates several businesses. Each business of the corporation is operated in the United States of America. Therefore, the corporation is embedded in the national economy. The national economy is a part of the international economy. If the corporation is referred to as a system, the components of the corporation that lie at a lower hierarchical level are called subsystems. Hence, the company is a subsystem of the corporation. Murdick and Ross (1971) suggest

> The system in the hierarchy that we are most interested in studying or controlling is usually called "**the system**." The business firm is viewed as "the system" or the "total system" when focus is on production, distribution of goods, and sources of profit and income.

The terms system and subsystem are relative expressions. When our attention is focused on the corporation, the corporation and the company are called the system and subsystem, respectively. However, when our aim is to study the company operations, the company and the manufacturing system are referred to as the system and subsystem, respectively. In a large system, a great many objects, attributes, and complex relationships are involved and the analysis becomes very difficult. Extremely large systems are called **supersystems**.

MORE ABOUT SYSTEMS

A system is usually represented by a block as in Figure 1.1. The outline of the block denotes the boundary of the system. The arrows indicate the inputs and outputs to the system. The system or processor operates on the inputs to yield the outputs. The transformation process that occurs within the system could be performed by various methods and by different means. Numerous activities that are interrelated in a complex manner might take place inside the system. These activities are grouped together in a single processor. The processor block is sometimes called a **black box**. The black box concept is useful when we do not wish to study the system in further detail or when we are unable to see what is going on within the system.

A **feedback system** has the ability to return a portion of its outputs into the inputs to influence the succeeding outputs of the system. Suppose that the output of a milling machine is inspected. Information on the quality defects is transmitted to the machinist. When the machine setup is adjusted, it might alter the quality of products manufactured in the following time periods. Figure 1.2 shows a feedback system. The principle of feedback is utilized to control the operations of an organization. By comparing the information on the actual performance with the predetermined standard, management can take corrective action to improve future operations.

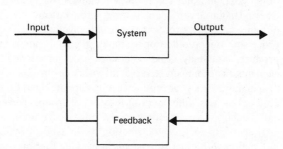

FIGURE 1.2   Graphical representation of a feedback system.

The condition of a system may change over time. The state of a system at any moment describes the condition of the system at the specified moment in time. We can define the state of a system at a given moment in time by the set of pertinent properties that the system has at that time. Likewise, the state of the system's environment at a given moment in time is defined by the set of pertinent properties that the environment has at that time.

SYSTEMS ANALYSIS

Systems analysis deals with the study of a system. The study of a system is not undertaken for its own sake, but is often aimed at determining effective ways of planning and allocating resources to achieve the desired goals. Systems analysis involves the use of analytical methods, which includes optimization techniques, economic concepts, and other mathematical methods. In systems analysis, the whole system under study is considered rather than any part of it. Solutions that would produce favorable results for the system as a whole are established. Most often the best solution for the system may not necessarily be the best for each component of the system. The overall result for the given system is the essence of the systems approach. Solomon and Weingart (1966, p. 16) describe it in this way:

> The systems approach, pioneered by this nation's aerospace industry, calls for a survey of the forest before the trees are examined in any detail. Look back for a moment at the list of elements in business data processing; think of these elements as the trees, and consider instead the whole forest—the complete flow of information needed to run your business efficiently. If you think of this whole package as a system (i.e., something that can be isolated, with a definable functional goal, and thus susceptible to analysis without extensive consideration of external factors), you are likely to see some startling relationships emerge. Instead of many unrelated processing operations, you may see several groups of related processing operations. For example, if your business involves distribution of products, you may visualize a sales order processing group, including such elements as accounts receivable, inventory control, production planning (if you also manufacture your products), and shipping or traffic control. For a different kind of business, different groups (or set of groups) will appear, but appear they will. Each group (or, more properly, subsystem) can then be isolated for further analysis of its goal, its contribution to the goal of the over-all system, and its interaction with other subsystems in meeting that over-all goal. This process of systems analysis then descends to the next level of detail, and so on down through the ultimate details of processing within the area under study, always bringing to bear the perspective developed at the higher levels.

Mathematical models are employed in systems analysis to represent the abstractions of real-world systems. Mathematical models are developed by utilizing the language of mathematics. The important objects, properties, and relationships are included in formulating the model while the unimportant ones are excluded. Sound judgment is necessary in selecting the aspects of the real-world system that should be incorporated in the model. Alternative policies are tried on the model and the resulting outcomes are observed. A large number of alternatives are usually available for complex problems. If only a few alternatives are generated, they may not include potential alternatives. However, generating a great many alternatives might require enormous evaluation effort. Hoag (1956) comments on this as follows:

> All one can say is that the analyst must be very careful to make his analysis neither too small nor too large, which is only to say that judgement is essential at this very early stage.

The alternative that satisfies the objective of the system is selected for implementation.

The two approaches of analyzing systems are process analysis and final outcome analysis. In the process analysis, the structure of the system is examined. The intermediate inputs and outputs are considered as a microscopic view. The system is studied through the analysis of subsystems. In the final process analysis, the macroscopic approach is adopted. The system is viewed as a black box and the relationships between the inputs and outputs are considered. Both the approaches may be utilized in the study of complex systems. The areas of applied science that are called systems analysis, operations research, management science, and systems approach are concerned with the application of objective scientific methods to the solution of management decision problems. Although differences between these areas exist, they are entirely semantical.

Many problems can be solved by using pencil and paper. However, there are problems that require extensive calculations. There are also other situations, where the same kind of calculations are performed frequently, perhaps on a daily or weekly basis. The manual approach using paper and pencil is time-consuming and impractical, if not impossible. Computers have the capacity for processing at a high speed many variables and masses of data the calculations for which are very accurate. The development of computers has enabled the systems analysts to attack complex problems and consider many more alternatives for the systems under study. Computer models enable the computer to perform the calculations according to mathematical models. The basic features of the FORTRAN language for building computer models are discussed in Chapter 2. The methods of developing computer models for various management decision situations are presented in Chapters 3 to 16.

CLASSIFICATION OF DECISION SITUATIONS

The decision making situations that are encountered in the study of systems can be classified into the three types:

Decision situation under certainty
Decision situation under risk
Decision situation under uncertainty

Decision situation under certainty occurs when complete information on the resulting outcome for each alternative or course of action is available to the decision maker. Each course of action is believed to result in only one outcome. It might appear to be a trivial type of decision situation, since we know exactly what will occur for each one of the alternative policies. It is, however, a difficult task because the availability of numerous alternatives means that a large computational effort is needed to evaluate them and select the appropriate one. Suppose that a foreman says, "I need 200 electric motors in order to assemble 200 refrigerators." This is a certainty situation. Analyses of certainty situations are presented in Chapters 3 to 9.

Decision situation under risk occurs when each course of action can lead to any one of several alternate outcomes. The decision maker could not say which one of them might occur. The probability of occurrence for the alternatives are known or can be estimated. Consider that the Giant Supply Corporation has the option of distributing either Product A or Product B in Canada. The possible outcomes and probability values are as follows:

ALTERNATIVE 1   Select Product A

Outcome	Large National Demand	Average National Demand	Limited National Demand
Sales value (in 10^6 dollars)	10	5	1
Probability	0.3	0.5	0.2

ALTERNATIVE 2   Select Product B

Outcome	Large National Demand	Average National Demand	Limited National Demand
Sales value (in 10^6 dollars)	10	5	1
Probability	0.1	0.7	0.2

The corporation wishes to select the product that yields the maximum expected sales. This is a decision situation under risk. The bulk of material presented in Chapters 10 to 15 deals with the decision situations under risk.

Decision situation under uncertainty prevails when any one of a set of outcomes might occur for each alternative and the decision maker cannot assign probabilities to the possible outcomes. For each alternative and outcome combination, the resulting payoff values to the decision maker are assumed to be known. In the Giant Supply Corporation illustration stated above, when the probability values are removed, we have a decision situation under uncertainty. Chapter 16 deals with the decision-making situations under uncertainty.

TYPES OF MODELS

One of the important aspects of systems analysis is to evaluate the alternative policies and predict the effects. The alternative plans cannot be tested on operating systems because it would be costly, unsafe, and time-consuming. The possibility of trying the alternatives on the proposed system is ruled out since this system is nonexistent. A model of the system under study is usually employed to test the alternatives and predict the resulting effects. A model is a representation or abstraction of the reality. A model cannot represent every aspect of reality because the model is only an approximation of the real object or situation. For a model to be representative of the reality, all significant features must be retained in the model. The models must be examined and validated repeatedly with observed data to obtain an acceptable model. Figure 1.3 shows the steps for the development of an acceptable model. Models can be classified into different types. Models can be categorized based on their structure into iconic, analogue, and symbolic models. *Iconic models* possess some of the physical properties of the things they represent. They are usually made on a different scale. Two-dimensional models such as photographs and engineering blueprints, and three-dimensional models such as

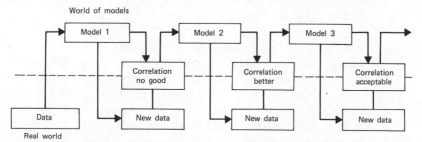

FIGURE 1.3 **Steps for the development of an acceptable model. (Source: E. S. Buffa,** *Operations Management: Problems and Models,* **Wiley, New York, 1972, p. 12).**

automobile, airplane, and physical facilities are examples of iconic models. Iconic models can be used to study the conditions that exist at a given moment in time. They cannot be utilized to study dynamic situations.

Analog models can be employed to represent dynamic situations. One set of properties is used in the analog models to describe another set of properties of the real objects and situations. A graph of sales by months, contour lines on a map, and flowcharts are examples of analog models. The length of lines represented on the graph of sales by months is analogous to the amount of sales. The contour lines on a map are analogous to elevation. The lines on the various kinds of flowcharts are analogous to material or information flow. Analog models are easier to manipulate than the iconic models and therefore, analogs are more suitable for many situations.

Symbolic models employ letters, numbers, and other kinds of symbols to describe real situations. The relationships between symbols exist, in general, mathematically. Symbolic models can be manipulated with less difficulty. However, the most difficult and expensive task might be the development of a symbolic model. One of the major advantages of symbolic models is that it is adaptable to manipulations by computers. It is often called a *mathematical model*. Our primary interest is in the mathematical models for the systems being studied. Mathematical models are structured utilizing two kinds of variables: controllable variables and uncontrollable variables. Controllable variables are under the control of the decision maker. Some of the controllable variables are units to be manufactured, selling price, and cities for factory location. Uncontrollable variables cannot be controlled by the decision maker. Some of the uncontrollable variables are supplier's price, competitor's price, competitor's advertising, and government actions. The objective of manipulating the mathematical model is generally to establish the values for the controllable variables so that some measure of effectiveness is optimized, such as maximize profit, minimize costs, and maximize sales.

Models can be categorized based on their generality of application into standard and custom-made models. *Standard models* have application in several functional areas of organizations. The techniques that are associated with operations research are described as standard models. The techniques include linear programming method, critical path method, and waiting line model. These methods can be employed in production, marketing, and data-processing operations. The proper values of the problem are inserted into the models, and the computational procedures are applied to determine the solutions.

A *custom-made model* is a specialized model intended to solve a unique problem. The concepts of various disciplines including mathematics are utilized to construct a specialized model suitable for the unique situation. The probabilistic bidding model and venture analysis model are examples of the custom-made model. The custom-made model is often the most difficult and expensive to construct.

SELECTED REFERENCES

Ackoff, R. L., and M. W. Sasieni, *Fundamentals of Operations Research*, Wiley, New York, 1968.

Buffa, E. S., *Operations Management—Problems and Models*, Wiley, New York, 1972.

Churchman, C. W., *Systems Approach*, Delacorte Press, New York, 1968.

De Neufville, R., and J. H. Stafford, *Systems Analysis for Engineers and Managers*, McGraw-Hill, New York, 1971.

Ellis, D. O., and F. J. Ludwig, *Systems Philosophy*, Prentice-Hall, Englewood Cliffs, N.J., 1962.

Hall, A. D., *A Methodology for Systems Engineering*, Van Nostrand, New York, 1962.

Hare, V. C. Jr., *Systems Analysis: A Diagnostic Approach*, Harcourt, Brace, and World, New York, 1967.

Hitch, C., *An Appreciation of Systems Analysis*, The RAND Corporation, Santa Monica, Calif., 1955.

Hoag, M. W., *An Introduction to Systems Analysis*, The RAND Corporation, Santa Monica, Calif., 1956.

Lee, A. M., *Systems Analysis Frameworks*, Wiley, New York, 1970.

McMillan, C., and R. F. Gonzalez, *Systems Analysis: A Computer Approach to Decision Models*, Richard D. Irwin, Homewood, Ill., 1973.

Murdick, R. G., and J. L. Ross, *Information Systems for Modern Management*, Prentice-Hall, Englewood Cliffs, N.J., 1971.

Nikoranov, S. P. *Systems Analysis: A Stage in the Development of the Methodology of Problem Solving in the USA*, in Optner, S. L. (ed), *Systems Analysis: Selected Readings*, Penguin Books, Middlesex, England, 1973.

Optner, S. L., *Systems Analysis for Business and Industrial Problem Solving*, Prentice-Hall, Englewood Cliffs, N.J., 1965.

Optner, S. L. (ed), *Systems Analysis: Selected Readings*, Penguin Books, Middlesex, England, 1973.

Solomon, I. I., and L. O. Weingart, *Management Uses of the Computer*, The New American Library, New York, 1966.

Thierauf, R. J., and R. A. Grosse, *Decision Making Through Operations Research*, Wiley, New York, 1970.

White, H. J., and S. Tauber, *Systems Analysis*, Saunders, Philadelphia, 1969.

CHAPTER 2 a review of computer programming

THE ELECTRONIC DIGITAL COMPUTER system has become a powerful tool for analyzing and solving problems in various fields such as business, government, science, education, and medicine. The wide usage of computer systems are attributed to their ability to process quickly a large amount of data, store information, perform complex operations, and produce errorfree results economically. But we cannot just tell the computer, "Do this problem for me" and obtain the desired results. A computer program designed for a particular problem will cause the computer system to perform a series of operations in a sequential form and then yield accurate results. A computer program consists of a set of instructions (also called "statements"). In this chapter, we discuss briefly the components of a computer system and then present the rules of writing instructions in FORTRAN IV language.

COMPONENTS OF A DIGITAL COMPUTER SYSTEM

The components of a digital computer system installed by a particular organization depend on the nature and amount of data processed as well as on the computer model used. A computer system will generally include the devices shown in Figure 2.1.

The standard way of preparing the program and data is to punch the already written program and data into punched

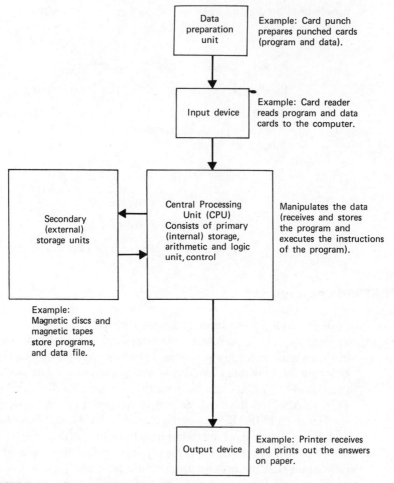

FIGURE 2.1   Elements of a computer system.

cards using a card punch. The card reader feeds the program and data into the storage of the processing unit by reading the punched cards. The card reader is an input device, and the punched card is the related input medium. Any errors in the program are detected, and a list of them is sent to the output device for recording. If the program is in acceptable form, the instructions are carried out. Data are provided, as required, in the execution phase. The results of the program go to the output device such as a printer, which records the solutions. Many types of input and output media are used with computer systems. These are the most widely used media and devices:

Input Media	Related Input Device
Punched cards	Card reader
Paper tape	Paper tape reader
Documents inscribed with magnetic characters (such as check)	Magnetic ink character reader

Output Media	Related Output Device
Paper	Printer
Punched card	Card punch
Paper tape	Paper tape punch

FORTRAN Language

FORTRAN is a programming language developed primarily for scientific applications. It is currently also employed for analyzing and solving business administration problems. The first version of FORTRAN, an acronym for FORmula TRANslation, was announced in 1957 and given the name FORTRAN I. Subsequently, it has been modified and improved. FORTRAN IV is the most recently developed, popular version of the already proven FORTRAN language. FORTRAN is machine-independent and, hence, the programmer writing in FORTRAN does not have to know how the computer works. Specifically, it is a procedure-oriented language; this means that the programmer should primarily be able to analyze the problem and come up with a computational procedure for writing a FORTRAN program.

STEPS IN WRITING A FORTRAN PROGRAM

Here are the various steps followed in solving a problem using FORTRAN language.

1. Analyze the problem.
2. Develop a computational procedure and prepare program flowchart.

3. Write the program statements and data on FORTRAN coding forms.
4. Keypunch the program and data into punched cards.
5. Submit the job (which consists of program cards and data check together with the job control cards).
6. Debug the program, if needed. Debugging is the process of detecting and correcting program errors. The compilation listing obtained from the computer is useful in the debugging process.
7. Execute the program after removing all errors and obtain the desired solution.

We shall illustrate these steps by a simple example. The Victor Parade Motors computes the weekly pay of each salesman as follows:

Weekly pay = Guaranteed base pay + bonus
Bonus   = 0 if the weekly sales is less than $1500.00
    and
Bonus   = 5 percent of weekly sales amount that is greater than $1500.00

The guaranteed base pay of any salesman depends on his (or her) past sales record and the number of years of service with the company. For each salesman, therefore, the base pay and the total weekly sales are the input data. The input values as well as the computed weekly pay are to be printed out.

The following variables are used in the program:

GBPAY   = guaranteed base pay
BONUS   = bonus earned by the salesman
WSALES = weekly sales amount
WPAY   = weekly pay

Suppose that the salesman 986 has made a total sales of $1360.50 and $3250.00 in the third and fourth week, respectively. His guaranteed base pay per week is $100.00. For the third week,

GBPAY   = 100.00 ⎫ Input data
WSALES = 1360.50 ⎭
BONUS   = 0.00   since   WSALES < 1500.00
WPAY   = GBPAY + BONUS
              = 100.00 + 0.00
              = 100.00

For the fourth week,

$$GBPAY = 100.00 \left.\right\} \text{Input data}$$
$$WSALES = 3250.00$$
$$BONUS = 0.05 (3250.00 - 1500.00)$$
$$= 87.50$$
$$WPAY = 100.00 + 87.50$$
$$= \underline{187.50}$$

Here is the computational procedure:

1. Obtain the values of GBPAY and WSALES.
2. Calculate BONUS.
3. Calculate WPAY.

The *flowchart* is a graphical representation of the logical sequence of steps to be performed in a program. A flowchart (also called a *block diagram*, or *logical flowchart*) serves as a blueprint in writing the program. A widely accepted set of symbols is used in flowcharting. The basic symbols are depicted below.

Symbol

Direction of flow. These flowlines and arrowheads show the flow of control from one program step to another. It links the symbols. It is also called the *program path line*.

Symbol

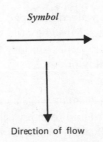

Direction of flow

Input or Output. This parallelogram shows the movement of data from a medium to the computer (such as input data being read from the punched cards by the computer) or the movement of results from the computer to an output device (such as printing the solutions from the computer storage on the printer).

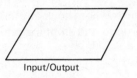

Input/Output

Processing. The rectangler refers to arithmetic operations (such as addition and multiplication) and the assignment of values to variables.

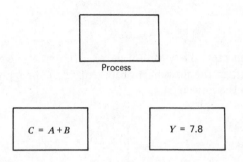

Process

$C = A + B$

$Y = 7.8$

Decision. The diamond-shaped box means testing to be done. Depending on the test result, the next program step would be performed. This symbol will have an outlet for each possible decision or choice. There will be *at least* two outlets from this symbol.

Decision

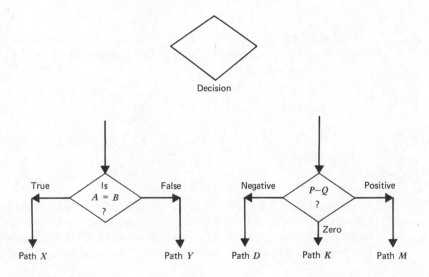

Terminal. The oval box shows any terminal point in the flowchart, such as the beginning point and stopping point of a flowchart.

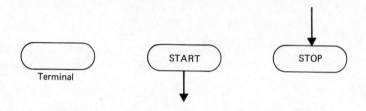

Terminal   START   STOP

Connector. This symbol shows an entry of the program path line from one part of the flowchart or an exit of the program path line to another part of the flowchart.

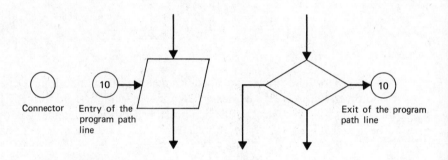

Connector   Entry of the program path line   Exit of the program path line

Predefined process. The special block represents an operation that is specified in detail elsewhere. Figure 2.2 shows the flowchart for the weekly pay computation problem.

Predefined process

The next step is writing the program statements and data on FORTRAN coding forms. This is called *coding*. In coding, the detailed computational procedure is transformed to a series of FORTRAN instructions. The coding makes it much easier to punch the program and data in the

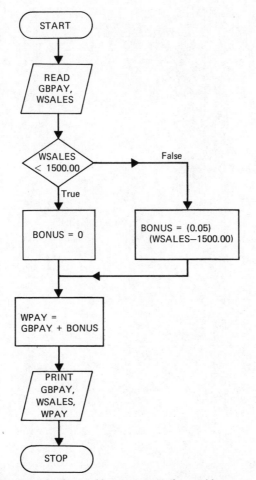

FIGURE 2.2   Flowchart for the weekly pay computation problem.

following step. The program statements are usually written on the coding form similar to that shown in Figure 2.3. There are 80 columns in the form. The columns are grouped into four fields in composing the instructions and used as follows:

Columns 1–5 are reserved for statement numbers. Statement numbers may be used for all the statements in the program. However, only those statements that are referred to elsewhere in the program need statement numbers. The numbers need not appear in sequence. The numbers must be positive integers containing from one to five decimal digits. A particular

```
IBM                                          FORTRAN Coding Form                              GX28-7327-6 U/M 050**
                                                                                                    Printed in U.S.A.

PROGRAM    VICTOR PARADE MOTORS - WEEKLY PAY                      PUNCHING  GRAPHIC              PAGE    OF
                                                                 INSTRUCTIONS PUNCH             CARD ELECTRO NUMBER*
PROGRAMMER  P - RAMALINGAM                          DATE

C    WEEKLY PAY PROBLEM
C
C    READ THE GIVEN VALUES
      READ 18, GBPAY, WSALES
C
C    COMPUTE BONUS
      IF ( WSALES .LT. 1500.00 ) GO TO 15
      BONUS = (.05) * ( WSALES - 1500.00 )
      GO TO 10
 15   BONUS = 0
C
C    COMPUTE WPAY
 10   WPAY = GBPAY + BONUS
C
C    PRINT OUT THE GIVEN AND COMPUTED VALUES
      PRINT 7, GBPAY, WSALES, WPAY
C
  7   FORMAT (5X,F7.2, 2X,F8.2,5X,5HWPAY =,F8.2)
 18   FORMAT ( 2F10.2 )
      STOP
      END

 100.00    3250.00            ← DATA
```

FIGURE 2.3   A FORTRAN program to determine the weekly pay.

statement number cannot be used to identify more than one statement in a program; that is, all numbers must be different. A number may be written anywhere in the five columns; the blanks are not interpreted as zeros by the computer, in contrast to the data cards. However, in order to insure readability and appearance, it is advisable to write the number right-justified, as shown in the statement number 15 in Figure 2.3.

Column 6 is reserved to specify continuation. When a statement does not fit on one line, it is continued to a second, third line, etc., by writing any one of the integers from 1 to 9, arbitrarily in column 6 of the continuation lines. Column 6 of the initial line that has continuation lines is usually left blank. A continuation line cannot have a statement number.

Columns 7–72 are used to write the actual FORTRAN statement. In each column of a line at most one letter, one number, or one special character is written. Blanks may be inserted in this field to improve readability. Blanks are ignored in all cases except in some FORMAT statements to be discussed later. Each statement is written on a new line in this field. However, if a statement requires more than the 66 spaces provided in this segment, it is continued on the successive line in the same field.

Columns 73–80 are the last eight columns, and they may be used for any identification purposes. The computer system ignores any information given in this field.

Column 1 is also used for a special purpose. A *C* in column 1 indicates that it is a comment. Following the *C* anything may be written from column 2 to 80. This line will not be processed by the computer. However, the comment line will be printed out in the program listing. It enables the programmer to describe the computational steps as shown in Figure 2.3.

Note that the above rules apply to writing FORTRAN statements only. All of the 80 columns may be used, however, for input data. The input data is written on the coding form according to the related instructions (FORMAT) in the program.

The coding form serves as an important means of communication in preparing the punched cards. If it is not neatly and legibly printed, many errors might occur in punching cards, and desired answers cannot be obtained. Hence, extreme care must be taken in coding. In each space, at most one character is printed. The coding is done only in printed capital letters. Since some characters* are easily confused with others,

---

* The letter "I" that occurs in the instructions of computer programs closely resembles the number "1". As a result, it is sometimes confusing to the beginning student. In this book the "1" used in the computer programs is represented by "1" and the letter "I" is represented by "I".

these "look-alikes" are differentiated as follows:

Letter	Number
Ø for the letter O	0
I	1 or 1
Z̶ for the letter Z	2
S	5
G	6

The next step is to punch the instructions printed on the coding sheet onto punched cards. The set of punched cards comprising the program is called the *source program deck*. The punching is done on the standard 80-column Hollerith card using a keypunch machine. Each line of the coding form is punched on a separate card. Each character in a line is punched into the corresponding column in the card. The characters are represented by one, two, or three punch holes at various rows in the particular column. Figure 2.4 shows a typical 80-column Hollerith card illustrating punched codes for the characters.

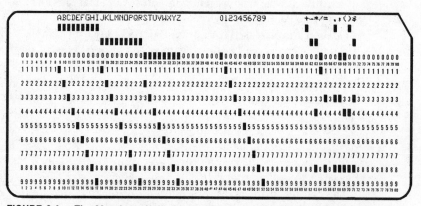

FIGURE 2.4   The 80-column Hollerith card illustrating punched codes for letters, numbers, and special characters.

The source program deck and the data deck for our weekly pay program are shown in Figure 2.5. The control cards are unique to the various computer installations, and this information can be obtained from the computer laboratory assistants. The job is then run on the computer system, and the printout is obtained. Figure 2.6 shows the resulting computer printout for our program. We note that the WPAY is $187.50

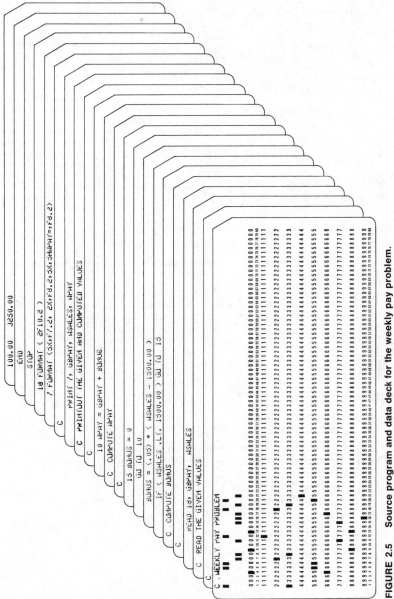

FIGURE 2.5   Source program and data deck for the weekly pay problem.

23

```
C   WEEKLY PAY PROBLEM
C
C   READ THE GIVEN VALUES
        READ 18, GBPAY,  WSALES
C
C   COMPUTE BONUS
        IF ( WSALES .LT. 1500.00 ) GO TO  15
        BONUS = (.05) * ( WSALES - 1500.00 )
        GO TO  10
    15 BONUS = 0
C
C   COMPUTE WPAY
    10 WPAY = GBPAY + BONUS
C
C   PRINTOUT THE GIVEN AND COMPUTED VALUES
        PRINT 7, GBPAY, WSALES, WPAY
C
     7 FORMAT (5X,F7.2,  2X,F8.2,5X,5HWPAY=,F8.2)
    18 FORMAT ( 2F10.2 )
        STOP
        END


        100.00   3250.00     WPAY=  187.50
```

FIGURE 2.6   Computer program and solution for the weekly pay problem.

for the fourth week, and it is the same as the manual computational solution. If the weekly pay for another salesman is to be calculated, the same general program can be used with the appropriate data card.

BASIC ELEMENTS OF FORTRAN STATEMENTS

We present the characters used in FORTRAN and also the rules governing the constants and variables in this section. The FORTRAN language is composed of the following character set:

1. The 26 capital letters of the alphabet:
   ABCDEFGHIJKLMNOPQRSTUVWXYZ
2. The 10 decimal digits:
   0 1 2 3 4 5 6 7 8 9
3. The 11 special characters:
   (a) Five for arithmetic signs:
       + plus
       − minus
       * asterisk
       / slash
       = equal

(b) One for spacing:
   Blank (abscence of any coding). We shall denote a blank by *b*
(c) Five for punctuation or editing:
   . Decimal point
   , Comma
   (Open parenthesis
) Closed parenthesis
   $ Dollar sign

FORTRAN Constants

The two kinds of constant values used commonly in FORTRAN are (1) integer, and (2) real. A constant is formed by the decimal digits, 0 to 9. An *integer constant* (also called "fixed-point quantity") is one that does not have a fractional part. In the integer constants, the decimal point does not appear. An integer may be preceded by a + or − sign. If an integer constant does not have a sign, it is assumed to be positive. Embedded commas are not permitted. The following are valid fixed-point quantities:

$$15$$

$$-763$$

$$0$$

$$-4079$$

$$+707$$

The following are invalid fixed-point quantities:

   10,798   (comma not allowed)

   20.0   (decimal point not allowed)

   72+   (sign must precede the number)

   $42   (special character not permitted)

The integer numbers would be useful in representing counting numbers. For example, we can say that there are 78 electric motors in the inventory.

A *real constant* (also called "floating-point quantity") may be an integer or may contain a fractional part. A real constant must include a decimal point. A floating-point quantity may be preceded by a + or − sign. If a real constant does not have a sign, it is assumed positive.

Embedded commas are not permitted. The following are valid floating-point quantities:

$$347.00$$

$$+9.$$

$$-.41$$

$$0.0$$

$$83.05$$

$$-0.00003$$

The following are invalid floating-point quantities:

-51	(decimal point omitted)
$+7,146.2$	(comma not allowed)
$763.89	(special character not allowed)
$18.2-$	(sign must precede the number)

The real numbers would be useful in expressing measuring numbers. For instance, a salesman's weekly pay can be $253.63 and it is recorded as 253.63.

FORTRAN Variables

Variables are used in FORTRAN to represent quantities. A variable is able to take on many values, in contrast to the constants, during the execution of a program. A variable can have, however, only one value at any specified time during the execution of the program. This value is called the current value of the variable. The two types of variables are (1) integer and (2) real. An integer variable (or fixed-point variable) can take any of the quantities that are said to be integer constants. Similarly, a real variable (or floating point variable) can take any of the quantities that are permitted to be real constants. The rules governing the formation of variable names are given below.

1. A variable name consists of 1 to 6 characters.
2. The name may contain only letters and numbers. Special characters are not permitted.
3. The first character of the name must be a letter.

(3a) Integer variable.
   If the first character of a variable is one of the letters *I*, *J*, *K*, *L*, *M*, *N*, it is an integer variable.
(3b) Real variable.
   If the first character of a variable is one of the letters other than *I*, *J*, *K*, *L*, *M*, *N*, it is a real variable. In other words, a floating point variable starts with one of the letters *A* to *H* or *O* through *Z*.

The following are valid variables:

Variable Name	Mode
K64	Integer
JBOOKS	Integer
M73A	Integer
X5	Floating point
D	Floating point
PAY7K	Floating point

The following are invalid variables:

K234567	(Too long)
9M	(Does not begin with a letter.)
ITEM*	(Special character not allowed.)
C J	(The intervening blank space is an invalid character.)

FUNDAMENTAL FEATURES OF FORTRAN STATEMENTS

The FORTRAN statements may be classified into the four categories:

1. Arithmetic statements
2. Input and output statements
3. Control statements
4. Specification statements

We shall now briefly introduce the statement types and, then, discuss the features of each one of them in detail.

Arithmetic Statements

These statements are used to direct the computer to perform arithmetic operations and also to assign the results to a variable name. The following instructions, shown in Figure 2.6 of our weekly pay program are arithmetic statements:

BONUS = (.05) * (WSALES − 1500.00)
> Bonus is calculated and the calculated value is assigned to BONUS.

15   BONUS = 0
> A zero value is assigned to BONUS.

10   WPAY = GBPAY + BONUS
> Weekly pay is calculated and the computed value is assigned to WPAY.

Input and Output Statements

The input instructions are employed to direct the computer to obtain data from a specified input device (such as a card reader and paper tape reader) and store data in the storage unit. The following instruction shown in Figure 2.6 is an input statement:

READ 18, GBPAY, WSALES
> This READ statement causes the computer to read the input data card containing the values of GBPAY and WSALES by a card reader and store them.

The output instructions are used to direct the computer to transmit the output solution from the storage unit to a specified output device (such as a printer and card punch). The following instruction shown in Figure 2.6 is an output statement:

PRINT 7, GBPAY, WSALES, WPAY
> This PRINT statement causes the computer to transfer the values of GBPAY, WSALES, and WPAY to the printer, and it prints these values on a sheet of printer paper.

Control Statements

These are used to control the order in which the program is to be executed. Usually, FORTRAN statements are executed in a one-after-the-other order as they appear in the source program. However, sometimes it is required to perform instructions in some order other than the normal sequence. In some cases, if the same program may be used with different data, we return to the beginning of the program. Depending upon the values of computed intermediate results, we may need to execute various sections of the program. In these cases, the control statements can be used to direct the computer to change the order of execution in the desired manner. The control instructions are also used to terminate program execution. The following three instructions shown in Figure 2.6 are control statements:

IF (WSALES .LT. 1500.00) GO TO 15
>This statement causes the computer to choose either statement 15 or the next statement depending on the test result.

GO TO 10
>This statement causes the computer to execute statement 10 next.

STOP
>This statement signals the logical end of the program. It brings the program execution to a final halt.

Specification Statements

These provide descriptive information that the computer will need to execute other statements. For example, one type of specification statement—called FORMAT instruction—describes the type, size, and location of data in the input and output media (such as punched cards and printed paper). The following two instructions shown in Figure 2.6 are the specification statements.

18   FORMAT (2F10.2)
>This statement describes the arrangement of data in the input data deck.

7   FORMAT (5X, F7.2, 2X, F8.2, 5X, 5HWPAY =, F8.2)
>This statement provides information for arranging the output data on the printed paper.

Arithmetic Statements

Arithmetic statements are one of the most commonly used instructions in most FORTRAN programs. The arithmetic statement must be of the following form:

$$v = e$$

where v represents any variable and e is any arithmetic expression. The expression may be any meaningful combination of constants, variables, arithmetic operators, and parentheses. For example, consider the arithmetic statement

$$A = B + C + 7 \cdot 5$$

In the expression, we have two variables (B and C) and a constant, 7.5. The two arithmetic operators indicate that these three items are added. The computer evaluates the expression on the right-hand side of the equal sign, and it replaces the value of the variable on the left-hand side by the evaluated value. The "=" sign signifies "is replaced by," and it does not imply the usual "equal to" sign used in conventional arithmetic. The variables and constants are the quantities to be operated upon in the expression. They are called operands. The following arithmetic operators are symbols employed to indicate the arithmetic operations:

Operation	Symbol
Addition	+
Subtraction	−
Multiplication	*
Division	/
Exponentiation	**

Some algebraic expressions and their FORTRAN equivalents are given below.

Algebraic Expression	FORTRAN Equivalent
$a + 7 \cdot 9$	A + 7.9
$a - f$	A − F
$a \cdot b$	A * B
c/h	C/H
t^3	T ** 3

Parentheses are employed in an expression to group terms and also to separate operators that would otherwise appear next to one another. They do not imply multiplication.

In an expression where parentheses are not used, the arithmetic operations are performed according to the hierarchy:

First level: Exponentiation
Second level: Multiplication and division
Third level: Addition and subtraction

If the expression contains more than one operation on the same level, the computations are performed from left to right. Consider the expression

$$6 - 8/2 * 3 + 2 ** 3$$

There are no parentheses in the expression. First, the exponentiation is performed. Hence,

$$6 - 8/2 * 3 + 2 \underset{\smile}{**} 3$$
$$= 8$$

The resulting expression is

$$6 - 8/2 * 3 + 8$$

Now, multiplications and divisions are done. We have one each of the two operators. Since they are on the same level they are calculated from left to right. Continuing the procedure, we get

$$6 - \underset{\smile}{8/2} * 3 + 8$$
$$= 4$$
$$6 - \underset{\smile}{4 * 3} + 8$$
$$= 12$$
$$6 - 12 + 8$$

Only addition and subtraction operators appear in the reduced expression. They are performed from left to right because they are on the same level. Hence,

$$\underset{\smile}{6 - 12} + 8 \rightarrow -6 + 8$$
$$= -6$$
$$= 2$$

Parentheses may be used to force certain calculations to be performed before others. When parentheses are used in an expression, the part of the expression written within the parentheses is always evaluated first. If the sets of parentheses are "nested," that is, pairs are used one within the other, calculation occurs first in the innermost grouping, then the next innermost, and so on to the outermost. Consider the expression,

$$4 * (5 - 3)$$

The calculations are performed as follows:

$$4 * \underbrace{(5 - 3)}_{= 2} \rightarrow 4 * 2$$

$$= 8$$

Consider the expression with nested parentheses:

$$(-1 * (1 + (2 - 1) * 3)) ** 2 + 1$$

Calculating the innermost parentheses

$$(-1 * (1 + \underbrace{(2 - 1)}_{= 1} * 3)) ** 2 + 1$$

we obtain

$$(-1 * (1 + 1 * 3)) ** 2 + 1$$

Continuing the method, we get

$$(-1 * 4) ** 2 + 1 \rightarrow -4 ** 2 + 1$$

$$= 17$$

The following rules must be observed in writing arithmetic statements.

1. Operation symbols must not appear in adjacent positions. For example, the expression d^{-4} cannot be written as D ** −4. It may be coded as D ** (−4).
2. No operation symbol may be assumed to be present. All operation symbols must be stated explicitly. For example, the expression $x \cdot y$ indicating x times y cannot be written as (X) (Y). It may be stated as X * Y or (X) * (Y).

3. Any variable used in the arithmetic expression must have been previously defined. A variable is said to be defined when the computer is able to find from its memory a numerical value for the variable.
4. If parentheses are not employed in the expression, the operations are performed as follows: all exponentiations are evaluated first, next all multiplications, and divisions and, finally, all additions and subtractions. If two or more operations of the same rank appear in the expression, these operations are performed from left to right. When parentheses are used, the segment of expression in the innermost parentheses is evaluated first, leading to the evaluation of the outermost expression. All equal levels of groupings are calculated before moving to the next innermost set. Parentheses may be inserted freely whenever there is any doubt about whether the computer will evaluate the operations in the desired manner. If the parentheses are unnecessary, they do no harm.
5. An arithmetic expression must have all the constants and variables in one type, either integer or real. The integer and real quantities cannot be mixed in any expression with the exception of exponentiation. The rule for forming acceptable exponentiation operation is given below

Exponent

Type of Quantity	Integer	Real
Base Integer	Valid	Invalid
Real	Valid	Valid

(A^B) ← Exponent

↑

Base

The calculations in an expression are performed according to its type, real-type calculations or integer calculations. In the real-type calculations, the evaluations are performed as in conventional algebra. However, in integer calculations, only the whole numbers are kept, dropping the fractional positions after each operation. Mixing integer and real quantities, which is not allowed in most computers, is called *mixed mode*. However, the variable indicated on the left-hand side of the equal sign and the expression given on the right-hand side of the equal sign may be in different modes or types. Thus, the three statements, A = B, J = X, and Z = K are valid. The statement in which the left-side variable and the right-side expression are in different type indicates to the computer to convert the value into the type of the variable. When the expression is in floating-point type and the variable is in integer type, real-type calculation of the expression is performed, and the computed value is assigned to the integer variable in the integer form after dropping the fractional part. The process of dropping the fractional part is termed *truncation*.

Arithmetic Statement   *Result*

1. K = 4.95   K = 4 (4.95 is truncated to 4).
2. A = 8   A = 8.0.
3. N = 9/10   9/10 = 0 (0.9 is truncated to 0). Thus, N = 0.
4. J = 18.0/10.0   18.0/10.0 = 1.8 in real-type calculation. J = 1 (1.8 is truncated to 1).
5. E = 12/10 + 8/10 12/10 = 1 (1.2 is truncated to 1). 8/10 = 0 (0.8 is truncated to 0). E = 1 + 0 = 1.0.
6. E = 12.0/10.0   12.0/10.0 = 1.2 in real-type calculation. Thus, E = 1.2.

6. Left-hand side of an arithmetic statement must be one unsigned variable only. An unsigned variable is not expressed with a + or − sign preceding it. For example, +ALPHA and −K are invalid unsigned variables, but L is a valid unsigned variable.

Input and Output Statements

The input and output statements provide information to the computer on transmitting data between the computer and the external devices. A READ statement is the input instruction, and it causes the transfer of data from an appropriate input device to the computer memory. The statement is of the form

READ (i, n) List

where

i represents an unsigned integer constant specifying the input device to be used.
n represents the statement number of the related FORMAT statement.
List represents the list of variable names for which the values are to be read in.

The related FORMAT statement specifies where and how the data can be found in the given input medium, such as a punched card.

Consider the input statement

```
    READ (5, 73) A, B, IOTA, D
73  FORMAT (F5.1, F7.3, I5, F6.2)
```

Many computer installations use $i = 5$ to specify the card reader as the input device. For any computer system, this information can be obtained from its manufacturer. In our illustration, the value 5 implies the use of a card reader in inputting the data. The data of the variables A, B, IOTA, and D are transferred in that order. The number 73 indicates that a FORMAT statement labeled with the statement number 73 will tell the computer where and how the four data values are presented in the input data card. We shall discuss the basic features of the READ and FORMAT statements subsequently.

The only input device used in the earlier versions of FORTRAN has been a card reader. Therefore, READ statements were written omitting the number indicating the input device. The above mentioned instruction may be coded as

```
    READ 73, A, B, IOTA, D
73  FORMAT (F5.1, F7.3, I5, F6.2)
```

This form of input statement is valid today in most computers. Therefore, we apply this type of input instructions for solving most of the problems in this book. Moreover, most batch processing computer systems have only one input device and, thus, the READ statements are written for them, without specifying the input device number, as stated above.

A WRITE statement is the output instruction and it causes the transfer of data from the computer memory to an appropriate output device. The output statement is of the form

```
WRITE (i, n) List
```

where

   i represents an unsigned integer constant specifying the output device to be used.
   n represents the statement number of the related FORMAT statement.
   List represents the list of variable names whose values are to be written.

The WRITE statements are similar in structure to the READ statements. The FORMAT statement related to the WRITE statement specifies where and how the data are to be arranged on the output medium. Consider the output instruction:

```
        WRITE (6, 58) ITEMS, PRICE, BILL
    58  FORMAT (I6, F5.2, F8.2)
```

The value of i denotes the output device and it can be obtained from the manufacturers of the computer systems. The most frequently used i value for a printer is 6. Hence, the above WRITE statement indicates that a printer will be used to output the values of the three variables ITEMS, PRICE, AND BILL in that order, according to the layout described in the FORMAT statement 58.

In the earlier versions of FORTRAN, printers and card punches were the output devices. Without explicitly stating the output device by any number, the output instructions were written as follows:

```
        PRINT 10, ALPHA, JIX, GAMMA
    10  FORMAT (F10.2, I5, F5.2)

        PUNCH 22, KOUNT, SALES, COST
    22  FORMAT (I5, F4.2, F7.2)
```

The PRINT statement uses a printer to output the data on the printer paper, and the PUNCH statement uses a card punch to punch the results on the standard 80-column cards. These punched output cards may serve as the input data deck to some succeeding programs at some future time. These types of output statements are acceptable in most computer installations including time-sharing systems. Therefore, the programs that are composed in this book use the simplified PRINT statements. Thus, the WRITE instruction mentioned above can also be written as

```
        PRINT 58, ITEMS, PRICE, BILL
    58  FORMAT (I6, F5.2, F8.2)
```

The computer reads and interprets the input data cards according to the associated FORMAT instructions; and also it prints out the results on the printer conforming to the descriptions stated in the related FORMAT statement. Hence, the programmer should be very careful in formulating the FORMAT instructions, in punching the data cards, and also in interpreting the computer printout results. We present the features of the following format specifications:

1. I specification
2. F specification
3. H specification
4. X specification

The total number of columns (or characters) reserved for the data of a variable in the input medium or the output medium is called the *field width*.

I Specification   This is used to describe an integer constant and it has the form

Iw

where

I specifies that the data is an integer quantity.
w denotes the field width.

The values are represented right-justified in the alloted field. For example, if the variables KOUNT and JDIST have the values 89 and ⁻963, respectively, they are represented as given below.

```
      READ 30, KOUNT, JDIST
30    FORMAT (I4, I5)
```

Actual input data fields are

```
      bb89b − 963
      PRINT 7, JDIST, KOUNT
7     FORMAT (I6, I2)
```

Actual output fields are

bb − 96389

F Specification   This is used to describe a floating-point quantity in the usual decimal form and it has the appearance

Fw.d

where

F specifies that the data is a floating-point quantity.

w indicates the total field width.

d specifies the number of digits reserved after the decimal point in the alloted field.

The floating-point value is right-justified in the given field on the output medium. If the decimal point is used in a floating point value in preparing the input medium, the value may be recorded in any location in the alloted field. On the other hand, if the decimal point is not recorded on the input medium, computer will interpret the value according to the F specification. If the values of the three floating-point variables, A, B, and C, are 10.76, $^-$998.2, and 0.003, respectively, they would be represented as follows:

```
    READ 59, A, B, C
59  FORMAT (F6.2, F6.1, F7.3)
```

Actual input data fields are

b10.76 − 998.2bbb.003

```
    PRINT 26, B, C, A
26  FORMAT (F8.1, F4.3, F8.2)
```

Actual output fields are

bb − 998.2.003bbb10.76

H Specification (or Hollerith Specification)   This is used to reproduce a given set of characters into the respective columns in its field on the output medium. The sets of characters may include letters, numbers, and special characters. The H specification can be employed to provide descriptive messages and headings in the output. It has the appearance

wH message

where

w denotes the width of the message field

H identifies that it is Hollerith field

message represents *w* columns of information containing the characters to be printed out. The characters are counted immediately following the notation H.

For example, if YVAL = 12.36

```
      PRINT 75, YVAL
 75   FORMAT (16HbTHEbANSWERbIS = b, F5.2)
```

will cause the computer to print out

THE ANSWER IS = 12.36

Titles may be composed in the output as follows:

```
      PRINT 83
 83   FORMAT (18H MARKETING PROBLEM)
```

This will cause the computer to print out

MARKETING PROBLEM

X Specification   This is used to skip spaces and, thus, it causes blanks for the specified fieldwidth. The general form of the specification is

wX

where

w is the field width.

If the X specification is used in the format statement related to a READ instruction, the data represented in the specified field, if any, will be ignored. The X specification used in a format statement related to a PRINT instruction provides blank spaces in the specified field. Consider the following statements

```
      D = 15.50
      JC = D * 100.0
      PRINT 19, D, JC
 19   FORMAT (IX, F5.2, 10HbDOLLARSb =, I5, 6HbCENTS)
```

This will cause the computer to print out

b15.50bDOLLARSb = b1550bCENTS

The title

$$\xleftarrow{\text{20 blanks}} \text{INVESTMENTbPROBLEM}$$

can be printed out by the instructions

```
     PRINT 60
60   FORMAT (20X, 18HINVESTMENTbPROBLEM)
```

In addition to the features stated above, the following rules are also applicable in formulating input and output instructions as well as the FORMAT statements:

1. All the 80 columns in the punched data card can be used in transmitting the data.
2. The printer paper on which the output is printed ordinarily has about 132 spaces. This depends upon the printer used in the computer installation.
3. Care should be taken to see that the first column is not used in outputting the results because it is reserved for special purposes. This topic is discussed in a subsequent section.
4. It is advisable to allocate sufficient field width for each data item, including one space for the algebraic sign whenever necessary.
5. Any number of READ and/or PRINT statements may be related to a particular FORMAT statement. A FORMAT statement may be written anywhere in the program, above or below its related READ or PRINT instruction.
6. If no algebraic sign is shown in the input data, the values are assumed positive.
7. In transmitting the input data, sometimes errors occur because the numbers are incorrectly punched or because they are positioned in the incorrect columns, especially when a large amount of input data is to be punched. Hence, it is recommended that the input data is also printed out to see if the computer-accepted input data is the same as the original data.

Control Statements

The control statements change the usual order of processing the statements and transfer execution to some executable statement other than the next statement. Executable statements tell the computer explicitly what to do. All FORTRAN instructions are executable statements with the exception of specification statements. We shall explore the following control statements and illustrate them by numerical examples:

1. Unconditional GO TO statement
2. Conditional GO TO statement
3. Arithmetic IF statement
4. Relational IF statement
5. The STOP, PAUSE, and END statements

Unconditional GO TO Statement This is an executable statement and it is of the form

GO TO sn

where

sn refers to a statement number

When the GO TO statement is encountered, the next statement to be executed is the one referred by sn. The referred statement can be above or below this statement, but it should never reference itself. After transfer has occurred, the computer continues with the statement following sn. Some examples of the GO TO statements are

GO TO 7
GO TO 204

Note that the unconditional transfer of control statement shown in Figure 2.6, GO TO 10, causes the computer to transfer control to the statement 10, which lies below the particular control instruction.

Conditional GO TO Statement This is used to transfer control to any one of several possible statements depending on the current value of an integer variable. The general form is

GO TO (sn_1, sn_2, sn_3 . . . , sn_m), iv

where

sn_1, sn_2, sn_3, . . . , sn_m is a list of statement numbers.

iv is an unsigned and unsubscripted integer variable.
If the value of the variable iv is 1, execution is transferred to statement sn_1; if iv equal 2, execution is transferred to sn_2 and so on. The value of the variable does not specify the statement number to which transfer is to be made but, rather, the position of the statement number in the list.

The variable may not take negative values and zero. The value of the variable may not also exceed the number of statement numbers in the list. The statement numbers may appear in any order in the list.

Consider the statement:

GO TO (39, 66, 5, 714), JOB

This will cause the computer to transfer control to

Statement number 39 if JOB = 1
Statement number 66 if JOB = 2
Statement number 5 if JOB = 3
Statement number 714 if JOB = 4

Arithmetic IF Statement   This is a conditional transfer of control statement and it is of the form

IF (expression) sn_1, sn_2, sn_3

where the expression may be any arithmetic expression and sn_1, sn_2, sn_3 are statement numbers. This statement transfers control to any one of the three statements depending on whether the value of the expression is less than zero, equal to zero, or more than zero (i.e., negative, zero, or positive).

Statement sn_1 will be executed if the value is negative.
Statement sn_2 will be executed if the value is zero.
Statement sn_3 will be executed if the value is positive.

Consider the statement

IF (A − 10.00)623, 2, 180

This will cause the computer to transfer control to

Statement number 623 if A = 5.60
Statement number 2 if A = 10.00
Statement number 180 if A = 12.09

Relational IF Statement   This is a conditional transfer of control statement and it has the general form

IF (e) statement 1
statement 2

where e is a relational expression and statement 1 and statement 2 are any executable instructions. Statement 1, however, may not be a DO statement (this topic is discussed in a later part in this chapter) or another relational IF statement. The relational expression consists of two arithmetic expressions, and it is written using any one of the relational operations. This statement transfers control to any one of the two statements depending on whether the relational expression is true or false. The decision for transferring control is made as follows:

1. The relational expression is evaluated and determined if it is true or false.
2. If true, control is transferred to statement 1. Consequently, if the statement 1 is a control statement, execution is shifted according to the statement 1. If statement 1 is not a control statement, it is performed and then the statement 2 is executed.
3. If false, control passes directly to statement 2 skipping statement 1.

The relational operators are

Operator	Meaning
.LT.	Less than
.LE.	Less than or equal to
.EQ.	Equal to
.NE.	Not equal to
.GT.	Greater than
.GE.	Greater than or equal to

Note that the operators are always enclosed within a pair of decimal points to differentiate them from variable names. Consider the following instructions

IF (X.EQ. 5.0) A = 3.0 B = −4.0	If X = 5.0, control is transfered to A = 3.0; otherwise B = −4.0 is executed.
If (A.LT. 15.7) GO TO 26 K = −10	If A has any value less than 15.7, control is transfered to GO TO 26, which in turn transfers execution to statement 26. If the value of A is 15.7 or more, K = −10 is executed.

Refer to the flowchart shown in Figure 2.2 and the related program in Figure 2.6 for an application of the relational IF statement.

The STOP, PAUSE, and END Statements   These are the three control instructions that indicate the termination of the program. The logical end of the program is specified by a STOP statement and it has the form

STOP

When a STOP statement is encountered, it terminates the program although it does not physically stop the computer. Several STOP statements may be included in a program, but in each program there should be at least one STOP statement. A PAUSE statement is used to bring the program execution to a temporary halt. The computer operator can inspect the results and, if desired, the execution is continued at the next statement by pressing the "START" button on the computer console. The form of the statement is

PAUSE

The main difference between the STOP and PAUSE instructions is that a final halt of program execution occurs on encountering the STOP, while a temporary halt is caused by the PAUSE statement. The END statement indicates the physical end of the program. This statement signals to the computer that all of the FORTRAN source program statements have been read. The last card of a program must be an END and the form of instruction is

END

ADDITIONAL FEATURES OF FORTRAN STATEMENTS

We have described some important features and basic rules of writing FORTRAN programs in the last section. In this section we shall continue the description.

Data Arrays

Suppose the amount of wages earned by each employee during a particular week is given as follows:

Employee 1 $117.62
Employee 2 $132.79
Employee 3 $180.00
Employee 4 $114.85
     ⋮   ⋮
Employee 200 $119.83

The process of reading and storing these values in the computer memory is a difficult task. We are to formulate 200 variable names, then the corresponding values can be inputted and stored in the respective computer memory locations by preparing a program. In ordinary mathematical notation, these amounts are represented by subscripted variables, such as $S_1, S_2, S_3, S_4, \ldots, S_{200}$. The variable name S refers to the data set, and the subscript denotes the individual element in the set. In our illustration, the data array S specifies the set of 200 quantities: S_1 for 117.62, S_2 for 132.79, and so on. A data array is a sequence of quantities arranged in a convenient order. A *one-dimensional array* is a simple list of quantities arranged in sequence. The one-dimensional array for the employee wages is shown in Figure 2.7.

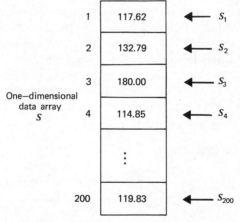

FIGURE 2.7   A one-dimensional array.

A *two-dimensional array* (also called a matrix) consists of a series of horizontal rows and vertical columns. Any element in a two-dimensional array is identified by two subscripts; the first subscript denotes the related row number and the second subscript indicates the related column number. Suppose the PQL company owns three stores. The daily sales values for each one of the stores specifying the breakdown figures for various departments are depicted in array C of Figure 2.8. This is a two-dimensional array with three rows and four columns. There are 12 elements in

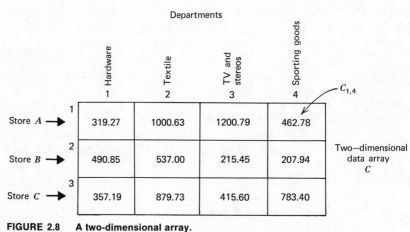

FIGURE 2.8   A two-dimensional array.

the table. The element $C_{1,4}$ lies at the intersection of row 1 and column 4. This represents the Store A sales amount by the sporting goods department.

A *two-dimensional array* (also called a matrix) consists of a series of arrays. The array D, shown in Figure 2.9, illustrates a three-dimensional array. For the PQL Company, suppose the relevant data for each day is presented in a different plane in sequence: the first day sales in the first plane, the second day sales in the second plane, and so on. Each element in the three-dimensional array is specified by three subscripts, the first subscript refers to the row number, the second to the column number, and the third to the plane number. Subscripted variables are used in FORTRAN programs to store and manipulate a large quantity of related data. The general form is

Name (Subscript)

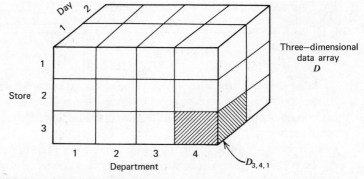

FIGURE 2.9   A three-dimensional array.

The rules for composing subscripted variable names are the same as those for the nonsubscripted (simple) variables. All of the elements of any subscripted variable must be of the same type: integer or real. The type of the quantities is established by the first character of the variable name as with simple variables. The value of the subscript is a positive integer value and zero is not allowed.

RATE (5) = the fifth element in the one-dimensional array RATE. The quantities in the array are of the floating-point type.

KOUNT (3) = the third value in the one-dimensional array KOUNT. The quantities are of the integer type.

CREDIT (4, 6) = the element that lies at the intersection of row 4 and column 6 in the two-dimensional array CREDIT. The values are of the floating-point type.

NBOOKS (4, 3, 10) = the element situated in row 4, column 3, and plane 10 in the three-dimensional array NBOOKS. The values are of the integer type.

The following arithmetic expressions are acceptable as subscripts:

General Form	Example
ic	PROFIT (5)
iv	NETPAY (J7)
$iv + ic$	NETREV (K + 10)
$iv - ic$	AMT (J1 − 3)
$ic * iv$	QTY (2 * L17)
$ic * iv + ic'$	SALES (7 * M + 2)
$ic * iv - ic'$	JUNITS (8 * N10 − 6)

where

ic and ic' are unsigned, nonsubscripted, and integer numbers greater than zero

iv denotes an unsigned, integer variable

DIMENSION Statement This is a specification statement, and it indicates to the computer the number of subscripts used for each subscripted variable and the maximum value for each individual subscript of each variable. It has the general form

DIMENSION name$_1$ (subscripts), name$_2$ (subscripts), . . .

The DIMENSION statements include all of the subscripted variables employed in the program. Consider the statement:

DIMENSION A(4, 5), K(8)

This instruction specifies to the computer that the variable A is two-dimensional, and it reserves 20 storage locations ($= 4 \times 5$) for the elements of A. Similarly, the statement denotes the existence of the one-dimensional data array K and the computer sets aside eight memory locations for the members of K. The DIMENSION statement must specify the maximum anticipated size for each data array to ascertain the reservation of adequate storage spaces. It is not required that all reserved storage spaces should be used every time a program is run. One DIMENSION statement may be employed to specify any number of data arrays. There may be any number of DIMENSION instructions in a program. The DIMENSION statement must always precede any statement that refers to the data array. It is a good practice, however, to locate the DIMENSION statement at the beginning of the program.

Consider that a small mail-order house sells five different items. Their unit cost values are given in the data array COST:

(1)	(2)	(3)	(4)	(5)
$5.00	$3.00	$10.00	$7.00	$4.95
Item 1	Item 2	Item 3	Item 4	Item 5

COST

The number of units of the items ordered by a customer is presented in the array KUNITS:

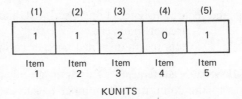

(1)	(2)	(3)	(4)	(5)
1	1	2	0	1
Item 1	Item 2	Item 3	Item 4	Item 5

KUNITS

The total cost of the order (TCOST) can be computed using the subscripted variables as shown below.

```
item 1: COST(1) · KUNITS(1) =  (5.00) · (1) =   5.00
item 2: COST(2) · KUNITS(2) =  (3.00) · (1) =   3.00
item 3: COST(3) · KUNITS(3) = (10.00) · (2) =  20.00
item 4: COST(4) · KUNITS(4) =  (7.00) · (0) =   0.00
item 5: COST(5) · KUNITS(5) =  (4.95) · (1) =   4.95
                                TCOST = $32.95
```

The following program could be used to perform the calculations:

```
        DIMENSION COST(5), KUNITS(5)
        READ 90, COST(1), COST(2), COST(3), COST(4), COST(5)
        READ 10, KUNITS(1), KUNITS(2), KUNITS(3),
       9KUNITS(4), KUNITS(5)
        PRINT 90, COST(1), COST(2), COST(3), COST(4), COST(5)
        PRINT 10, KUNITS(1), KUNITS(2), KUNITS(3),
       8KUNITS(4), KUNITS(5)
C
        TCOST = 0
        K     = 1
C
   60   UNITS = KUNITS(K)
        TCOST = TCOST + COST(K) * UNITS
        K     = K + 1
        IF (K .LE. 5) GO TO 60
C
        PRINT 8, TCOST
   10   FORMAT (I3, I3, I3, I3, I3)
   90   FORMAT (F5.2, F5.2, F5.2, F5.2, F5.2)
    8   FORMAT (5X, 6HTCOST =, F8.2)
        STOP
        END
```

When subscripted variables are used in a program, their location must be identified by their subscripts. If the entire array is read or printed out, however, the array name only is specified, omitting the subscripts. The READ and PRINT statements related to transmitting the values of the subscripted variables can be written as follows:

```
READ 90, COST
READ 90, KUNITS
PRINT 90, COST
PRINT 90, KUNITS
```

The effect of these are the same as those stated in the above mail order house program. In the case of a one-dimensional array, if only the array name is written in the input/output statements, the elements are transmitted in the usual order, element 1 first, element 2 second, and so on. If a two- or three-dimensional array name is encountered in the READ or PRINT statements without specifying the subscripts, the transmission starts with the element whose subscripts are 1's and continuing so that the first subscript changes most rapidly, the second subscript changes next most rapidly, and the third subscript, if it is a three-dimensional array, changes least rapidly. If the values of the two-dimensional array C shown in Figure 2.8 are stored in the computer memory, the quantities are printed out as follows on only listing C in the PRINT statement:

Printed Value	Element Identification
319.27	$C(1, 1)$
490.85	$C(2, 1)$
357.19	$C(3, 1)$
1000.63	$C(1, 2)$
537.00	$C(2, 2)$
879.73	$C(3, 2)$
1200.79	$C(1, 3)$
215.45	$C(2, 3)$
415.60	$C(3, 3)$
462.78	$C(1, 4)$
207.94	$C(2, 4)$
783.40	$C(3, 4)$

The DO Statements

A set of instructions are repeated in many computational methods many times using different values for one or more variables each time. This process of performing such an iterative or recursive procedure with a computer program is termed *looping*. Consider the following segment of the mail-order house program presented under Data Array:

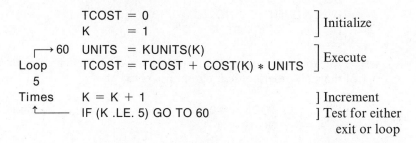

The IF statement has been employed for looping. We can also accomplish the execution of looping by DO statements. The DO statement is one of the most powerful features in the FORTRAN language. The DO statement is an executable statement with the following general form:

DO sn iv = m_1, m_2, m_3

where

sn   is the statement number of the last statement in the loop.

iv   is a nonsubscripted integer variable written without a sign. Its value is initialized at the beginning of the first loop and is incremented at the beginning of each successive loop. It is also called *the index*.

m_1 is the *initial value* of iv

m_2 is the maximum value that iv may assume. It must be greater than or equal to m_1 and it is also called the *test value*.

m_3 is the increment by which sn is increased. If $m_3 = 1$, m_3 may be omitted. It is also called the *increment value*.

m_1, m_2, m_3 are unsigned integer constants or unsigned nonsubscripted integer variables, whose values are defined before the DO statement is executed.

The DO statement operates as follows: The set of instructions after the DO statement through the one with the statement number sn are executed many times. This repetitive process continues as required starting with iv = m_1 and being incremented by m_3 before starting each succeeding loop. If the iv value is less than or equal to m_2, the repeated execution continues. If the iv value is greater than m_2, control is transferred automatically to the statement following the statement with the statement number sn. Examples of DO statements are

```
DO 763 J = 7, 35, 2
DO 5 KOUNT = 1, NODE
DO 60 I = J, KLAST, K
```

The following segment indicates the use of DO statement in the portion of the mail-order house program presented above:

$$
\left.
\begin{array}{l}
\quad\quad\quad \text{TCOST} = 0 \\
\quad\quad\quad \text{DO 19 K} = 1, 5 \\
\text{Loop} \\
\quad 5 \quad\quad\quad \text{UNITS} = \text{KUNITS(K)} \\
\text{Times} \\
\quad\quad 19 \quad \text{TCOST} = \text{TCOST} + \text{COST(K)} * \text{UNITS}
\end{array}
\right]
\begin{array}{l}
\text{Initialize} \\[2em]
\text{Do loop}
\end{array}
$$

The variable TCOST is initialized with zero before encountering the DO statement. The DO statement tells that the set of statements following the DO statements, through statement 19 are repeated 5 times, with K = 1, 2, 3, 4, and 5. In the first iteration K = 1 and UNITS is assigned with the KUNITS (1) value. The TCOST value is the sum of the initialized TCOST value and the cost of the items 1. In the second iteration K = 2. Since 2 is less than 5, the iteration is continued. Now, UNITS equals KUNITS (2). The new TCOST value is the sum of the current TCOST value and the cost of the items 2. Thus, TCOST is the total cost of items 1 and 2. Then, the third iteration (K = 3), the fourth iteration (K = 4), and the fifth iteration are executed in the similar manner. After this, K = 6, and it is greater than the test value (5). Hence, the looping process is terminated and the control is transferred to the statement immediately following statement 19. Note that the value of m_3 does not appear in the DO statement and it indicates that $m_3 = 1$.

The following rules govern the use of DO statements:

1. The first statement following the DO statement must be an executable statement. It must not be a specification instruction such as the FORMAT or DIMENSION statements.
2. The last statement in the DO range should not be a transfer statement. If a transfer statement is encountered as the final instruction in the DO range, a CONTINUE statement may be included to fulfill the rule.

We note in Figure 2.10a that when A is less than 100.0, control is passed to the DO statement. The IF statement acts as the last

```
10   DO 57 I = 1, 20, 2              DO 57 I = 1, 20, 2
     READ 26, A                      READ 26, A
26   FORMAT (F10.2)          26      FORMAT (F10.2)
     IF (A.LT.100.0) GO TO 10        IF (A.LT.100.0) GO TO 57
57   GO TO 20                        GO TO 20
20   A = A − 30.6            57      CONTINUE
       ⋮
                             20      A = A − 30.6
                                       ⋮
```

(*a*) Invalid DO Loop   (*b*) Valid DO Loop

FIGURE 2.10   DO loop termination.

statement in the DO loop. This is not permitted according to rule 2. Also, the final statement in the DO range is a transfer statement. This is also not allowed. The segment of a program shown in Figure 2.10*b* is valid and performs the same calculations.

The *CONTINUE statement* is a "dummy statement." The CONTINUE statement does not alter the sequence of execution. When a CONTINUE instruction is encountered, it produces no action other than continuing to the next statement. It may be placed anywhere in the program.

3. The statements placed within the DO loop must not change the current value of the index variable (iv), as well as those of the integer variables denoting the initial value (m_1), test value (m_2), and increment value (m_3).

4. A DO loop can contain DO loops within its range. This is called the *nesting of DO loops*. All statements of an inner DO loop must be entirely contained within the outer DO loop. It is allowed that the inner DO loop and the outer DO loop may end with the same statement. The set of statements shown in Figure 2.11 could be used to read and store the values of matrix *C* shown in Figure 2.8.

```
     DIMENSION C(3, 4)
     DO 40 I = 1, 3
     DO 57 J = 1, 4
     READ 34, C(I, J)
57   CONTINUE
40   CONTINUE
34   FORMAT (F10.2)
```

FIGURE 2.11   Valid nested DO loops.

Another way of writing these instructions is displayed in Figure 2.12.

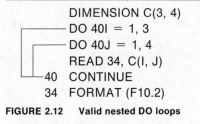

```
         DIMENSION C(3, 4)
         DO 40I = 1, 3
         DO 40J = 1, 4
         READ 34, C(I, J)
      40 CONTINUE
      34 FORMAT (F10.2)
```
FIGURE 2.12   Valid nested DO loops

The DO loops shown in Figure 2.13, however, are invalid because all of the inner loop instructions are not contained within the outer loop.

```
         DIMENSION C(3, 4)
         DO 40I = 1, 3
         DO 57J = 1, 4
         READ 34, C(I, J)
      40 CONTINUE
      57 CONTINUE
      34 FORMAT (F10.2)
```
FIGURE 2.13   Invalid nested DO loops.

There is no specific limitation on the number of DO loops that may be nested. When nested DO loops are encountered, the inner loop is repeated the specified number of times for each iteration of the outer loop.

5. (a) Control may be transferred out of the range of a DO loop from any statement within its range. Figure 2.10*b* illustrates an acceptable transfer of control.

   (b) Control may not be transferred into the range of a DO loop from any statement that lies outside its range. (See Figures 2.14 *a* and *d*.)

   (c) In nested DO loops, control may be transferred from any statement situated within the inner DO loop range to outside the range of the inner loop and within the range of the outer loop. (See Figure 2.14*b*.) Control may be transferred within the range of a DO loop (Figure 2.14*c*).

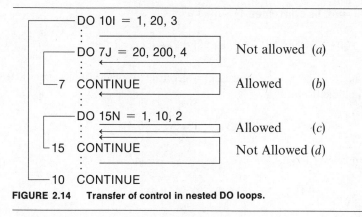

FIGURE 2.14   Transfer of control in nested DO loops.

Input and Output Statements

We have discussed the fundamental features of input and output in the last section. The additional input and output features that follow will simplify the formulation of certain READ and PRINT instructions.

Multiple Specifications   When several variables in consecutive order in the READ and PRINT statements have the same specifications in the related FORMAT statements, the multiple specifications can be used. This repetition of the same specifications can be specified by placing an integer number in front of the specification to indicate the number of times it is to be used.

        READ 79, A, D, C, XVAL
    79 FORMAT (5X, F10.2, F8.1, F8.1, F6.4)

We note that the specification F8.1 is repeated twice in consecutive order. Hence, the FORMAT statement may be written as

    79 FORMAT (5X, F10.2, 2F8.1, F6.4)

Consider the following FORMAT statement:

    700 FORMAT (F8.2, F8.2, 5X, F8.2, F8.2)

This can be conveniently stated as

```
700   FORMAT (2F8.2, 5X, 2F8.2)
```

It is also possible to represent a group of repeated field specifications in an efficient manner.

```
5   FORMAT (I7, 5X, I3, 5X, I3, 5X, F10.4)
```

can be written as

```
5   FORMAT (I7, 2(5X, I3), 5X, F10.4)
```

The term 2(5X,I3) indicates that the group (5X, I3) is repeated twice as 5X, I3, 5X, I3.

Implied Loops   Data arrays are used frequently in the computer programs. The elements of a data array can be transmitted to or from the computer in several ways. They can be handled individually as follows:

```
    PRINT 20, B(1), B(2), B(3), B(4)
20   FORMAT (4F8.2)
```

The same process can be performed by an implied DO loop:

```
    PRINT 20, (B(K), K = 1, 4)
20   FORMAT (4F8.2)
```

This will cause the computer to print out the first four values of the data array B. If the eight values of the array B are to be read starting with the B(11) value and ending with the B(18) value, we could write

```
     READ 200, (B(J), J = 11, 18)
200   FORMAT (8F10.2)
```

A further example of the implied loop is the following:

```
    PRINT 30, ( (D(L, K), K = 1, 2), L = 1, 3)
30   FORMAT (6F5.2)
```

This would print the six values of array D in one line according to FORMAT 30 in the order:

D(1, 1)   D(1, 2)   D(2, 1)   D(2, 2)   D(3, 1)   D(3, 2)

These six values can be printed out in six lines, one value per line, by these instructions:

```
    PRINT 35, ( (D(L, K), K = 1, 2), L = 1, 3)
35  FORMAT (F5.2)
```

The equivalent program segment to execute the output is:

```
    DO 12 L = 1, 3
    DO 12 K = 1, 2
    PRINT 37, D(L, K)
12  CONTINUE
37  FORMAT (F5.2)
```

This will cause the computer to print out the values in the following order:

D(1, 1)
D(1, 2)
D(2, 1)
D(2, 2)
D(3, 1)
D(3, 2)

Note that in each one of the three segments for printing the array D, the innermost loop has the index K and it changes the most rapidly. The outermost loop has the index L and it changes the least rapidly as in nested DO loops.

Control Specifications   For controlling line spacing in printing out the results, the control specifications are used. A slash (/) may be used in the input and output statements. A slash included in the FORMAT statement related to a READ statement indicates that the items following the slash sign are to be read from the subsequent card if the input device is a card reader. For example, consider this statement:

        READ 1006, X, Y, COST
    1006   FORMAT (F5.2, /, F6.3, F10.2)

When this read statement is encountered, X would be read from one punched card, and Y as well as COST would be read from the next card. A slash included in the FORMAT statement related to a PRINT statement indicates that the items following the slash sign are to be transmitted on the next line.

        PRINT 789, I, A, B
    789   FORMAT (I7, F7.2, /, F10.3)

Upon execution of the PRINT instruction, I and A would be printed out on a line and B on the next line. Any number of slashes may be used as appropriate. If K slashes are written at the beginning or end of a FORMAT instruction, K cards or lines are skipped depending on whether the FORMAT statement is associated with a READ or PRINT instruction. If K slashes are written in the middle of a FORMAT, (K-1) cards or lines are skipped in transmitting the data.

FORTRAN *Statement*	*Result*
1. READ 10, A, B, C, I 10 FORMAT (///, 3F10.2, I5)	Skips the first three input cards and reads the four values given on the fourth card.
2. READ 7, I, K, Q 7 FORMAT (2I10, ///, F10.3)	Reads the values of I and K from the first card, skips the following two cards and reads the value of Q from the fourth card.
3. PRINT 9, X, Y, Z 9 FORMAT (2F10.3, /, F5.2)	Prints the values of X and Y on the first line, skips to the second line, and prints the value of Z in the first five spaces of the second line.
4. PRINT 6, A, B, L, N 6 FORMAT (//, 2F5.1, 2I7, /)	Skips the first two lines, prints the values of A, B, L, and N on the third line, and skips the fourth line

The vertical spacing is also controlled by the *carriage control character*. The carriage control character is the character that is contained in the first position of the output. It is not printed in the output. The printing device interprets the first character as follows:

First Character	*Vertical Spacing Before Printing*
+	No advance
Blank	One line
0	Two lines
1	First line of next page

The carriage control characters are transmitted to the computer easily by the one-character Hollerith specifications. These are illustrated by the incomplete FORMAT instructions:

FORMAT (1H + , ...)   indicates no advance
FORMAT (1H, ...)   indicates one line advance
FORMAT (1H0, ...)   indicates two lines advance
FORMAT (1H1, ...)   indicates skipping the rest of the page and
                                printing in the first line of the next page

SUBPROGRAMS

A subprogram is a group of computer instructions that perform a specific task as a part of the main program. It is intended to be used as a part of the main program rather than by itself. When a program segment is used several times in different places in a program, we can write a subprogram only once and thus save time and effort. Whenever a subprogram is called in the main program or another subprogram, the associated group of instructions are executed and the calculated results are transmitted to the main program or the related subprogram from where it has been branced earlier. The subprograms simplify debugging since they can be tested individually. Long and complex computer programs are composed of a series of subprograms. We can use this technique to assign several programmers to one problem and, thus, reduce the time needed to complete the task. There are four forms of subprograms:

1. FORTRAN-Supplied functions
2. One-statement functions
3. Function subprograms
4. Subroutine subprograms

FORTRAN-Supplied Functions

Several standard computations are encountered often in solving problems. The function routines supplied by the computer manufacturer are used to simplify the writing of programs.

A function represents the mathematical relationship by which the value of one variable, *dependent variable*, is established for the given values of one or more *independent variables*. The value of the dependent variable is the function value. The independent variables are called the *arguments*. A programmer does not have to write the complete function instructions. The programmer is only required to use the appropriate function name in the instructions followed by a set of arguments in parentheses.

The set of supplied functions available in a computer system depends on the needs of the people utilizing the system and also on the computer memory space. In almost every computer installation, the following built-in functions are provided.

Function	Description	Algebraic Formula
SIN(X)	Sine of X radians	$\sin x$
COS(X)	Cosine of X radians	$\cos x$
ATAN(X)	Arc tangent of X radians	$\arctan x$
EXP(X)	Exponentiate e to the power of X	e^x
ALOG(X)	Natural logarithm of X (log to base e)	$\log_e x$
ALOG10(X)	Logarithm of X to base 10	$\log_{10} x$
ABS(X)	Absolute value of X	$|x|$
SQRT(X)	Square root of X	$\sqrt{x}$

An argument is an arithmetic expression. In formulating the arguments, we find that the rules governing the arithmetic expression are applicable. The function is evaluated using the current values of the variables specified in the argument. The evaluation gives a single value and is transferred to the location occupied by the function in the FORTRAN statement. Here are some examples of FORTRAN supplied functions:

```
D = SQRT(A)
Z = SQRT(P + 77.3)
J = Y + ALOG(SQRT(B)) + SIN(0.5)
GAMMA = ABS(E/F) + 100.0
```

One-Statement Functions

The one-statement function contains, as the name implies, a single statement to define the function and it is placed in the main program. This single statement defines the mathematical relationship between the independent variables (arguments) and the dependent variable (function name). The defining statement is of the form

$f(Arg_1, Arg_2, \ldots) = exp$

where

f is the function name and it is formed in the same way as a variable name. If the first character of the variable name is any one of the letters I through N, the function value is an integer quantity. Otherwise, it is a floating-point quantity.

$Arg_1, Arg_2, \ldots$ are the dummy arguments used in the function. They are any set of variables chosen by the programmer. They indicate how many arguments are required when the subprogram is to be executed.

exp is any arithmetic expression and it must include all of the arguments specified in the defining statement ($Arg_1, Arg_2, \ldots$). Subscripted variables are not allowed. The expression indicates the operation to be executed when this function is called by the main program.

This defining statement is not executed. To utilize the subprogram, the function name together with the actual variables are written in parentheses in the appropriate places in the program as needed. The actual variables and the dummy arguments must correspond in number, mode, and sequence. On encountering the function name with the actual variables, the function is executed as specified in the defining statement using the current values of the actual variables. The resulting value is transferred into the related statement as the value of the function. Here are some examples of one-statement functions:

(a) Defining statement:

DELTA(X, Y, Z) = X ** 2 + Y ** 3 + Z

Using the function:

A = ALPHA + DELTA(P, F, C)

These statements will cause the computer to calculate A as follows:

A = ALPHA + (P ** 2 + F ** 3 + C)

(b) Defining statement:

K706(I, J) = I * 3 + J

Using the function:

BETA = K706(J, L)/2

These statements will cause the computer to calculate BETA as follows:

BETA = (J * 3 + L)/2

The two requirements that there must be only one defining statement and also only one resulting value must be returned to the main program make it very restrictive. The function subprogram eliminates the first restriction.

Function Subprograms

A subprogram is written as a complete program. It can be tested separately, but it must be executed with the main program in solving problems. A function subprogram is always a subsidiary program to another program, termed the *calling program*. When the function subprogram is called, control transfers to the subprogram, executes the subprogram, and returns a single value to the calling program. The function subprogram may contain several statements. It has this appearance:

FUNCTION Name (Arg$_1$, Arg$_2$, . . .)
$\vdots$
Name = expression
$\vdots$
RETURN
END

where

FUNCTION Name (Arg_1, Arg_2, . . .) is the function statement (first statement in the function subprogram).

Name   is a variable name and it is used as the unique name to identify the subprogram. The name must be used at least once as a nonsubscripted variable name on the left-hand side of an arithmetic statement in the subprogram. On encountering the function name in the calling program, the function subprogram is executed, the current value of the subprogram is assigned to the variable name, and control is returned back to the calling program. The first character of the name indicates the mode of the subprogram value.

Arg_1, Arg_2, . . .   are the nonsubscripted variables used as dummy arguments in the subprogram.

The function subprograms are called in the main program in the same way as the one-statement functions. It is also required that the actual variables and the dummy arguments must agree in number, mode, and sequence.

Here are the specifications of a subprogram:

1. A subprogram can call another subprogram. When it happens, the calling subprogram can be considered as the main program in applying the rules for the calling subprogram.
2. When any one of the dummy arguments is a subscripted variable, only its name is stated in the first statement of the subprogram. The related DIMENSION statements must immediately follow the function statement.
3. When data arrays, if any, are used in the arguments, they must have the same size in the main program and the subprogram.

4. A RETURN statement is used to indicate the logical terminal point of the subprogram. When a RETURN statement is encountered, the execution of the subprogram is terminated, control is transferred back to the calling program, and the FUNCTION value is returned into the calling statement. Any number of RETURN statements may be used in a subprogram.
5. The END statement is placed as the last physical statement in the subprogram, and it signals the computer that the particular subprogram has been compiled completely.
6. A subprogram is executed separately from the main program. Therefore, the same statement numbers and variable names may be used in the subprogram and the main program. For example, there may be an instruction with the statement number 506 in a main program and also an instruction with the same statement number 506 in the associated subprogram.

Consider the following example:

Main Program

```
DIMENSION SALES(100), UNITS(100), X(100)
    ⋮
KSIZE = 100
TSALES = 75.0 + SUM(SALES, KSIZE)

JSIZE = 65
UOLD = 15
TUNITS = UOLD + SUM(UNITS, JSIZE)
    ⋮
STOP
END
```

Function subprogram

```
    FUNCTION SUM(X, K)
    DIMENSION X(100)
    SUM = 0
    DO 70 J = 1, K
    SUM = SUM + X(J)
70  CONTINUE
    RETURN
    END
```

The main program calls the subprogram to compute the values of TSALES and TUNITS. When the subprogram SUM is called for the first time, the data array SALES is transferred to the array X in the subprogram, and KSIZE replaces K in the subprogram. The value of SUM is returned to the main program, and 75.0 is added to SUM giving the value of TSALES. When the subprogram is called for the second time, the data array UNITS is transferred to the array X in the subprogram and JSIZE replaces K in the subprogram. The resulting value of SUM is returned to the main program. UOLD and SUM are added, and the resulting value is assigned to TUNITS. The function subprogram can only return a single value to the calling program. The subroutine subprogram removes this serious restriction.

Subroutine Subprograms

A subroutine subprogram, like the function subprogram, is written as a separate program. It usually contains a series of statements. It can return, in contrast to the function subprogram, any number of values to the calling program. The general form of the subprogram is as follows:

SUBROUTINE Name (Arg$_1$, Arg$_2$, . . .)
:
RETURN
END

where

Name   is the name of the subprogram supplied by the pro-
                  grammer. The rules governing the formulation of
                  variable names are applicable in structuring the
                  subroutine name. The name must be unique. The
                  first character of the name does not imply the mode
                  of the data because the subroutine name is not em-
                  ployed for returning a single value to the calling
                  program.

Arg$_1$, Arg$_2$, . . . are the dummy arguments used in defining the sub-
                  routine. The arguments may be nonsubscripted vari-
                  ables, array names, or dummy function or subroutine
                  names.

A subroutine is called in the main program by a CALL statement and it is of the form

CALL Name (Arg$_1$, Arg$_2$, . . .)

where

Name refers to the subroutine name
Arg$_1$, Arg$_2$, . . . are the arguments to be used in the subroutine

The arguments specified in the SUBROUTINE statement and the CALL statement must agree in number, mode, and sequence. The specifications of a subprogram presented under "Function subprograms" section are also pertinent to the formulation of subroutine subprograms. The following is an example of the use of the SUBROUTINE subprogram:

Main program

```
DIMENSION PTS(10, 5), X(10, 5), Y(10)
DIMENSION PTSROW(10)
    :
M = 7
N = 5
CALL SUM (X, M, Y, N)
    :
STOP
END
```

Subprogram

```
      SUBROUTINE SUM(PTS, J, PTSROW, L)
      DIMENSION PTS(10, 5), PTSROW(10)
      DO 89 K = 1, J
      TOTAL = 0
      DO 66 I = 1, L
      TOTAL = TOTAL + PTS(K, I)
 66   CONTINUE
 89   PTSROW(K) = TOTAL
      RETURN
      END
```

The CALL statement in the main program causes the computer to store the values of the X array, M, and N in the PTS array, J, and L, respectively. The subprogram calculates the row totals for the first seven rows of the matrix X and stores them in the one-dimensional array PTSROW. This is transferred to the corresponding argument, namely, to array Y.

PROBLEMS

2.1 Write a computer program to calculate and print out the values of a, b, c, d, e, and j, where

$$a = x + y$$
$$b = x - y$$
$$c = x \cdot y$$
$$d = x/y$$
$$e = x^y$$
$$j = x/y$$

(Note j denotes truncated value.)

Assume that $x = 3$ and $y = 4$. The values of x and y are given to the computer by the assignment statements

$$X = 3$$
$$Y = 4$$

2.2 Prepare a computer program that will cause the computer to sum the first 20 positive integers beginning with 1: $1 + 2 + 3 + \cdots + 20$, and print out the result.

2.3 Temperatures can be recorded in degrees of Celsius (Centigrade) or degrees of Fahrenheit. Write a computer program that will generate a table showing the degree Celsius values (one through 100 degrees in steps of 1 degree, namely 1, 2, 3, ... etc.) and the corresponding Fahrenheit values expressed to 3 decimal places. The formula for converting Celsius to Fahrenheit is

$$F = \frac{9C + 160}{5}$$

where

$C =$ temperature in degrees of Celsius

$F =$ temperature in degrees of Fahrenheit

2.4 A small company is planning to manufacture steel boxes. The owner feels that any number of boxes up to 25 boxes can be sold. The estimated values are

$$\text{Fixed cost } (F) \qquad = \$30$$
$$\text{Variable cost per box } (V) = \$\ 5$$
$$\text{Revenue per box } (R) \qquad = \$\ 8$$

The relationship between the number of boxes produced (Q), F, V, R and the total profit (TP) can be stated by the equations

$$TC = F + V \cdot Q$$
$$TR = R \cdot Q$$
$$TP = TR - TC$$

where

TC = total cost for manufacturing Q boxes

TR = total revenue earned by sellqng Q boxes

TP = total profit

When TP is a negative value, it indicates that the total effect is the corresponding loss. Write a computer program that will cause the computer to calculate and print out the values of Q, TC, TR, and TP for the 25 possibilities, starting with $Q = 1$, through $Q = 25$ boxes.

2.5 The unit price of an item is high when the order size is small. The unit price drops when a larger order is placed. Here is the price schedule:

Order Size (Units)	Unit Price
1 to 999	$10.00
1000 to 4999	$ 9.50
5000 and above	$ 9.05

Write a computer program that will enable the computer to read the order size from a card, calculate the total cost of the order, and print out the order size, selected unit price, and the total cost of the order. Prepare a data deck for an order size of 1100 units.

2.6 The income tax is calculated for the employees of a mail-order house according to the following table

Annual Income	Tax
Less than $2500	No tax
$2500 or over but less than $6000	5% of the amount over $2500
$6000 or over	$175 plus 8% of the amount over $6000

Write a computer program that will cause the computer to read the annual income of an employee, compute the tax, and print out the input value and the tax. The computer must perform these operations for four employees. Design a data deck for the annual incomes of $7200.00, $2406.75, $3000.00, and $5000.00.

2.7 A set of 10 floating-point values are given. Write a program that will enable the computer to read the values, establish the smallest value among them and print out the input data as well as the solution. Assume that the given values are

$$107.1, 200.9, -109.6, 83.4, 0.8, 79.8, 1572.3, -94.9, 8.0, 15.9$$

2.8 Write a computer program to calculate the total value for a set of 10 integer values. The 10 values are to be punched in one card. The values are

$$55 \quad 32 \quad -46 \quad 35 \quad 35 \quad 74 \quad -33 \quad 63 \quad 69 \quad 71$$

Prepare the input data deck for the given values.

2.9 John Doe goes to livestock market with $200 to buy 100 heads of stock. The following are the prices: calves, $30 each; pigs, $2 each; chickens, $0.50 each. He buys 100 heads for his $200. Write a computer program that will determine and print out for each one of the possibilities the number of each he can buy.

2.10 Write a computer program that will cause the computer to read the values of A, B, and C of a quadratic equation with the general form $AX^2 + BX + C = 0$; calculate and print out its roots considering all of the three possibilities:

When $A = 0$, the equation becomes a linear relationship of the form $BX + C = 0$, thus, $X = -C/B$

When $(B^2 - 4AC)$ is negative, print "THE ROOTS ARE IMAGINARY." If the above two situations are not encountered, the two roots $(X_1$ and $X_2)$ are determined applying the expressions:

$$X_1 = \frac{-B + \sqrt{B^2 - 4AC}}{2A}$$

$$X_2 = \frac{-B - \sqrt{B^2 - 4AC}}{2A}$$

The program should be able to solve the four sets of data automatically.

A	B	C
2	6	5
1	-7	12
0	8	20
8	15	3

2.11 Write a computer program that will read a one-dimensional array of 10 floating-point values from a single card, arrange these values in ascending order, and print out the input values in the given order and the end result of sorted numbers in ascending order. Assume that the given data array is

59.2 99.3 22.8 78.4 6.0 70.5 24.0 −29.7 43.9 31.9

2.12 Consider Problem 2.11. Modify the program so that the values will be arranged and printed in descending order instead of in ascending order. Use the same data array given in Problem 2.11.

─────────────── SELECTED REFERENCES ───────────────

Davis, G. B., *Introduction to Electronic Computers*, McGraw-Hill, New York, Second Edition, 1971.

Emerick, P. L., and J. W. Wilkinson, *Computer Programming for Business and Social Science*, Richard D. Irwin, Homewood, Ill. 1970.

Ford, D. H., *Basic FORTRAN IV Programming*, Richard D. Irwin, Homewood, Ill., 1971.

Lee, R. M., *A Short Course in FORTRAN IV Programming*, McGraw-Hill, New York, 1967.

McCameron, F. A., *FORTRAN Logic and Programming*, Richard D. Irwin, Homewood, Ill., 1968.

McCracken, D. D., *A Guide to FORTRAN IV Programming*, Wiley, New York, Second Edition, 1972.

Mullish, H., *Modern Programming: FORTRAN IV*, Blaisdell, Waltham, Mass. 1968.

Murrill, P. W., and C. L. Smith, *FORTRAN IV Programming for Engineers and Scientists*, International Textbook Scranton, Penn. 1968.

Nydegger, A. C., *An Introduction to Computer Programming with an emphasis on FORTRAN IV*, Addison-Wesley, Reading, Mass., 1968.

Raun, D. L., *An Introduction to FORTRAN Computer Programming for Business Analysis*, Dickenson, Belmont, Calif., 1968.

Rule, W. P., *FORTRAN IV Programming*, Weber & Schmidt, Boston, Mass., 1968.

Schick, W., and C. J. Merz, Jr., *FORTRAN for Engineering*, McGraw-Hill, New York, 1972.

Schmidt, R. N., and W. E. Meyers, *Introduction to Computer Science and Data Processing*, Holt, Rinehart and Winston, New York, Second Edition, 1970.

Smith, R. E., *The Bases of FORTRAN*, Control Data Corporation, Minneapolis, Minn., 1967.

Vazsonyi, A., *Introduction to Electronic Data Processing*, Richard D. Irwin, Homewood, Ill., 1973.

CHAPTER 3

forecasting for planning

A FORECAST is an estimate of the level of activity in some period of time in the future. Forecasting is the process of determining the future estimate for the given activity. Forecasting plays a vital role in establishing plans for the future. Forecasts are the critical inputs for diverse managerial decisions, and these decision areas include the following:

Location and expansion of plants
Product planning
Purchase of raw materials
Weekly production schedules
Hiring and training sales force
Capital and cash budgets
Scheduling items into production

A good forecast is characterized by a minimal deviation from the actual activity level in the planned period. If we could obtain better forecasts, we could acquire the required resources such as men, machines, and money, allocate the resources adequately, and operate the system efficiently. Forecasting, however, is guess work. As Reinfield (1959) points out, "... the makeup of a good forecast is rather like a good Martini: One part experience; two parts science, with a good bit of judgment tossed in it." The forecasts obtained through the development and proper application of mathematical techniques provide more reliable estimates than intuitive forecasts. In the course of determining a forecast, the appropriate methodology is selected, and the relevant information is gathered. It is necessary to know the planning period over which the forecast will be made. The planning period can be classified into the three groups:

Long-range plans: For plant location, plant expansion, product development, and capital planning, time spans of five years or more are considered.

Intermediate-range plans: For setting production and inventory budgets, sales planning, production planning of seasonal products and manpower, equipment, money and materials planning, time horizons of one to two years are considered.

Short-range plans: The order sizes and the time of ordering from outside suppliers, lot sizes and the time of production, leveling work force and operating on overtime are determined on daily or weekly basis or in advance of three to six weeks.

APPROACHES FOR FORECASTING

Forecasts are obtained in a number of ways. They can be determined based on:

Estimates of salesmen and retailers
Customer opinions
Consumer opinions
Executive opinions
Marketing trials
Market research
Economic data such as the gross national product, personal and farm income, and retail and wholesale indices.
Industry data such as the annual production and estimated dollar sales of various industries.
Time series analysis
A combination of approaches

The demand of a particular item or activity for the past periods at regular intervals of time is collected in the time-series analysis. It may be a series of weekly sales, a series of monthly freight car loadings, and so on. The time series data is examined for observable regularities or patterns, and the observed patterns are assumed to prevail in the future. Based upon this assumption, the forecast for the planning period is determined.

Time Series Analysis

In time series analysis we look back at the past data, analyze them, and then project or extrapolate the historical pattern into the planning period. The components of observed demand are the average demand, trend, cyclical, and seasonal elements, and random variation. The average demand is the central tendency of the time series at a given time. Trend element is the regular movement of a series over a long period of time. Seasonal variations are regularly repeating patterns in the time series. They repeat in most situations on a yearly basis. The up and down movements in demand that differ from seasonal fluctuations and extend over longer periods of time are said to be cyclic variations. The cyclic fluctuations may be due to recurring periods of prosperity, recession, political elections, or war. Cyclic variations are more complex to determine. Since they are beyond the scope of our book, we will not discuss them. The random variations are the fluctuations resulting from natural chances.

The time series of product P5 is shown in Figure 3.1. Note that basically the actual demand increases with time, and the time series exhibits a growth trend. The monthly demand for product P6 is shown in Figure 3.2. The higher demand in each year occurs in May, June, and July, and lower demand is encountered in January, February, and December. The pattern repeats on a yearly basis. We conclude that seasonal vairations and random variations prevail in the time series. Figure 3.3 displays the demand of product P7. No pattern seems to exist in the series. The movements due to trend and seasonal components are not prevailing. The

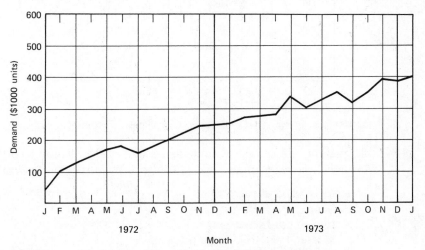

FIGURE 3.1   Time series with growth trend and random variations.

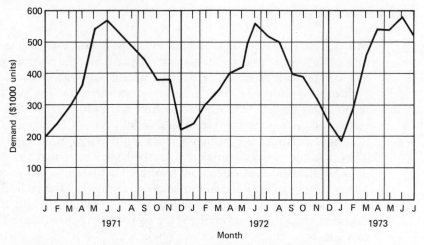

FIGURE 3.2 Time series with seasonal and random variations.

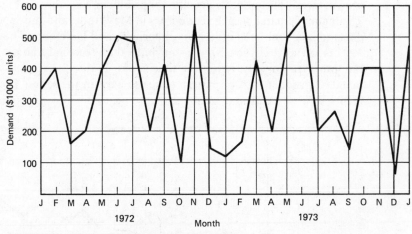

FIGURE 3.3 Time series with random fluctuations.

fluctuations are caused by random variations. We shall discuss the following effective techniques and illustrate them with numerical examples:

1. Least squares regression
2. Simple moving average
3. Exponential smoothing
4. Seasonal variations with trend effect

LEAST SQUARES REGRESSION

When the past data can be described best by a straight line, the method of least squares regression line can be employed. Here is an example that illustrates the technique.

Example 3.1

The Miller Company has examined the records of its hand tool division for the past five years and it has gathered the following sales data:

Year	Sales ($1000)
1	40
2	50
3	55
4	72
5	81

The company wishes to determine the sales forecast for next year (year 6). Figure 3.4 shows the data plotted on a graph. We observe that the past data values appear to fall on a straight line. The method of least squares regression line provides a mathematical approach to determine the line of best fit. The general equation of any straight line can be represented as follows:

$$Y = a + bX \tag{3.1}$$

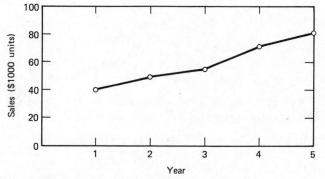

FIGURE 3.4 **Sales data for regression analysis.**

where

Y = value of the dependent variable. This denotes the forecast at time X

X = value of the independent variable; it specifies the time period

a = Y intercept of the regression line

b = amount of change in Y for an increase of one unit in the value of X

We can determine the line of best fit once the values of the two unknowns, a and b, are established. The formulas to find the two unknowns are:

$$a = \frac{\sum X^2 \sum Y - \sum X \sum XY}{n \sum X^2 - (\sum X)^2} \tag{3.2}$$

$$b = \frac{n \sum XY - \sum X \sum Y}{n \sum X^2 - (\sum X)^2} \tag{3.3}$$

where

$\sum$ = summation symbol

n = number of paired data points given

Thus, for determining the equation for the regression line, the values of $\sum X$, $\sum Y$, $\sum X^2$, and $\sum XY$ are calculated and substituted in expressions 3.2 and 3.3.

Year (X)	Sales (Y)	X²	XY
1	40	1	40
2	50	4	100
3	55	9	165
4	72	16	288
5	81	25	405
$\sum = 15$	298	55	998

If we substitute the total values and $n(=5)$ in formulas 3.2 and 3.3, we obtain values for a and b.

$$a = \frac{(55)(298) - (15)(998)}{5(55) - (15)^2} = 28.4$$

$$b = \frac{5(998) - 15(298)}{5(55) - (15)^2} = 10.4$$

Thus, the equation of the regression line from (3.1) becomes

$$Y = 28.4 + 10.4X \qquad (3.4)$$

The sales forecasts (Y_f) for the past years computed by employing the regression line are the points on the line for the corresponding X values. The calculations of the forecast are performed by applying equation 3.4:

X	Calculation $(Y_f = 28.4 + 10.4X)$	Sales Forecast (Y_f)
1	$Y_f = 28.4 + 10.4(1)$	38.80
2	$Y_f = 28.4 + 10.4(2)$	49.20
3	$Y_f = 28.4 + 10.4(3)$	59.60
4	$Y_f = 28.4 + 10.4(4)$	70.00
5	$Y_f = 28.4 + 10.4(5)$	80.40

Likewise, the forecast for next year is established when $X = 6$, $Y_6 = 28.4 + 10.4(6) = \underline{90.80.}$ The regression line and the sales forecast for next year are displayed in Figure 3.5. The least squares line has the two

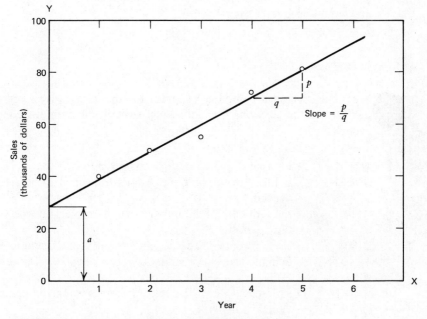

FIGURE 3.5   Yearly sales and the resulting regression line.

characteristics:

1. The sum of deviations between the actual data points (Y_a) and the points (Y_f) on the line of best fit for the corresponding X values is zero.

$$\sum(Y_a - Y_f) = 0 \qquad (3.5)$$

2. The sum of squares of the deviations is the minimum.

$$\sum(Y_a - Y_f)^2 = \text{minimum} \qquad (3.6)$$

X	Y_a	Y_f	$(Y_a - Y_f)$	$(Y_a - Y_f)^2$
1	40	38.80	1.2	1.44
2	50	49.20	0.8	0.64
3	55	59.60	−4.6	21.16
4	72	70.00	2.0	4.00
5	81	80.40	0.6	0.36
			0	27.60

Note that the sum of deviations is zero. The sum of squared deviations is 27.60. This signifies that the sum of squared deviations for any other line would be greater than 27.60.

A Computer Model

We shall develop a computer program for the least squares regression method in this section. The following variables are employed in the program:

NYRS = number of paired data points given (n)

JNEED = number of future time periods for which forecasts are needed

X(K) = the Kth data of the independent variable. It denotes the time period

Y(K) = the Kth data of the dependent variable. It denotes the demand (sales) occurred in X(K)th period (Y_a for period K)

XX(K) = squared value of the Kth element among the given X values

XY(K) = value obtained by multiplying X(K) and Y(K)

FCAST(K) = forecast for the Kth period (Y_f for period K)
DELTA(K) = the deviation value between the actual data and the forecast for the Kth period ($Y_a - Y_f$)
SUMX   = sum of X values ($\sum X$)
SUMY   = sum of Y values ($\sum Y$)
SUMXX = sum of X^2 values ($\sum X^2$)
SUMXY = sum of XY values ($\sum XY$)
A   = Y intercept of the regression line (a)
B   = slope of the regression line (b)
SDELTA = sum of deviations [$\sum (Y_a - Y_f)$]
SFCAST = sum of forecast values ($\sum Y_f$)
LSLF   = this subprogram calculates the values of A, B, and SDELTA

The variables X( ), Y( ), XX( ), XY( ), FCAST( ), and DELTA( ) are one-dimensional arrays of size NYRS elements. The flowcharts for the main program and subprogram LSLF are displayed in Figures 3.6 and 3.7, respectively.

The input data are the values of NYRS, JNEED, and NYRS sets of X and Y. These values might be read into the computer and printed out by the instructions:

```
      READ 15, NYRS, JNEED
      PRINT 20, NYRS, JNEED
      DO 500 K = 1, NYRS
      READ 25, X(K), Y(K)
      PRINT 30, X(K), Y(K)
  500 CONTINUE
   15 FORMAT ( )
   20 FORMAT ( )
   25 FORMAT ( )
   30 FORMAT ( )
```

Next, the values of A, B, and SDELTA are calculated and printed out. These operations can be performed by including the statements:

```
      CALL LSLF (X, Y, NYRS, XX, XY, FCAST, DELTA,
    1 SUMX, SUMY, SUMXX, SUMXY, SDELTA, SFCAST, A, B)
      PRINT 35, A, B
   35 FORMAT ( )
```

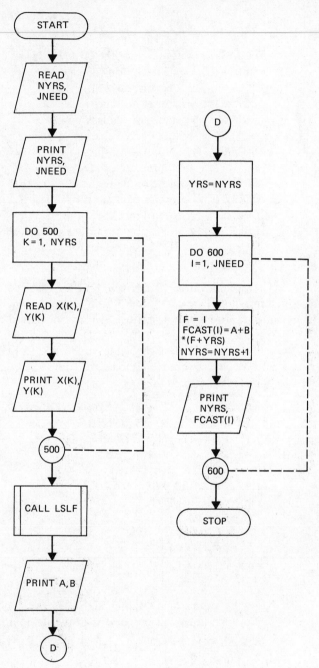

FIGURE 3.6   Flowchart for the main program of the least squares regression technique.

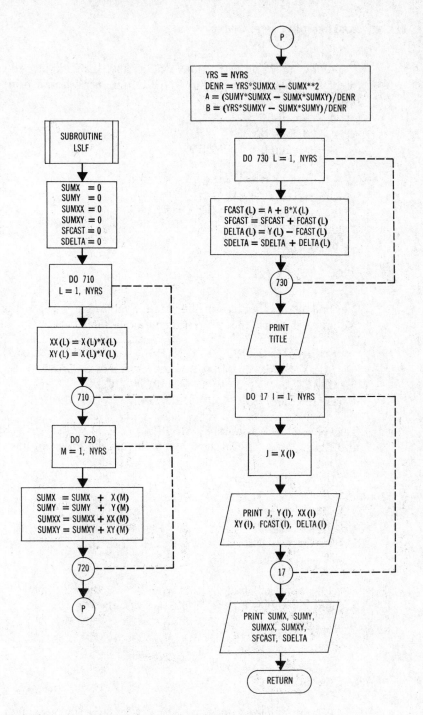

FIGURE 3.7   Flowchart for SUBROUTINE LSLF.

The following set of instructions would cause the computer to calculate forecasts for each of the future JNEED time periods and print out the results:

```
      YRS = NYRS
      DO 600 I = 1, JNEED
      F = I
      FCAST(I) = A + B * (F + YRS)
      NYRS = NYRS + 1
      PRINT 40, NYRS, FCAST(I)
  600 CONTINUE
   40 FORMAT ( )
      STOP
      END
```

We shall discuss the steps included in the subprogram LSLF (Figure 3.7). We might initialize the subroutine by these statements:

```
      SUBROUTINE LSLF (X, Y, NYRS, XX, XY, FCAST, DELTA,
    1 SUMX, SUMY, SUMXX, SUMXY, SDELTA, SFCAST, A, B)
      DIMENSION X(10), Y(10), XX(10), XY(10)
      DIMENSION FCAST(10), DELTA(10)
```

The values of the variables SUMX, SUMY, SUMXX, SUMXY, SFCAST, and SDELTA are set at zero by employing the following instructions:

```
      SUMX = 0
      SUMY = 0
      SUMXX = 0
      SUMXY = 0
      SFCAST = 0
      SDELTA = 0
```

The following set of statements would cause the computer to calculate XX and XY values for the NYRS sets of data:

```
      DO 710 L = 1, NYRS
      XX(L) = X(L) * X(L)
      XY(L) = X(L) * Y(L)
  710 CONTINUE
```

For determining the values of $\sum X$, $\sum Y$, $\sum X^2$, and $\sum XY$, we might employ these statements:

```
      DO 720 M = 1, NYRS
      SUMX = SUMX + X(M)
      SUMY = SUMY + Y(M)
      SUMXX = SUMXX + XX(M)
      SUMXY = SUMXY + XY(M)
720   CONTINUE
```

The values of A and B are established by utilizing the instructions below:

```
YRS = NYRS
DENR = YRS * SUMXX − SUMX ** 2
A = (SUMY * SUMXX − SUMX * SUMXY)/DENR
B = (YRS * SUMXY − SUMX * SUMY)/DENR
```

Next, the data arrays FCAST and DELTA are determined. The values of SFCAST and SDELTA are also calculated using FCAST and DELTA. These operations are performed by the following statements:

```
      DO 730 L = 1, NYRS
      FCAST(L) = A + B * X(L)
      SFCAST = SFCAST + FCAST(L)
      DELTA(L) = Y(L) − FCAST(L)
      SDELTA = SDELTA + DELTA(L)
730   CONTINUE
```

Subsequently, the computed values are printed out, and control is transferred back to the main program.

```
      DO 17 I − 1, NYRS
      J = X(I)
      PRINT 300, J, Y(I), XX(I), XY(I), FCAST(I),
    1 DELTA(I)
  17  CONTINUE
      PRINT 400, SUMX, SUMY, SUMXX, SUMXY, SFCAST,
    1 SDELTA
300   FORMAT ( )
400   FORMAT ( )
      RETURN
      END
```

A complete computer program for the least squares regression method composed of our previously developed program segments is shown in Figure 3.8. This program can be employed to solve problems up to 10 paired data points ($n \leq 10$). For solving Example 3.1 by the computer

```
      DIMENSION X(10),Y(10),XX(10),XY(10)
      DIMENSION FCAST(10),DELTA(10)
C  READ AND PRINT OUT INPUT DATA.
      READ 15, NYRS,JNEED
      PRINT 20, NYRS,JNEED
C
      DO 500 K=1,NYRS
      READ 25, X(K),Y(K)
      PRINT 30, X(K),Y(K)
  500 CONTINUE
C
      CALL LSLF(X,Y,NYRS,XX,XY,FCAST,DELTA,
     1SUMX,SUMY,SUMXX,SUMXY,SDELTA,SFCAST,A,B)
      PRINT 35, A,B
      YRS=NYRS
C
C  CALCULATE AND PRINT OUT THE FORECAST.
C
      DO 600 I=1,JNEED
      F=I
      FCAST(I)=A+B*(F+YRS)
      NYRS=NYRS+1
      PRINT 40, NYRS,FCAST(I)
  600 CONTINUE
C
   15 FORMAT(2I1)
   20 FORMAT(1H1,1X,2I1)
   25 FORMAT(2F5.2)
   30 FORMAT(1X,2(F5.2,1X))
   35 FORMAT(11X,17HTHE VALUE OF A IS,
     1F6.2,1X,8HAND B IS, F6.2,//)
   40 FORMAT(11X,27HTHE SALES FORECAST FOR YEAR,
     1I3,1X,4HIS  ,F7.2,//)
      STOP
      END



      SUBROUTINE LSLF(X,Y,NYRS,XX,XY,FCAST,DELTA,
     1SUMX,SUMY,SUMXX,SUMXY,SDELTA,SFCAST,A,B)
C
C  THIS IS THE LEAST SQUARES LINE
C  FITTING.
C
      DIMENSION X(10),Y(10),XX(10),XY(10)
      DIMENSION FCAST(10),DELTA(10)
```

FIGURE 3.8 Computer program for the least squares regression method (illustrated for Example 3.1).

```
C
C   INITIALIZE SUM VALUE
        SUMX=0
        SUMY=0
        SUMXX=0
        SUMXY=0
        SFCAST=0
        SDELTA=0
C
C   FOR EACH YEAR, PERFORM THE CALCULATIONS.
C
        DO 710 L=1,NYRS
        XX(L)=X(L)*X(L)
        XY(L)=X(L)*Y(L)
  710 CONTINUE
C
        DO 720 M=1,NYRS
        SUMX=SUMX+X(M)
        SUMY=SUMY+Y(M)
        SUMXX=SUMXX+XX(M)
        SUMXY=SUMXY+XY(M)
  720 CONTINUE
C
        YRS=NYRS
        DENR=YRS*SUMXX-SUMX**2
        A=(SUMY*SUMXX-SUMX*SUMXY)/DENR
        B=(YRS*SUMXY-SUMX*SUMY)/DENR
C
        DO 730 L=1,NYRS
        FCAST(L)=A+B*X(L)
        SFCAST=SFCAST+FCAST(L)
        DELTA(L)=Y(L)-FCAST(L)
        SDELTA=SDELTA+DELTA(L)
  730 CONTINUE
C
        PRINT 200
C
        DO 17 I=1,NYRS
        J=X(I)
        PRINT 300, J,Y(I),XX(I),XY(I),FCAST(I),
       1DELTA(I)
   17 CONTINUE
C
        PRINT 400, SUMX,SUMY,SUMXX,SUMXY,SFCAST,
       1SDELTA
C
  200 FORMAT(//,6X,4HYEAR,5X,5HSALES,7X,2HXX,
       17X,2HXY,6X,8HFORECAST,4X,5HDELTA,/)
  300 FORMAT(I9,1X,5F10.2)
  400 FORMAT(/,6F10.2,//)
        RETURN
        END
```

FIGURE 3.8 (*Continued*)

program, the input data is prepared according to FORMAT statements 15 and 25 (see Figure 3.8). The data deck has the following appearance:

```
5 I

I·0040·00

2·0050·00

3·0055·00

4·0072·00

5·0081·00
```

The resulting computer printout is shown in Figure 3.9.

```
51
1.00 40.00
2.00 50.00
3.00 55.00
4.00 72.00
5.00 81.00
```

YEAR	SALES	XX	XY	FORECAST	DELTA
1	40.00	1.00	40.00	38.80	1.20
2	50.00	4.00	100.00	49.20	.80
3	55.00	9.00	165.00	59.60	-4.60
4	72.00	16.00	288.00	70.00	2.00
5	81.00	25.00	405.00	80.40	.60
15.00	298.00	55.00	998.00	298.00	.00

THE VALUE OF A IS 28.40 AND B IS 10.40

THE SALES FORECAST FOR YEAR 6 IS 90.80

FIGURE 3.9 Computer printout for the Miller Company forecasting problem.

SIMPLE MOVING AVERAGE

In the simple moving average method, the demand values that have occurred in the more recent periods are considered. The forecast for any period is calculated by adding the actual demands for the desired number of immediate past periods and dividing the sum by the number of periods considered. A forecast for each period is obtained by adding the more recent demand and dropping the oldest demand. Mathematically, the n-period moving average is expressed as follows:

$$F_K = \frac{D_{K-1} + D_{K-2} + \cdots + D_{K-n}}{n}$$

where

F_K = forecast for period K

D_T = actual demand (sales) in period T

Thus, the forecast for August in the three-month moving average method is the average of the three month demands, May, June, and July. Likewise the forecast for September is the average of the three month demands, June, July, and August. Note that the more recent demand (demand in August) is added, and the oldest data (demand in May) is dropped. The actual sales for chemicals (in tons) at Taylor Enterprises and three-period and six-period moving average forecasts are shown in Table 3.1 and Figure 3.10.

For the three-period moving average,

$$F_4 = \frac{D_1 + D_2 + D_3}{3} = \frac{500 + 520 + 570}{3}$$

$$= 530.00 \text{ tons}$$

and

$$F_5 = \frac{D_2 + D_3 + D_4}{3} = \frac{520 + 570 + 530}{3}$$

$$= 540.00 \text{ tons}$$

TABLE 3.1   Actual Sales of Chemicals and a Three-Period and a Six-Period Moving Average

Period	Actual Sales (Tons)	Three-Period Moving Average (Tons)	Six-Period Moving Average (Tons)
1	500.00		
2	520.00		
3	570.00		
4	530.00	530.00	
5	590.00	540.00	
6	580.00	563.33	
7	480.00	566.67	548.33
8	520.00	550.00	545.00
9	520.00	526.67	545.00
10		506.67	536.67

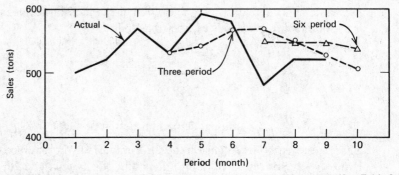

FIGURE 3.10   Actual sales of chemicals and moving average forecasts. (See Table 3.1.)

The forecast for next month (period 10) is 506.67 tons by the three-month moving average and 536.67 tons by the six-month moving average. When a forecast is established by using the moving average technique, the forecast is too small if the time series has an upward trend. On the other hand, the forecast is too large if the time series has a downward trend. We note from Figure 3.10 that the forecasts of three-period moving average respond to the variations in demand and follow the trend quickly. Conversely, the forecasts of six-period moving average do not react to the movements of demand readily and are much more stable. Various n values (number of periods included in the moving average) should be tried, and the particular n value whose forecasts respond to bona fide movements in trend is selected to determine the future forecasts.

A Computer Model

In this section, we shall design a computer program for the moving average technique and illustrate the use of the program by the Taylor Enterprises forecasting problem. Let us establish the following list of variables:

MAXT	= number of past demand (sales) values given
KTIMES	= number of periods considered in calculating moving average (n)
NMAX	= number of periods considered for calculating forecasts (MAXT + 1)
JROUT	= number of periods for which forecasts will be actually calculated
JUPTO	= index specifying the most recent period that is included in the sum of sales. JUPTO = 5 for F_6

SUM   = sum of more recent n demand values (sales)
MONTH(K) = Kth period number
SALES(K)   = actual demand (sales) occurred in MONTH(K) period
FCAST(K)   = forecast for MONTH(K) period
DELTA(K)   = difference between forecast and actual demand (sales)
                for MONTH(K)
                = FCAST(K) − SALES(K)

The variables SALES( ) and DELTA( ) are one-dimensional arrays with MAXT elements, and the variables MONTH( ) and FCAST( ) have NMAX elements. Figure 3.11 displays the flowchart for the moving average technique.

We might transmit the input data to the computer memory by the statements:

```
        READ 100, MAXT, KTIMES
        NMAX = MAXT + 1
        READ 110, (MONTH(I), I = 1, NMAX)
        READ 120, (SALES(K), K = 1, MAXT)
100   FORMAT ( )
110   FORMAT ( )
120   FORMAT ( )
```

The data arrays FCAST and DELTA are initialized by the instructions:

```
        DO 300 J = 1, MAXT
        FCAST(J) = 0
        DELTA(J) = 0
300   CONTINUE
```

The value of JROUT might be established by the statement:

```
        JROUT = MAXT − KTIMES + 1
```

The FCAST and DELTA values for each of the JROUT periods are determined by including the following statements:

```
        DO 320 J = 1, JROUT
        SUM = 0
        JUPTO = J + KTIMES − 1
        DO 330 K = J, JUPTO
        SUM = SUM + SALES(K)
330   CONTINUE
```

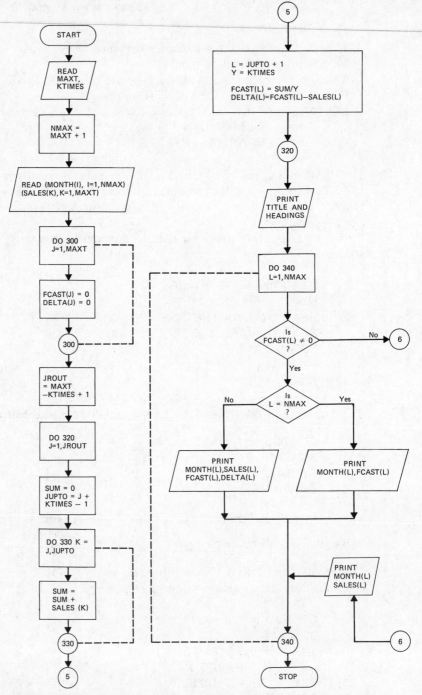

FIGURE 3.11   Flowchart for the main program of the moving average technique.

```
      L = JUPTO + 1
      Y = KTIMES
      FCAST(L) = SUM/Y
      DELTA(L) = FCAST(L) - SALES(L)
320   CONTINUE
```

The input data as well as the calculated values are transmitted to the printer by the following instructions:

```
      DO 340 L = 1, NMAX
      IF (FCAST(L).NE.O) GO TO 360
      PRINT 205, MONTH(L), SALES(L)
      GO TO 340
      IF (L.EQ. NMAX) GO TO 380
360   PRINT 210, MONTH(L), SALES(L), FCAST(L), DELTA(L)
      GO TO 340
380   PRINT 215, MONTH(L), FCAST(L)
340   CONTINUE
205   FORMAT ( )
210   FORMAT ( )
215   FORMAT ( )
```

Figure 3.12 shows the complete computer program composed of our previously developed program segments.

The input data for the Taylor Enterprises problem are the following:

```
MAXT   = 9
KTIMES = 3 (for three-period moving average)
```

MONTH(1)	= 1	SALES(1)	= 500.0
MONTH(2)	= 2	SALES(2)	= 520.0
MONTH(3)	= 3	SALES(3)	= 570.0
MONTH(4)	= 4	SALES(4)	= 530.0
MONTH(5)	= 5	SALES(5)	= 590.0
MONTH(6)	= 6	SALES(6)	= 580.0
MONTH(7)	= 7	SALES(7)	= 480.0
MONTH(8)	= 8	SALES(8)	= 520.0
MONTH(9)	= 9	SALES(9)	= 520.0
MONTH(10)	= 10		

```
            DIMENSION MONTH(10),SALES(9),FCAST(10),DELTA(10)
C
C   READ INPUT DATA
            READ 100,MAXT,KTIMES
            NMAX=MAXT+1
            READ 110,(MONTH(I),I=1,NMAX)
            READ 120,(SALES(K),K=1,MAXT)
C
C   INITIALIZE FCAST AND DELTA VALUES
            DO 300 J=1,MAXT
            FCAST(J)=0
            DELTA(J)=0
        300 CONTINUE
C
C   CALCULATE FCAST AND DELTA VALUES
            JROUT=MAXT-KTIMES+1
C
            DO 320 J=1,JROUT
            SUM=0
            JUPTO=J+KTIMES-1
C
            DO 330 K=J,JUPTO
            SUM=SUM+SALES(K)
        330 CONTINUE
C
            L=JUPTO+1
            Y=KTIMES
            FCAST(L)=SUM/Y
            DELTA(L)=FCAST(L)-SALES(L)
        320 CONTINUE
C
            PRINT 150
            PRINT 200
            PRINT 201
            PRINT 202
C
C   PRINTOUT THE CALCULATED VALUES
            DO 340 L=1,NMAX
            IF(FCAST(L).NE.0)GO TO 360
            PRINT 205,MONTH(L),SALES(L)
            GO TO 340
        360 IF(L.EQ.NMAX)GO TO 380
            PRINT 210,MONTH(L),SALES(L),FCAST(L),DELTA(L)
            GO TO 340
        380 PRINT 215,MONTH(L),FCAST(L)
        340 CONTINUE


C
        100 FORMAT(2I2)
        110 FORMAT(10I2)
        120 FORMAT(9F5.1)
        150 FORMAT(1H1)
        200 FORMAT(5X,39HFORCASTING BY THE MOVING AVERAGE METHOD)
        201 FORMAT(7X,4HTIME,3X,6HACTUAL,2X,7HFORCAST,1X,10HDIFFERENCE)
        202 FORMAT(6X,6HPERIOD,2X,5HSALES,4X, 7H(UNITS),3X,7H(UNITS))
        205 FORMAT(5X,I5,3X,F7.2)
        210 FORMAT(5X,I5,3X,F7.2,1X,F7.2,2X,F8.2)
        215 FORMAT(5X,I5,11X,F7.2)
            STOP
            END
```

FIGURE 3.12   Computer program for the simple moving average method (illustrated for the Taylor Enterprises problem).

The data deck is prepared according to the FORMAT statements 100, 110, and 120; it has the following appearance:

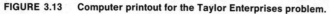

The resulting computer printout is displayed in Figure 3.13.

```
FORCASTING BY THE MOVING AVERAGE METHOD
 TIME      ACTUAL    FORCAST  DIFFERENCE
PERIOD     SALES     (UNITS)    (UNITS)
   1       500.00
   2       520.00
   3       570.00
   4       530.00    530.00         0
   5       590.00    540.00     -50.00
   6       580.00    563.33     -16.67
   7       480.00    566.67      86.67
   8       520.00    550.00      30.00
   9       520.00    526.67       6.67
  10                 506.67
```

FIGURE 3.13   Computer printout for the Taylor Enterprises problem.

EXPONENTIAL SMOOTHING

A more reliable forecast can be determined by emphasizing recent data that portray the future better and considering all data so that fluctuations in the given demand pattern are smoothed out. Exponential smoothing technique is an efficient forecasting method, and it readily lends itself to computer modeling.

The calculation of a single exponential smoothing forecast is simply the latest forecast plus a fraction (known as the smoothing constant) of the difference between the actual demand in the current period and the latest forecast. The smoothing constant is frequently called alpha (α), and it must be between 0 and 1. The equation for the exponentially smoothed forecast is written as:

$$F_k = F_{k-1} + \alpha(D_k - F_{k-1}) \qquad (3.8)$$

where

F_k = exponentially smoothed forecast determined this period. This estimate is used to forecast the demand for the next period.

F_{k-1} = exponentially smoothed forecast established last period; this is the latest forecast

α = smoothing constant $(0 \leqq \alpha \leqq 1)$

The equation 3.8 can be rearranged as

$$F_k = \alpha D_k + (1 - \alpha)F_{k-1} \tag{3.9}$$

Of this, F_{k-1} can be written according to the equation 3.9 as

$$F_{k-1} = \alpha D_{k-1} + (1 - \alpha)F_{k-2}$$

Substituting this for F_{k-1} in (4.9), we obtain

$$F_k = \alpha D_k + (1 - \alpha)[\alpha D_{k-1} + (1 - \alpha)F_{k-2}]$$
$$= \alpha D_k + (1 - \alpha)D_{k-1} + (1 - \alpha)^2 F_{k-2}$$

By continuing successively the substituting process for $F_{k-2}, F_{k-3}, F_{k-4}$, etc., the following series has resulted:

$$F_k = \alpha D_k + \alpha(1 - \alpha)D_{k-1} + \alpha(1 - \alpha)^2 D_{k-2}$$
$$+ \alpha(1 - \alpha)^3 D_{k-3} + \alpha(1 - \alpha)^4 D_{k-4} + \cdots \tag{3.10}$$

When $\alpha = 0.10$, the series becomes

$$F_k = 0.1D_k + 0.09D_{k-1} + 0.081D_{k-2}$$
$$+ 0.073D_{k-3} + 0.065D_{k-4} + \cdots \tag{3.11}$$

The function (3.11) indicates that in computing the forecast for the $(k + 1)$th period, more weight is given to recent data and less weight to older data. However, if we employ this function to calculate the forecast, a large amount of computer memory space is needed to store the past demand data. The operations can be performed with smaller memory space by employing the equation 3.8.

Example 3.2

The following figures represent the monthly revenue of Sunshine Moving Company from shipping timber (in thousands of dollars) for the past 24 months:

Month	Revenue ($1000)	Month	Revenue ($1000)
1	5	13	35
2	15	14	38
3	12	15	39
4	18	16	33
5	14	17	45
6	20	18	48
7	24	19	50
8	18	20	47
9	25	21	55
10	25	22	55
11	32	23	51
12	20	24	60

The forecast for next month (month 25) can be computed by exponentially smoothing method, as described next.

We note from expression 3.8 that for calculating the forecast for any month (F_k), data on the actual demand and the forecast for the previous period, D_k and F_{k-1}, are needed. Hence, the computations are done successively starting with month 1. No forecasts are available for months before month 1. However, if demand data values for months earlier than month 1 are available, the average of any desired demand values is computed and set to be the F_1 value. If this information is not available, as in our example, any reasonable value or the D_1 itself is assigned to be the F_1 value. In our problem let us assign D_1 to be the F_1 value. Also, we shall assume an alpha of 0.10, and illustrate the calculations.

$$F_1 = 5$$

From (3.8)

$$F_2 = 5 + 0.10(15 - 5)$$
$$= 6$$

$$F_3 = 6 + 0.10(12 - 6)$$
$$= 6.6$$

The calculated forecasts for the 24 months are depicted in Table 3.2. The monthly revenue and the forecasts are shown on the graph in Figure 3.14. The revenue data plotted in Figure 3.14 signify the existence of growth trend. The exponential smoothing forecasts give lower values than the actual revenue. For a time series with a downward trend, the forecasts would give larger values. In those situations, the forecasts are said to lag the trend. When a trend movement prevails, improved forecasts are determined by correcting for lag due to trend.

**TABLE 3.2   Monthly Revenue of Sunshine Moving Company from
Shipping Timber and Exponential Smoothed Forecasts
(in Thousands of Dollars)**

Month (k)	Revenue (D_k)	Forecast (F_k)	Month (k)	Revenue (D_k)	Forecast (F_k)
1	5	(5.00)	13	35	18.25
2	15	6.00	14	38	20.23
3	12	6.60	15	39	22.10
4	18	7.74	16	33	23.19
5	14	8.37	17	45	25.37
6	20	9.53	18	48	27.64
7	24	10.98	19	50	29.87
8	18	11.68	20	47	31.59
9	25	13.01	21	55	33.93
10	25	14.21	22	55	36.03
11	32	15.99	23	51	37.53
12	20	16.39	24	60	39.78

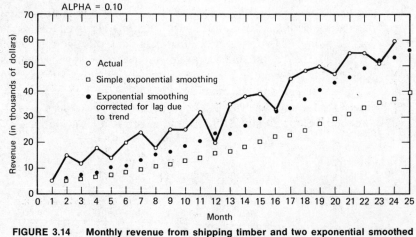

**FIGURE 3.14   Monthly revenue from shipping timber and two exponential smoothed
forecasts.**

Correcting for Lag Due to Trend

When a time series has an observable trend component, exponentially
smoothing forecasts would lag the trend. We shall discuss a technique
for correcting lag due to trend. The apparent trend in any period k is
calculated by expression 3.12:

$$t_k = F_k - F_{k-1} \qquad (3.12)$$

where
$$t_k = \text{apparent trend in period } k$$

Next, we determine the exponentially smoothed trend (T_k) by employing expression 3.13:

$$T_k = T_{k-1} + \alpha(t_k - T_{k-1}) \tag{3.13}$$

Subsequently, expected demand for the current period (ED_k) is computed.

$$ED_k = F_k + \left(\frac{1-\alpha}{\alpha}\right) T_k \tag{3.14}$$

Then, the forecast for the $(k + n)$th period is calculated using ED_k value by extrapolation as follows:

$$F_{k,k+n} = ED_k + nT_k \tag{3.15}$$

where

$F_{k,k+n} = $ forecast for $(k + n)$th period by exponential smoothing corrected for lag due to trend

Therefore, the forecast for next period ($F_{k,k+1}$) can be determined from equation 3.15 in the following way:

$$F_{k,k+1} = ED_k + T_k \tag{3.16}$$

The computations are performed successively beginning with month 1. Again, the difficulty is that F_0 and T_0 are not known. Frequently, F_0 is assigned the value of D_1, and T_0 is set at zero.

Assume

$$F_0 = D_1 = 5$$

$$T_0 = 0$$

when $k = 1$

$$t_1 = 5 - 5 = 0$$

$$T_1 = 0 + 0.10(0 - 0) = 0$$

$$ED_1 = 5 + \left(\frac{1 - 0.10}{0.10}\right) 0 = 5$$

The forecast for month 2, made in month 1, is

$$F_{1,2} = 5 + 0 = 5$$

when

$$k = 2$$

$$t_2 = 6 - 5 = 1$$

$$T_2 = 0 + 0.10(1 - 0) = 0.10$$

$$ED_2 = 6 + \left(\frac{1 - 0.10}{0.10}\right) 0.10 = 6.90$$

TABLE 3.3   Forecasts by Exponential Smoothing Corrected for Lag due to Trend (Alpha = 0.10)

Month k	Revenue D_k	F_k	t_k	T_k	ED_k	$F_{k,k+1}$
1	5	5.00	0	0	5.00	5.00
2	15	6.00	1.00	0.10	6.90	7.00
3	12	6.60	0.60	0.15	7.95	8.10
4	18	7.74	1.14	0.25	9.98	10.23
5	14	8.37	0.63	0.29	10.95	11.23
6	20	9.53	1.16	0.37	12.90	13.27
7	24	10.98	1.45	0.48	15.31	15.79
8	18	11.68	0.70	0.50	16.21	16.72
9	25	13.01	1.33	0.59	18.29	18.88
10	25	14.21	1.20	0.65	20.04	20.69
11	32	15.99	1.78	0.76	22.84	23.60
12	20	16.39	0.40	0.72	22.91	23.64
13	35	18.25	1.86	0.84	25.80	26.64
14	38	20.23	1.97	0.95	28.80	29.75
15	39	22.10	1.88	1.04	31.51	32.55
16	33	23.19	1.09	1.05	32.64	33.68
17	45	25.37	2.18	1.16	35.83	37.00
18	48	27.64	2.26	1.27	39.09	40.36
19	50	29.87	2.24	1.37	42.19	43.56
20	47	31.59	1.71	1.40	44.21	45.62
21	55	33.93	2.34	1.50	47.40	48.90
22	55	36.03	2.11	1.56	50.06	51.61
23	51	37.53	1.50	1.55	51.50	53.05
24	60	39.78	2.25	1.62	54.37	55.99

The forecast for month 3, made in month 1, is

$$F_{2,3} = 6.90 + 0.10 = 7.00$$

Likewise, the subsequent forecasts are determined. The corrected exponential smoothing forecasts are shown in Table 3.3 and Figure 3.14. We have developed a method for calculating the forecasts by exponential smoothing corrected for lag due to trend. For the arbitrarily selected smoothing constant value, the computations were performed. The next question is how to determine the best value for the smoothing constant? The best value of alpha can be established by using the following approaches:

METHOD I. Take an alpha value, calculate the corrected forecast values, $(F_{k,k+1})$, and plot the forecasts on the graph. Continue the process for different alpha values. Select the one that tracks the actual data points best.

METHOD II. Take an alpha value, compute the corrected forecast values $(F_{k,k+1})$, and determine the sum of the absolute values of forecasting errors. Repeat the calculations for various alpha values. The best value of alpha yields the smallest sum. It is used to forecast the next period's level of activity. The forecast error in each period is the difference between the actual demand (D_k) and the exponentially smoothed forecast $(F_{k-1,k})$ for that period.

$$\text{SAVFE} = \sum_{k=1}^{k=N} \left| D_k - F_{k-1,k} \right| \tag{3.17}$$

where

$\quad$ SAVFE $=$ sum of absolute values of forecasting errors

$\quad$ $N =$ number of past data points given

This method removes any subjective judgment in determining which alpha tracts the actual data best. Also, it eliminates the sketching effort that is required in Method I. Computer systems can be utilized to calculate SAVFE values for different alpha values, and to select the best alpha value. In calculating SAVFE values, a reasonable value for $F_{0,1}$ is required. Usually, the value of $F_{0,1}$ is set equal to D_1. We shall design a computer program and determine the best alpha value for our example problem in the next section.

A Computer Program

We have discussed the solution procedure for the exponential smoothing model for time series with trend and random movements. We shall now develop a computer program for the solution procedure in this section. Let us establish the following list of variables:

$\quad$ KDATA $\quad=$ number of past data points given (N)

$\quad$ ALPHA $\quad\;\;=$ smoothing constant (α)

$\quad$ SELECT $\quad=$ index that tells the computer if the computed values are to be printed out

$\qquad\qquad\qquad$ SELECT $= 0$, when the best alpha is not established and the computed values are not to be printed out

$\qquad\qquad\qquad$ SELECT $= 1$, when the best alpha is established and the calculated values are to be printed out

$\quad$ TABLE $\quad\;\;=$ this subprogram is employed to calculate the SAVFE value for any given alpha value

EXA   = exponentially smoothed forecast determined this period (F_k)

EXANEW = the recently calculated EXA value

AT   = apparent trend in kth period (t_k)

ATNEW   = the recently calculated AT value

ET   = exponential smoothed trend (T_k)

ETNEW   = the recently calculated ET value

FCASTN = expected demand for current period (ED_k)

FCAST   = forecast for next period corrected for lag due to trend $(F_{k,k+1})$

JAD(K)   = actual demand (sales) for period K

AD   = actual demand (sales) that occurred in period L

SDELTA = sum of absolute values of forecasting errors (SAVFE)

The variable JAD( ) is a one-dimensional array with KDATA elements. Figures 3.15 and 3.16 display the flowcharts for the main program and subprogram TABLE, respectively.

We might initialize the program and transmit the input data to the computer memory by the following statements:

```
    KDATA = 24
    READ 100, (JAD(K), K = 1, KDATA)
100 FORMAT
```

For the 100 alpha values from 0.01 to 1.00 in steps of 0.01, SDELTA values can be computed and tested to determine which one of them yields the smallest SDELTA value, by including the instructions:

```
    SMALL = 999999.99
    DO 200 N = 1, 100
    X = N
    ALPHA = X * 0.01
    SELECT = 0
    CALL TABLE (JAD,SELECT,KDATA,
  3 SDELTA, L. EXA, EXANEW, AT, ATNEW, ET, ETNEW,
  4 FCAST, FCASTN, ALPHA)
    IF (SDELTA .GE. SMALL) GOTO 200
    SMALL = SDELTA
    S = ALPHA
200 CONTINUE
```

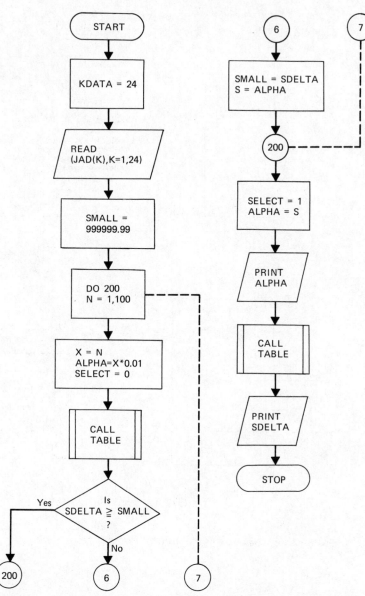

FIGURE 3.15 Flowchart for the main program of the exponential smoothing model.

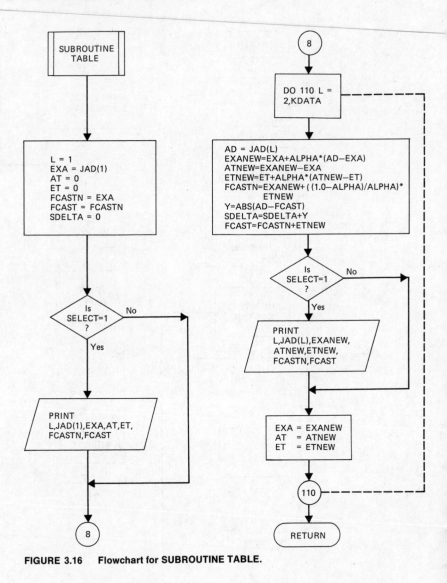

FIGURE 3.16   Flowchart for SUBROUTINE TABLE.

The value of S represents the best alpha value. The value of selected alpha, and the related FCAST as well as SDELTA values can be generated and printed out by the following set of statements:

```
        SELECT = 1
        ALPHA = S
        PRINT 120, ALPHA
        CALL TABLE (JAD, SELECT, KDATA,
      3 SDELTA, L, EXA, EXANEW, AT, ATNEW, ET, ETNEW,
      4 FCAST, FCASTN, ALPHA)
        PRINT 110, SDELTA
  110   FORMAT ( )
  120   FORMAT ( )
```

The subprogram TABLE might be initialized by the statements:

```
    SUBROUTINE TABLE (JAD, SELECT, KDATA,
  3 SDELTA, L, EXA, EXANEW, AT, ATNEW, ET, ETNEW,
  4 FCAST, FCASTN, ALPHA)
    DIMENSION JAD ( )
    L = 1
    EXA = JAD (1)
    AT = 0
    ET = 0
    FCASTN = EXA
    FCAST = FCASTN
    SDELTA = 0
```

The following instructions would cause the computer to print out the given and calculated values for period 1 for the selected alpha value:

```
        IF (SELECT .NE. 1) GO TO 100
        PRINT 700, L, JAD(1), EXA, AT, ET, FCASTN, FCAST
  700   FORMAT ( )
```

Next, the values of F_k, t_k, T_k, ED_k, and $F_{k,k+1}$ are calculated for the successive past periods. Then the given and calculated values are printed out for the selected alpha value only. The following set of statements would enable the computer to perform the operations and subsequently transfer the control to the main program:

```
100   DO 110 L = 2, KDATA
      AD = JAD (L)
      EXANEW = EXA + ALPHA * (AD − EXA)
      ATNEW = EXANEW − EXA
      ETNEW = ET + ALPHA * (ATNEW − ET)
      FCASTN = EXANEW + ((1.0 − ALPHA)/ALPHA) * ETNEW
      Y = ABS(AD − FCAST)
      SDELTA = SDELTA + Y
      FCAST = FCASTN + ETNEW
      IF (SELECT .NE. 1) GO TO 120
      PRINT 700, L, JAD(L), EXANEW, ATNEW,
    7 ETNEW, FCASTN, FCAST
120   EXA = EXANEW
      AT = ATNEW
      ET = ETNEW
110   CONTINUE
      RETURN
      END
```

Figure 3.17 gives the complete computer program composed of our previously designed program segments. There are 24 past data points in Example 3.2, and they are inputted according to FORMAT statement 100. The input data deck has the following appearance:

The resulting computer printout is shown in Figure 3.18. It indicates that the best alpha value is 0.27, and based on this selected alpha, the forecast for next months' revenue is $60,410. Figure 3.19 displays the actual revenue and forecasted revenue values for the selected smoothing constant. Buffa and Taubert (1972) and Shore (1973) have developed various exponential smoothing models for different time series.

```
        DIMENSION JAD ( 30 )
C
C   INITIALIZE KDATA
        KDATA = 24
C
C   INPUT THE ACTUAL DEMAND VALUES
        READ 100, ( JAD(K), K = 1,12 )
        READ 100, ( JAD(K), K = 13,24 )
C
C   CALCULATE SDELTA FOR ALPHA VALUES
C   CHOOSE ALPHA WITH THE SMALLEST SDELTA
C
        SMALL = 999999.99
C
        DO 200 N = 1,100
        X = N
        ALPHA = X * 0.01
        SELECT = 0
C
        CALL        TABLE ( JAD,SELECT, KDATA,
       3SDELTA, L, EXA,EXANEW,AT, ATNEW, ET, ETNEW,
       4FCAST, FCASTN, ALPHA   )
C
        IF ( SDELTA .GE. SMALL ) GO TO 200
        SMALL = SDELTA
        S = ALPHA
    200 CONTINUE
C
        PRINT 90
C
C   PRINT THE TABLE FOR SELECTED ALPHA
        SELECT = 1
        ALPHA = S
        PRINT 120, ALPHA
        PRINT 130
C
        CALL        TABLE ( JAD,SELECT, KDATA,
       3SDELTA, L, EXA,EXANEW,AT, ATNEW, ET, ETNEW,
       4FCAST, FCASTN, ALPHA   )
C
        PRINT 110, SDELTA
C
     90 FORMAT (1H1 )
    100 FORMAT ( 12(I2, 1X ) )
    110 FORMAT ( /,5X, 8HSAVFE =   , F9.3 )
    120 FORMAT ( 10X, 7HALPHA = , F4.2 , / )
    130 FORMAT ( 9X, 1HK,3X,3HJAD,5X,3HEXA,10X,
       72HAT, 8X, 2HET, 4X, 6HFCASTN, 5X, 5HFCAST ,/)
        STOP
        END
```

FIGURE 3.17   Computer program for determining forecasts by the exponential smoothing method (illustrated for Example 3.2).

```
        SUBROUTINE TABLE ( JAD,SELECT, KDATA,
      3SDELTA, L, EXA,EXANEW,AT, ATNEW, ET, ETNEW,
      4FCAST, FCASTN, ALPHA  )
C
        DIMENSION JAD ( 30 )
C  FIND SDELTA FOR THE GIVEN ALPHA VALUE
C
C  INITIALIZE L,EXA,AT,ET,FCASTN,
C     FCAST, SDELTA VALUES
C
        L = 1
        EXA = JAD (1)
        AT = 0
        ET = 0
        FCASTN = EXA
        FCAST = FCASTN
        SDELTA = 0
        IF ( SELECT .NE. 1 ) GO TO 100
        PRINT 700, L, JAD(1), EXA, AT, ET, FCASTN, FCAST
C
    100 DO 110 L = 2, KDATA
        AD = JAD (L)
        EXANEW = EXA + ALPHA * ( AD - EXA )
        ATNEW = EXANEW - EXA
        ETNEW = ET + ALPHA * ( ATNEW - ET )
        FCASTN = EXANEW + ((1.0-ALPHA) / ALPHA) * ETNEW
        Y = ABS (AD-FCAST)
        SDELTA = SDELTA + Y
        FCAST = FCASTN + ETNEW
        IF ( SELECT .NE. 1 ) GO TO 120
        PRINT 700, L, JAD(L), EXANEW, ATNEW,
      7ETNEW, FCASTN, FCAST
    120 EXA = EXANEW
        AT = ATNEW
        ET = ETNEW
    110 CONTINUE
C
    700 FORMAT ( 5X, 2I6, 6F10.2 )
        RETURN
        END
```

FIGURE 3.17 (*Continued*)

ALPHA = .27

K	JAD	EXA	AT	ET	FCASTN	FCAST
1	5	5.00	0	0	5.00	5.00
2	15	7.70	2.70	.73	9.67	10.40
3	12	8.86	1.16	.85	11.15	11.99
4	18	11.33	2.47	1.28	14.80	16.08
5	14	12.05	.72	1.13	15.11	16.24
6	20	14.20	2.15	1.41	18.00	19.40
7	24	16.84	2.65	1.74	21.55	23.29
8	18	17.16	.31	1.36	20.82	22.17
9	25	19.27	2.12	1.56	23.49	25.06
10	25	20.82	1.55	1.56	25.03	26.59
11	32	23.84	3.02	1.95	29.12	31.07
12	20	22.80	-1.04	1.14	25.90	27.04
13	35	26.10	3.29	1.73	30.76	32.48
14	38	29.31	3.21	2.13	35.06	37.19
15	39	31.93	2.62	2.26	38.03	40.29
16	33	32.22	.29	1.73	36.89	38.61
17	45	35.67	3.45	2.19	41.60	43.79
18	48	39.00	3.33	2.50	45.76	48.26
19	50	41.97	2.97	2.63	49.07	51.70
20	47	43.33	1.36	2.28	49.50	51.79
21	55	46.48	3.15	2.52	53.29	55.81
22	55	48.78	2.30	2.46	55.43	57.89
23	51	49.38	.60	1.96	54.67	56.63
24	60	52.25	2.87	2.20	58.20	60.41

SAVEE = 106.593

FIGURE 3.18 Computer printout for Example 3.2.

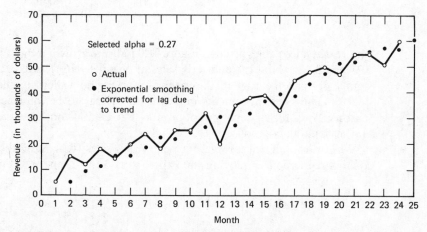

FIGURE 3.19 Monthly revenue from shipping timber and exponential smoothed forecasts for the selected alpha value.

SEASONAL VARIATIONS WITH TREND EFFECT

The demand for some items or activities exhibits a growth or decay trend movement, and heavy or light demands occur in particular seasons or months of each year. The products or activities that are subject to trend and seasonal effects might include the demand for skiing equipment, winter coats, boats, automobiles, offering courses on skiing, and construction work. To illustrate the solution procedure for determining forecasts for seasonal products or tasks, let us consider Example 3.3.

Example 3.3

The management of Teleota, Inc., has collected the quarterly sales, shown in Table 3.4, of product $P76$ (in $1000 units) for the past five years.

TABLE 3.4   Quarterly Sales of Product P76 for Teleota, Inc. ($1000 units)

Year	Quarter				Yearly Sales
	Winter	Spring	Summer	Fall	
(X)	(1)	(2)	(3)	(4)	(Y)
1	50	75	67	60	252
2	54	82	73	65	274
3	59	86	76	77	298
4	64	81	80	68	293
5	63	92	81	81	317

The yearly sales values given in Figure 3.20 indicate growth trend. Figures 3.21 and 3.22 exhibit the quarterly sales of product $P76$. The data points signify peaks appearing in the second quarter and valleys in the first quarter. The sales movements in each year appear to be influenced by seasons. Thus, the sales pattern of product $P76$ exhibits both trend effects and seasonal fluctuations.

We can establish a forecast for next year's sales (year 6) by the least squares regression employing the expressions 3.1, 3.2, and 3.3. For our illustration,

$$\sum X = 15 \qquad \sum X^2 = 55$$
$$\sum Y = 1434 \qquad \sum XY = 4451$$
$$n = 5$$

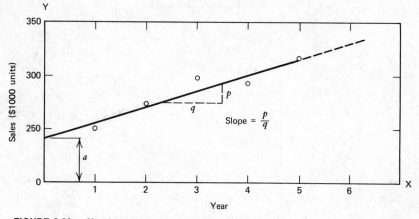

FIGURE 3.20   Yearly sales and trend line for Product P76.

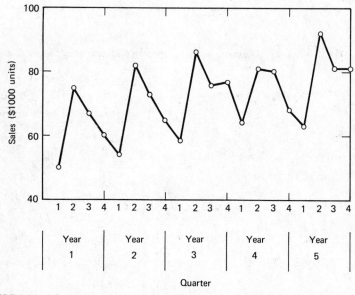

FIGURE 3.21   Quarterly sales for Product P76.

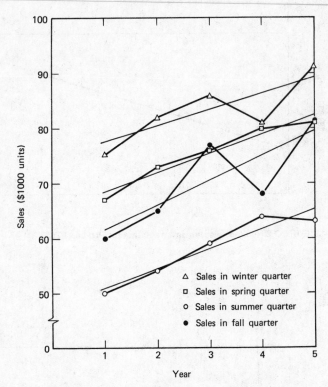

FIGURE 3.22 Quarterly sales for Product P76 indicating seasonal and trend movements.

From (3.2)

$$a = \frac{\sum X^2 \sum Y - \sum X \sum XY}{n\sum X^2 - (\sum X)^2} = \frac{(55)(1434) - (15)(4451)}{5(55) - 15^2}$$

$$= 242.10$$

From (3.3)

$$b = \frac{n\sum XY - \sum X \sum Y}{n\sum X^2 - (\sum X)^2} = \frac{5(4451) - 15(1434)}{5(55) - 15^2}$$

$$= 14.90$$

The equation of the trend line is obtained from (3.1).

$$Y = 242.10 + 14.90(X)$$

The trend line is shown in Figure 3.20. When $X = 6$

$$Y = 242.10 + 14.90(6) = 331.50$$

Hence, the forecast for next year's sales is \$331,500. The seasonal forecasts for next year are determined based on the seasonal sales pattern prevailing in the past. Here are the computational steps:

STEP 1. Calculate the total sales of a particular quarter (such as winter) over the given years. Repeat the process for each of the four quarters:

$$TS_j = Q_{1,j} + Q_{2,j} + \cdots + Q_{n,j} \qquad \text{for } j = 1, 2, 3, 4 \qquad (3.18)$$

where

$$Q_{i,j} = \text{sales in quarter } j \text{ for year } i$$

$$j = 1 \text{ for winter quarter}$$

$$j = 2 \text{ for spring quarter}$$

$$j = 3 \text{ for summer quarter}$$

$$j = 4 \text{ for fall quarter}$$

$$TS_j = \text{total sales in quarter } j$$

STEP 2. Determine the average sales for each of the four quarters

$$AS_j = \frac{TS_j}{n} \qquad (3.19)$$

where

$$AS_j = \text{average sales for quarter } j.$$

STEP 3. Calculate the average of quarterly sales of all quarters.

$$AQS = \frac{AQ_1 + AQ_2 + AQ_3 + AQ_4}{4} \qquad (3.20)$$

STEP 4. where AQS = average of quarterly sales of all quarters.

Compute the quarterly indexes (QI_j) for each of the four quarters.

$$QI_j = \frac{AS_j}{AQS} \qquad (3.21)$$

STEP 5. Determine the forecasts of quarterly sales by multiplying the average quarterly forecast by the corresponding quarterly index.

$$FQ_j = \frac{Y_f}{4} (QI_j) \qquad (3.22)$$

where

$$FQ_j = \text{forecast of sales for quarter } j \text{ in next year}$$

$$Y_f = \text{forecasts of sales for next year}$$

The calculations required for determining QI_j values are shown in Table 3.5.

TABLE 3.5 Calculations for Determining QI_j Values

	Quarter j			
Year i	Winter $j = 1$	Spring $j = 2$	Summer $j = 3$	Fall $j = 4$
1	50	75	67	60
2	54	82	73	65
3	59	86	76	77
4	64	81	80	68
5	63	92	81	81
TS_j	290	416	377	351
AS_j	58	83.2	75.4	70.2
QI_j	0.809	1.160	1.052	0.979

$$AQS = 71.7$$

$$FQ_1 = \frac{331.50}{4} \times 0.809 = 67.04 \rightarrow \$67,040$$

$$FQ_2 = \frac{331.50}{4} \times 1.160 = 96.17 \rightarrow \$96,170$$

$$FQ_3 = \frac{331.50}{4} \times 1.052 = 87.15 \rightarrow \$87,150$$

$$FQ_4 = \frac{331.50}{4} \times 0.979 = \underline{81.14} \rightarrow \$81,140$$

Forecast of sales for next year $= 331.50 \rightarrow \$331,500$

Note that the lines of best fit drawn for the quarterly sales in Figure 3.22 are approximately parrallel. In order to use the seasonal index approach, this condition must be satisfied.

A Computer Program

Most of the variables utilized in the computer model for the least squares regression (Figure 3.8) are employed in this computer program. The following new variables are used:

TOTAL(K)	= yearly demand (sales) for the year $X(K)$
DEMAND(K,L)	= value that specifies the demand (sales) occurred in year $X(K)$ for the quarter L
SUM(J)	= total sales in quarter $j(TS_j)$
XMEAN(J)	= average sales for quarter $j(AS_j)$
AVE	= average of quarterly sales of all quarters (AQS)
Q(L)	= quarterly forecast for quarter L
LSLF	= this subprogram computes A, B, and SDELTA for the least squares regression model (see Figures 4.7 and 4.8)

The variables SUM( ), XMEAN( ), AND Q( ) are one-dimensional arrays with four elements. The variable TOTAL( ) is a one-dimensional array with NYRS elements. The two-dimensional array DEMAND(,) is of the size NYRS by four. We might transmit the input data to the computer memory and print out the data by the following statements:

```
      READ 15, NYRS, JNEED
      PRINT 17, NYRS, JNEED
      N = NYRS
      DO 100 I = 1, NYRS
      READ 20, (DEMAND(I, J), J = 1, 4)
      PRINT 25, (DEMAND(I, J), J = 1, 4)
      X(I) = I
100   CONTINUE
 15   FORMAT ( )
 17   FORMAT ( )
 20   FORMAT ( )
 25   FORMAT ( )
```

The following statements would cause the computer to calculate the yearly demand (sales) values:

```
      DO 110 I = 1, NYRS
      T = 0
      DO 120 J = 1, 4
120   T = T + DEMAND(I, J)
      TOTAL(I) = T
110   CONTINUE
```

Next TS_j and AS_j values are computed by employing the statements:

```
        Y = NYRS
        DO 130 J = 1, 4
        S = 0
        DO 140 I = 1, NYRS
  140   S = S + DEMAND(I, J)
        SUM(J) = S
        XMEAN(J) = S/Y
  130   CONTINUE
```

The subprogram LSLF depicted in Figure 3.8 can be used to determine the values of A, B, and SDELTA of the linear regression model for the yearly sales. The value of AQS can be established employing the expression 3.20. The values of A and B can be transmitted to the printer. These operations might be performed by the instructions:

```
        CALL LSLF(X, TOTAL, NYRS, XX, XY, FCAST, DELTA,
      1 SUMX, SUMY, SUMXX, SUMXY, SDELTA, SFCAST, A, B)
        CXMEAN = 0
        DO 150 M = 1, 4
  150   CXMEAN = CXMEAN + XMEAN(M)
        AVER = CXMEAN/4.0
        PRINT 30, A, B
   30   FORMAT ( )
```

The sales forecast for future years can be determined by using the mathematical relationship (3.1). Subsequently, the quarterly forecasts can be established applying the equation 3.22. The following set of statements would cause the computer to calculate these values and print out the computer values:

```
        FN = NYRS
        DO 160 IJ = 1, JNEED
        F = IJ
        FCAST(IJ) = A + B * (FN + F)
        N = N + 1
        PRINT 50, N, FCAST(IJ)
        DO 160 L = 1, 4
        Q(L) = (XMEAN(L)/AVER) * (FCAST(IJ)/4.0)
        PRINT 60, L, Q(L)
  160   CONTINUE
```

```
50  FORMAT ( )
60  FORMAT ( )
```

Figure 3.23 shows the complete computer program that includes the main program composed of our previously developed program segments and the subprogram LSLF discussed in the least squares regression section of this chapter.

```
          DIMENSION DEMAND(10,4),TOTAL(10),SUM(4)
          DIMENSION XMEAN(4),Q(4),X(10),XX(10)
          DIMENSION XY(10),FCAST(10),DELTA(10),Y(10)
C
C   READ AND PRINT OUT INPUT DATA.
          READ 15, NYRS,JNEED
          PRINT 17, NYRS,JNEED
          N=NYRS
C
          DO 100 I=1,NYRS
          READ 20, (DEMAND(I,J),J=1,4)
          PRINT 25, (DEMAND(I,J),J=1,4)
          X(I)=I
    100 CONTINUE
C
          DO 110 I=1,NYRS
          T=0
C
          DO 120 J=1,4
    120 T=T+DEMAND(I,J)
C
          TOTAL(I)=T
    110 CONTINUE
C
          Y=NYRS
C
          DO 130 J=1,4
          S=0
C
C   FOR EACH TIME PERIOD, PERFORM THE
C   CALCULATIONS.
C
          DO 140 I=1,NYRS
    140 S=S+DEMAND(I,J)
C
          SUM(J)=S
          XMEAN(J)=S/Y
    130 CONTINUE
C
          CALL LSLF(X,TOTAL,NYRS,XX,XY,FCAST,DELTA,
         1SUMX,SUMY,SUMXX,SUMXY,SDELTA,SFCAST,A,B)
          CXMEAN=0
C
          DO 150 M=1,4
    150 CXMEAN=CXMEAN+XMEAN(M)
C
```

FIGURE 3.23   Computer program for the forecasting model with seasonal variations and trend effect (illustrated for Example 3.3).

```
            AVER=CXMEAN/4.0
            PRINT 30, A,B
            FN=NYRS
C
C   CALCULATE THE FORECAST AND PRINT OUT.
C
            DO 160 IJ=1,JNEED
            F=IJ
            FCAST(IJ)=A+B*(FN+F)
            N=N+1
            PRINT 50, N,FCAST(IJ)
C
C   FOR EACH QUARTER, CALCULATE THE
C   FORECAST AND PRINT OUT.
C
            DO 160 L=1,4
            Q(L)=(XMEAN(L)/AVER)*(FCAST(IJ)/4.0)
            PRINT 60, L,Q(L)
       160 CONTINUE
C
            PRINT 40
C
        15 FORMAT(2I1)
        17 FORMAT(1H1,1X,2I1)
        20 FORMAT(4F5.1)
        25 FORMAT(1X,4(F5.1,1X))
        30 FORMAT(6X,17HTHE VALUE OF A IS,F7.2,2X,
           18HAND B IS,F7.2,//)
        40 FORMAT(1H0)
        50 FORMAT(6X,27HTHE SALES FORECAST FOR YEAR,
           1I3,1X,4HIS  ,F7.2)
        60 FORMAT(6X,I5,F10.2)
            STOP
            END



            SUBROUTINE LSLF(X,Y,NYRS,XX,XY,FCAST,DELTA,
           1SUMX,SUMY,SUMXX,SUMXY,SDELTA,SFCAST,A,B)
C
C   THIS IS THE LEAST SQUARES LINE
C   FITTING.
C
            DIMENSION X(10),Y(10),XX(10),XY(10)
            DIMENSION FCAST(10),DELTA(10)
C
C   INITIALIZE SUM VALUE
            SUMX=0
            SUMY=0
```

FIGURE 3.23 (Continued)

```
      SUMXX=0
      SUMXY=0
      SFCAST=0
      SDELTA=0
C
C   FOR EACH YEAR, PERFORM THE CALCULATIONS.
C
      DO 710 L=1,NYRS
      XX(L)=X(L)*X(L)
      XY(L)=X(L)*Y(L)
  710 CONTINUE
C
      DO 720 M=1,NYRS
      SUMX=SUMX+X(M)
      SUMY=SUMY+Y(M)
      SUMXX=SUMXX+XX(M)
      SUMXY=SUMXY+XY(M)
  720 CONTINUE
C
      YRS=NYRS
      DENR=YRS*SUMXX-SUMX**2
      A=(SUMY*SUMXX-SUMX*SUMXY)/DENR
      B=(YRS*SUMXY-SUMX*SUMY)/DENR
C
      DO 730 L=1,NYRS
      FCAST(L)=A+B*X(L)
      SFCAST=SFCAST+FCAST(L)
      DELTA(L)=Y(L)-FCAST(L)
      SDELTA=SDELTA+DELTA(L)
  730 CONTINUE
C
      PRINT 200
C
      DO 17 I=1,NYRS
      J=X(I)
      PRINT 300, J,Y(I),XX(I),XY(I),FCAST(I),
     1DELTA(I)
   17 CONTINUE
C
      PRINT 400, SUMX,SUMY,SUMXX,SUMXY,SFCAST,
     1SDELTA
C
  200 FORMAT(//,6X,4HYEAR,5X,5HSALES,7X,2HXX,
     17X,2HXY,6X,8HFORECAST,4X,5HDELTA,/)
  300 FORMAT(I9,1X,5F10.2)
  400 FORMAT(/,6F10.2,//)
      RETURN
      END
```

FIGURE 3.23 (*Continued*)

Let us employ the computer program to solve our Example 3.3. We have

NYRS = 5
JNEED = 1

The input data deck is designed according to the FORMAT statements 15 and 20. The data deck has the following appearance:

5 /				
50.0	75.0	67.0	60.0
54.0	82.0	73.0	65.0
59.0	86.0	76.0	77.0
64.0	81.0	80.0	68.0
63.0	92.0	81.0	81.0

The resulting computer printout is shown in Figure 3.24.

```
51
50.0   75.0   67.0   60.0
54.0   82.0   73.0   65.0
59.0   86.0   76.0   77.0
64.0   81.0   80.0   68.0
63.0   92.0   81.0   81.0
```

YEAR	SALES	XX	XY	FORECAST	DELTA
1	252.00	1.00	252.00	257.00	-5.00
2	274.00	4.00	548.00	271.90	2.10
3	298.00	9.00	894.00	286.80	11.20
4	293.00	16.00	1172.00	301.70	-8.70
5	317.00	25.00	1585.00	316.60	.40
15.00	1434.00	55.00	4451.00	1434.00	0

THE VALUE OF A IS 242.10   AND B IS   14.90

THE SALES FORECAST FOR YEAR   6 IS   331.50
```
1     67.04
2     96.17
3     87.15
4     81.14
```

FIGURE 3.24   Computer printout for Example 3.3.

PROBLEMS

3.1   The historical demand for bicycles for the past 12 months were as follows:

Month	Demand	Month	Demand
January	140	July	158
February	150	August	168
March	150	September	165
April	154	October	168
May	150	November	174
June	160	December	182

(a) Using least squares regression method, what would you estimate demand to be for the next three months, January through March?

(b) Construct a graph showing the past demand data and the three forecasts.

3.2   Consider Problem 3.1. Write a FORTRAN program to determine the forecasts for next year, January through December, by least squares regression.

3.3   Given the following one-year data of monthly demand for a product:

Month	Demand	Month	Demand
January	190	July	210
February	180	August	214
March	190	September	214
April	218	October	220
May	200	November	230
June	195	December	230

(a) Calculate a four-month moving average for the time series.

(b) Draw a graph showing the demand data and the forecast values.

3.4   Monthly sales in thousands of dollars for the past two years are given in the next page.

Month	Two Years Ago	One Year Ago
January	150	250
February	180	242
March	208	240
April	250	212
May	300	330
June	344	250
July	250	360
August	186	360
September	212	460
October	315	380
November	260	310
December	260	280

Assume $F_0 = D_1$ and $T_0 = 0$

(a) Determine the forecast for next January using the simple exponential smoothing for $\alpha = 0.10$.

(b) Determine the forecast for next January using the exponential smoothing corrected for lag due to trend for $\alpha = 0.10$.

(c) Plot the sales data and the two sets of forecasts determined in (a) and (b) on a graph. What conclusions can you make?

3.5 Consider Problem 3.4. Write a FORTRAN program to cause the computer to calculate the sum of absolute values (SAVFE) of forecasting errors using exponential smoothing corrected for lag due to trend, and to print out the SAVFE values for the 100 alpha values from 0.01 through 1.00 in steps of 0.01 and the best alpha value. The forecasts for the selected alpha must be also printed out.

3.6 The following figures are the number of students registered in ABC Secretarial School:

Quarter	Year 1	Year 2	Year 3	Year 4	Year 5	Year 6
Winter	83	84	96	99	103	112
Spring	62	68	70	77	62	75
Summer	105	110	120	126	132	145
Fall	60	66	72	74	68	82

(a) Plot the data on a graph and examine if trend and seasonal effects are significant.

(b) Calculate the forecasts of the number of students registered in year 7 and year 8 using the least squares regression method.
(c) Determine the forecasts for each of the four quarters in year 7 and year 8.

SELECTED REFERENCES

Biegel, J. E., *Production Control: A Quantitative Approach*, Prentice-Hall, Englewood Cliffs, N.J., Second Edition, 1971.

Brown, R. G., *Smoothing, Forecasting and Prediction*, Prentice-Hall, Englewood Cliffs, N.J., 1963.

Buffa, E. S., and W. H. Taubert, *Production-Inventory Systems: Planning and Control*, Richard D. Irwin, Homewood, Ill., Revised Edition, 1972.

Chambers, J. C., S. K. Mullick, and D. D. Smith, "How to Choose the Right Forecasting Technique," *Harvard Business Review*, July–August 1971, pp. 45–74.

Geoffrion, A. M., "A Summary of Exponential Smoothing," *Journal of Industrial Engineering*, Vol. XIII, No. 4, July–August, 1962.

Groff, G. E., and J. F. Muth., *Operations Management: Analysis for Decisions*, Richard D. Irwin, Homewood, Ill., 1972.

Levin, R. I., McLaughlin, R. P. Lamone, and J. F. Koltas, *Production/Operations Management: Contemporary Policy for Managing Operating Systems*, McGraw-Hill, New York, 1972.

Magee, J. F., and D. M. Boodman, *Production Planning and Inventory Control*, McGraw-Hill, New York, Second Edition, 1967.

Mayer, R. R., *Production Management*, McGraw-Hill, New York, Second Edition, 1968.

Optner, S. L., *Systems Analysis for Business Management*, Prentice-Hall, Englewood Cliffs, N.J., Second Edition, 1968.

Parkey, G. G. C., and E. L. Segura, "How to Get a Better Forecast," *Harvard Business Review*, March–April, 1971, pp. 99–109.

Reinfeld, N. V., *Production Control*, Prentice-Hall, Englewood Cliffs, N.J., 1959.

Riggs, J. L., *Production Systems: Planning, Analysis, and Control*, Wiley, New York, 1970.

Shore, B., *Operations Management*, McGraw-Hill, New York, 1973.

Winters, P. R., "Forecasting Sales by Exponentially Weighted Moving Averages," *Management Science*, Vol. 6, No. 3, April 1960, pp. 324–342.

matrix models

CHAPTER 4

M ATRICES ARE EXTREMELY USEFUL in the formulation and analysis of many systems problems. Matrices offer a convenient notational device for solving problems whose solution would otherwise be clumsy and inefficient. Types of matrices and the matrix operations are presented in the first part of this chapter. Next, the Gauss reduction method used for solving a system of linear equations is discussed. Finally, the application of matrices is illustrated by two examples—the input-output analysis and the production requirements planning problem.

TYPES OF MATRICES

We are familiar with data that are arranged in the tabular form or in rectangular arrays. A manufacturer might prepare the summary of weekly shipments as follows:

Geographical Area	Product			
	1	2	3	4
Area 1	60	45	40	39
Area 2	70	20	52	28
Area 3	40	36	70	60

A matrix is a rectangular array of elements such as the summary of of shipments shown above. If the above matrix is designated S, the element in the second row and fourth column is referred to as s_{24}. Thus,

$$s_{24} = 28$$

Likewise, s_{ij} is the element in the ith row and jth column.

The matrix S has 3 rows and 4 columns. There is a total of 12 elements ($=3 \times 4$). The *order* (*size*) of a matrix indicates the number of rows and columns in the matrix. The size of the matrix S is 3×4 (3 by 4). If a matrix A has m rows and n columns, it will have $(m)(n)$ elements and it is of the size $m \times n$. It is represented as follows:

$$A = \begin{bmatrix} a_{11} & a_{12} & a_{13} & \cdots & a_{1n} \\ a_{21} & a_{22} & a_{23} & \cdots & a_{2n} \\ \vdots & & & & \vdots \\ a_{m1} & a_{m2} & a_{m3} & \cdots & a_{mn} \end{bmatrix}$$

A matrix does not have a numerical value; it represents the arrangement of numbers in an orderly manner. Matrices are designated by special names, when the number of rows (m) and number of columns (n) have certain values or when the elements in a matrix have particular values.

Row Vector A row vector is a matrix with one row and two or more columns. For example,

$$B = \begin{bmatrix} 7 & 0 & 9 & 2 \end{bmatrix}$$

$$C = \begin{bmatrix} 66 & -9 & -7 \end{bmatrix}$$

are row vectors of size (1×4) and (1×3), respectively.

Column Vector A column vector is a matrix with two or more rows and one column. For example,

$$D = \begin{bmatrix} 4 \\ 5 \\ 89 \end{bmatrix} \qquad E = \begin{bmatrix} 98 \\ 0 \\ -7 \\ 43 \end{bmatrix}$$

are column vectors of size (3×1) and (4×1), respectively.

Square Matrix A square matrix is a matrix in which $m = n$. For example,

$$F = \begin{bmatrix} 4 & 5 \\ 7 & 9 \end{bmatrix} \qquad G = \begin{bmatrix} 1 & 2 & 3 \\ 4 & 5 & 6 \\ 7 & 8 & 9 \end{bmatrix}$$

are the square matrices of size (2×2) and (3×3), respectively.

Zero Matrix A matrix in which each element is zero is called a zero matrix. It is also known as a *null matrix*. For example,

$$H = \begin{bmatrix} 0 & 0 & 0 \\ 0 & 0 & 0 \end{bmatrix} \qquad I = \begin{bmatrix} 0 & 0 & 0 \\ 0 & 0 & 0 \\ 0 & 0 & 0 \end{bmatrix}$$

are the null matrices of size (2×3) and (3×3), respectively.

Diagonal Matrix A square matrix in which all nondiagonal elements are zero is called a diagonal matrix. For example, among the two square matrices,

$$J = \begin{bmatrix} 1 & 0 & 0 \\ 0 & 6 & 0 \\ 0 & 0 & 3 \end{bmatrix} \qquad K = \begin{bmatrix} 9 & 0 & 3 \\ 0 & 5 & 0 \\ 0 & 0 & 8 \end{bmatrix}$$

J is a diagonal matrix. However, K is not a diagonal matrix because the nondiagonal element k_{13} is not a zero.

Identity Matrix An identity matrix is a square matrix in which all the diagonal elements are one and all nondiagonal elements are zero. It is also called a *unit matrix*. For example,

$$L = \begin{bmatrix} 1 & 0 & 0 \\ 0 & 1 & 0 \\ 0 & 0 & 1 \end{bmatrix} \qquad M = \begin{bmatrix} 1 & 0 & 0 & 0 \\ 0 & 1 & 0 & 0 \\ 0 & 0 & 1 & 0 \\ 0 & 0 & 0 & 1 \end{bmatrix}$$

are unit matrices of size (3×3) and (4×4), respectively.

Symmetric Matrix A symmetric matrix A is a square matrix in which $A_{ij} = A_{ji}$ for all i and j. For example,

$$N = \begin{bmatrix} 1 & 7 & 5 \\ 7 & 6 & -3 \\ 5 & -3 & 4 \end{bmatrix}$$

is a symmetric matrix of size (3×3).

Transpose of a Matrix The transpose of a matrix A is obtained by interchanging the rows and columns of A such that the first row of A becomes the first column of the transposed matrix, the second row of A becomes the second column of the transposed matrix, and so on. For example, the transpose of the matrices,

$$P = \begin{bmatrix} 3 & 2 & 6 \end{bmatrix} \qquad Q = \begin{bmatrix} 7 & 4 \\ 1 & 0 \\ 8 & 2 \end{bmatrix}$$

are R and S, respectively.

$$R = \begin{bmatrix} 3 \\ 2 \\ 6 \end{bmatrix} \qquad S = \begin{bmatrix} 7 & 1 & 8 \\ 4 & 0 & 2 \end{bmatrix}$$

Note that the transpose of a symmetric matrix is the same matrix.

Equal Matrices Two matrices, A and B, are said to be equal if these two conditions are satisfied:

1. A and B have the same size.
2. Each element of A is equal to the corresponding element of B for all elements.

For example, $A = B$ where

$$A = \begin{bmatrix} 3 & 4 & 1 \\ 6 & 5 & 9 \end{bmatrix} \qquad B = \begin{bmatrix} 3 & 4 & 1 \\ 6 & 5 & 9 \end{bmatrix}$$

MATRIX OPERATIONS

In matrix algebra, matrices can be added, subtracted, and multiplied, but they cannot be divided.

Matrix Addition and Subtraction

Two matrices can be added (subtracted) only if they are of the same size. When the two matrices, A and B, are added, the elements of the new matrix C are formed by adding the corresponding elements of A and B. When B is subtracted from A, the elements of the new matrix C are determined by subtracting the B elements from the corresponding A elements. The size of the new matrix C is the same as the size of matrix A or matrix B.

Example for Matrix Addition Suppose that a small company sells three products through two salesmen. The sales volumes of the salesmen are tabulated each week indicating the breakdown figures for the three products as shown below.

Summary of sales for the preceding week (first week) is depicted in matrix IA.

Matrix IA
Product

	1	2	3
Salesman 1	400	100	1200
Salesman 2	1050	50	800

The company wishes to determine the cumulative sales values at the end of each consecutive week. The sales for the second week are given in matrix IB.

Matrix IB
Product

	1	2	3
Salesman 1	600	50	700
Salesman 2	800	200	1000

The cumulative sales volume at the end of the second week can be determined by adding IA and IB. Let

IC = The cumulative sales values at the end of the second week

Thus,

$$IA + IB = IC$$

$$\begin{bmatrix} 400 & 100 & 1200 \\ 1050 & 50 & 800 \end{bmatrix} + \begin{bmatrix} 600 & 50 & 700 \\ 800 & 200 & 1000 \end{bmatrix}$$

$$= \begin{bmatrix} 400 + 600 & 100 + 50 & 1200 + 700 \\ 1050 + 800 & 50 + 200 & 800 + 1000 \end{bmatrix}$$

$$= \begin{bmatrix} 1000 & 150 & 1900 \\ 1850 & 250 & 1800 \end{bmatrix}$$

From the IC values, we can determine the cumulative sales of the salesmen by adding the row numbers. The cumulative sales of the products can be established by adding the column numbers of matrix *IC*.

Total sales of salesman 1 through the second week = 1000 + 150
$$+ 1900 = \$3050$$
Total sales of salesman 2 through the second week = 1850 + 250
$$+ 1800 = \$3900$$
Total sales of product 1 through the second week = 1000 + 1850
$$= \$2850$$
Total sales of product 2 through the second week = 150 + 250
$$= \$400$$
Total sales of product 3 through the second week = 1900 + 1800
$$= \$3700$$

To calculate the cumulative sales figures through the third week, we may use the same approach. Matrix *IA* is imagined to be the total sales through the preceding week (second week) and matrix *IB* is visualized as the sales of the third week. Then, *IC* is obtained by adding *IA* and *IB*. *IC* gives the cumulative values through the third week. The elements of the updated matrix *IC* can be computed by the program depicted in Figure 4.1.

```
      DIMENSION IA(2,3),IB(2,3),IC(2,3)
C
      DO 10 K = 1,2
      READ 70,(IA(K,N), N=1,3)
   10 PRINT 70,(IA(K,N), N=1,3)
C
      PRINT 60
      DO 20 K = 1,2
      READ 70,(IB(K,N),N=1,3)
   20 PRINT 70,(IB(K,N),N=1,3)
C
      DO 30 L = 1,2
      DO 30 M = 1,3
   30 IC(L,M) = IA(L,M)+IB(L,M)
C
      PRINT 60
      DO 40 N= 1,2
   40 PRINT 70,(IC(N,K),K=1,3)
C
   60 FORMAT (1H0)
   70 FORMAT (3I6)
      STOP
      END
```

FIGURE 4.1 Computer program for the cumulative sales calculation problem.

Example of Matrix Subtraction The Ronald Electric Company operates many stores in New Orleans. The products are supplied to the stores by three warehouses. The inventory for two products available at the beginning of a particular week is displayed in matrix IA as follows:

Matrix IA
Warehouse

	1	2	3
Product 1	450	720	860
Product 2	150	250	260

The number of units of each product supplied by the warehouses to the stores for this particular week is presented in the matrix IB.

Matrix IB
Warehouse

	1	2	3
Product 1	100	500	350
Product 2	70	120	200

The total number of units available in inventory at the end of the week in each warehouse (before considering the receipt of shipments, if any) can be determined by the expression

$$IC = IA - IB \qquad (4.1)$$

For our illustration, the calculations are

$$IC = \begin{bmatrix} 450 - 100 & 720 - 500 & 860 - 350 \\ 150 - 70 & 250 - 120 & 260 - 200 \end{bmatrix}$$
$$= \begin{bmatrix} 350 & 220 & 510 \\ 80 & 130 & 60 \end{bmatrix}$$

The column totals of matrix IC indicate the total units available in the corresponding warehouses. The row totals of matrix IC denote the total units of the corresponding products available in the three warehouses. Figure 4.1 can be employed after making one change in it to compute and printout the elements of the new matrix IC. The modification is that the statement 30 should be altered as follows:

```
30 IC(L, M) = IA(L, M) − IB(L, M)
```

Matrix Multiplication

Matrix A can be multiplied by matrix B, if, and only if, the number of columns of A is equal to the number of rows of B.

$$\underset{\text{(size } m \times r)}{A} \times \underset{\text{(size } s \times n)}{B} = \underset{\text{(size } m \times n)}{C} \tag{4.2}$$

In other words, it is required that $r = s$. The resulting matrix C would be of the size $m \times n$. If r is not equal to s, the two matrices cannot be multiplied. In the matrix product A times B, matrix A is called the *pre-multiplier*, and matrix B is called the *postmultiplier*. Each element of C, namely C_{ij}, is determined by adding the values that are obtained by multiplying the ith row elements of A by the corresponding jth column elements of B.

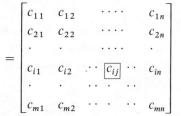

When

$$\underset{\text{(size } m \times r)}{A} \times \underset{\text{(size } r \times n)}{B} = \underset{\text{(size } m \times n)}{C}$$

then,

$$C_{ij} = a_{i1} \cdot b_{1j} + a_{i2} \cdot b_{2j} + \cdots + a_{ir} \cdot b_{rj} \tag{4.3}$$

Multiplication of a Row Vector by a Column Vector   A truck is loaded with four products. Their quantities in boxes and freight charges per box are given in the vectors Q and F.

Row Vector Q
Quantities in Boxes

Product

P1	P2	P3	P4

12	15	20	10

(Size 1 × 4)

Column Vector F
Freight Charges per Box

P1	$2.00
P2	$1.00
P3	$1.50
P4	$3.00

(Size 4 × 1)

For product $P1$, freight charge = 12(2.00) =   24.00
For product $P2$, freight charge = 15(1.00) =   15.00
For product $P3$, freight charge = 20(1.50) =   30.00
For product $P4$, freight charge = 10(3.00) =   30.00
Total freight charge = $99.00

The matrix product QF can be found from expression 4.2 and the result C would be a single number. The resulting C value indicates the total freight charge.

$$C = 12(2.00) + 15(1.00) + 20(1.50) + 10(3.00)$$
$$= \$99.00$$

Multiplication of a Matrix by a Column Vector   Three alloys are manufactured out of four metals according to the specifications shown in matrix A. The cost per pound of the metals is given in matrix B.

Matrix A
Amounts of Metals Needed
to Produce One Pound of
the Alloys

Metal

	M1	M2	M3	M4
Alloy 1	0.3	0.5	0.1	0.1
Alloy 2	0.2	0.3	0.4	0.1
Alloy 3	0.2	0.4	0.3	0.1

(Size 3 × 4)

Matrix B
Cost per Pound
of the Metals

M1	$4
M2	$10
M3	$8
M4	$20

(Size 4 × 1)

The total cost of metals for producing one pound of each alloy can be established by the matrix product $(A)(B)$.

Let matrix $C = (A)(B)$

Then,

$$C = \begin{bmatrix} 0.3 & 0.5 & 0.1 & 0.1 \\ 0.2 & 0.3 & 0.4 & 0.1 \\ 0.2 & 0.4 & 0.3 & 0.1 \end{bmatrix} \times \begin{bmatrix} 4 \\ 10 \\ 8 \\ 20 \end{bmatrix}$$

$$= \begin{bmatrix} (0.3)(4) + (0.5)(10) + (0.1)(8) + (0.1)(20) \\ (0.2)(4) + (0.3)(10) + (0.4)(8) + (0.1)(20) \\ (0.2)(4) + (0.4)(10) + (0.3)(8) + (0.1)(20) \end{bmatrix}$$

$$= \begin{bmatrix} 9.00 \\ 9.00 \\ 9.20 \end{bmatrix}$$

The total raw material cost for one pound of alloy 1, alloy 2, and alloy 3 are \$9.00, \$9.00, and \$9.20, respectively. Recall that the matrix multiplication is possible only if the number of columns of A is equal to the number of rows of B.

Multiplication of a Matrix by a Matrix   The Simmons Mailorder House sells four products. For each customer order, the company wishes to determine the total value of the order and the net weight of the package. The order quantities for three customers are given in matrix A. The unit price as well as the weight for each product are shown in matrix B.

Matrix A
Customer's Orders in Units

Customer	Product			
	P1	P2	P3	P4
C_1	4	1	5	2
C_2	0	2	3	0
C_3	3	0	4	0

(Size 3 × 4)

Matrix B
Unit Price and Weight
of the Products

	Cost (\$)	Weight (pounds)
P_1	2	1.5
P_2	3	2
P_3	1	1
P_4	4	2

(Size 4 × 2)

The matrix product C gives the desired values. Since A is of the size 3×4 and B is 4×2, C would be of the size 3×2.

$$C = \begin{bmatrix} 4 & 1 & 5 & 2 \\ 0 & 2 & 3 & 0 \\ 3 & 0 & 4 & 0 \end{bmatrix} \times \begin{bmatrix} 2 & 1.5 \\ 3 & 2 \\ 1 & 1 \\ 4 & 2 \end{bmatrix}$$

$$= \begin{matrix} (4)(2) + (1)(3) + (5)(1) + (2)(4) & (4)(1.5) + (1)(2) + (5)(1) + (2)(2) \\ (0)(2) + (2)(3) + (3)(1) + (0)(4) & (0)(1.5) + (2)(2) + (3)(1) + (0)(2) \\ (3)(2) + (0)(3) + (4)(1) + (0)(4) & (3)(1.5) + (0)(2) + (4)(1) + (0)(2) \end{matrix}$$

	Total Cost ($)	Net Weight (pounds)
C_1	24	17
$C = C_2$	9	7
C_3	10	8.5

The matrix C denotes the total cost and the net weight of each order for the three customers.

A Computer Program for the Matrix Multiplication   We shall formulate a computer program in this section that will cause the computer to calculate and print out the input data and the computed values. The number of rows and number of columns of the premultiplier (A) and the postmultiplier (B) cannot exceed 10 for using the computer model. Matrix C designates the solution of the matrix multiplication. The following notations are used in the program.

M = number of rows in matrix A
JR = number of columns in matrix A
KR = number of rows in matrix B
N = number of columns in matrix B

Figure 4.2 displays a flowchart for the model, and Figure 4.3 shows a general computer program for the matrix multiplication problem. Recall that JR must be equal to KR to perform the matrix product as given in expression 4.3. In the program, when JR is not equal to KR, the execution is terminated after printing A AND B *cannot be multiplied.* Otherwise, the input data and the resulting matrix C are printed out.

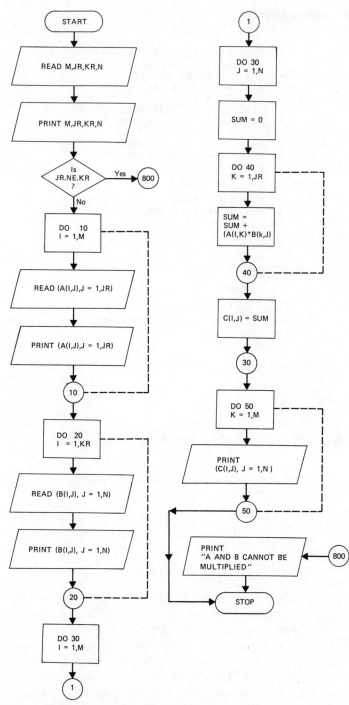

FIGURE 4.2 **Flowchart for matrix multiplication operations.**

```
      DIMENSION A(10,10), B(10,10), C(10,10)
      READ 110, M,JR,KR,N
      PRINT 110, M,JR,KR,N
      IF (JR .NE. KR) GO TO 800
      DO 10 I = 1,M
      READ 120, (A(I,J),J=1,JR)
   10 PRINT 120, (A(I,J),J=1,JR)
      DO 20 I = 1,KR
      READ 130, (B(I,J),J=1,N)
   20 PRINT 130, (B(I,J),J=1,N)
      DO 30 I = 1,M
      DO 30 J = 1,N
      SUM = 0
      DO 40 K = 1,JR
   40 SUM=SUM+A(I,K)*B(K,J)
   30 C(I,J) = SUM
      DO 50 K=1,M
   50 PRINT 140, (C(K,J),J=1,N)
      GO TO 810
  800 PRINT 150
  110 FORMAT (4I3)
  120 FORMAT (4F7.1)
  130 FORMAT (3F4.1)
  140 FORMAT (/,4F10.2)
  150 FORMAT (5X, 28HA AND B CANNOT BE MULTIPLIED)
  810 STOP
      END
```

FIGURE 4.3   **Computer program for matrix multiplication operations.**

SYSTEMS OF N LINEAR EQUATIONS IN N UNKNOWNS

A system of n linear equations in n unknowns can be represented in the form

$$a_{11}x_1 + a_{12}x_2 + \cdots + a_{1n}x_n = b_1$$
$$a_{21}x_1 + a_{22}x_2 + \cdots + a_{2n}x_n = b_2$$
$$\dots\dots\dots\dots\dots\dots\dots\dots\dots\dots\dots\dots\dots$$
$$a_{n1}x_1 + a_{n2}x_2 + \cdots + a_{nn}x_n = b_n \qquad (4.4)$$

We present in this section the algebraic reduction procedure and the Gauss reduction method for solving systems of linear equations. By solving systems of linear equations, we establish the values of the n unknowns, $x_1, x_2, \ldots, x_n$, using the n equations. A computer model is developed subsequently for the Gauss reduction technique. The number of equations in a system is equal to the number of variables. For example, to solve for 10 unknowns, 10 linear equations are required.

Algebraic Reduction Procedure

Consider the two equations

$$5X_1 + 10X_2 = 250 \tag{4.5}$$

$$10X_1 + 5X_2 = 200 \tag{4.6}$$

An equation may be multiplied or divided by the same nonzero value on both sides of the equation and the equality is not altered (Rule 1). For example, (4.5) may be rewritten in the form

$$10X_1 + 20X_2 = 500 \tag{4.7}$$

On both sides of an equation, equal amounts may be added or subtracted and the equality is unchanged (Rule 2).

We can determine the values of x_1 and x_2 from the equations 4.5 and 4.6 as follows: Subtracting (4.6) from (4.7), we get

$$(10x_1 + 20x_2) - (10x_1 + 5x_2) = 500 - 200$$

$$(10 - 10)x_1 + (20 - 5)x_2 = 300$$

$$15x_2 = 300$$

$$x_2 = \frac{300}{15} = 20$$

Substituting $x_2 = 20$ in (4.5), it reduces to

$$5x_1 + 10(20) = 250$$

$$5x_1 = 250 - 200 = 50$$

$$x_1 = \frac{50}{5} = 10$$

Therefore, the solution is

$$x_1 = 10$$

$$x_2 = 20$$

These two rules are applied to solve a system of linear equations. The procedure is illustrated for a problem with three unknowns. The zero coefficients are carried through the end of the solution so that there will be three coefficients in each equation at all times. Here are the three equations:

$$2x_1 + 4x_2 + 2x_3 = 20 \tag{4.8}$$

$$3x_1 + 4x_2 + 1x_3 = 18 \tag{4.9}$$

$$5x_1 + 3x_2 + 6x_3 = 48 \tag{4.10}$$

Multiplying (4.9) by 2 and subtracting (4.8) from it, the equations obtained are:

$$(4.9) \times 2 \qquad 6x_1 + 8x_2 + 2x_3 = 36 \tag{4.11}$$

$$(4.8) \qquad 2x_1 + 4x_2 + 2x_3 = 20$$

$$(4.11) - (4.8) \qquad 4x_1 + 4x_2 + 0x_3 = 16 \tag{4.12}$$

Equation 4.8 can be multiplied by 3 and from it (4.10) can be subtracted. This operation will not change the solution.

$$(4.8) \times 3 \qquad 6x_1 + 12x_2 + 6x_3 = 60 \tag{4.13}$$

$$(4.10) \qquad 5x_1 + 3x_2 + 6x_3 = 48$$

$$(4.13) - (4.10) \qquad 1x_1 + 9x_2 + 0x_3 = 12 \tag{4.14}$$

Equation 4.14 can be multiplied by 4 and from it (4.12) can be subtracted. This operation will not alter the solution.

$$(4.14) \times 4 \qquad 4x_1 + 36x_2 + 0x_3 = 48 \tag{4.15}$$

$$(4.12) \qquad 4x_1 + 4x_2 + 0x_3 = 16$$

$$(4.15) - (4.12) \qquad 0x_1 + 32x_2 + 0x_3 = 32 \tag{4.16}$$

Dividing (4.16) by 32, we get

$$(4.16)/32 \qquad 0x_1 + 1x_2 + 0x_3 = 1 \tag{4.17}$$

Therefore,

$$x_2 = 1$$

Subtracting 4 times (4.17) from (4.12), we obtain

$$(4.12) - (4.17) \times 4 \qquad 4x_1 + 0x_2 + 0x_3 = 12 \tag{4.18}$$

Equation 4.18 can be written as

$$(4.18)/4 \qquad 1x_1 + 0x_2 + 0x_3 = 3 \tag{4.19}$$

Hence,

$$x_1 = 3$$

Subtracting 3 times (4.19) and 4 times (4.17) from (4.9), we get

$$(4.9) - (4.19) \times 3 - (4.17) \times 4 \qquad 0x_1 + 0x_2 + 1x_3 = 5 \tag{4.20}$$

Therefore,

$$x_3 = 5$$

Equations 4.19, 4.17, and 4.20 are

$$1x_1 + 0x_2 + 0x_3 = 3$$
$$0x_1 + 1x_2 + 0x_3 = 1$$
$$0x_1 + 0x_2 + 1x_3 = 5 \tag{4.21}$$

This indicates that the values of x_1, x_2, and x_3 are 3, 1, and 5, respectively. The validity of the solution may be checked by substituting the values in the original equations.

$$(4.8) \qquad 2(3) + 4(1) + 2(5) = 20$$

$$(4.9) \qquad 3(3) + 4(1) + 1(5) = 18$$

$$(4.10) \qquad 5(3) + 3(1) + 6(5) = 48$$

Since all of the three equations are satisfied, this indicates that we have the valid solution. Note that the square matrix containing the coefficients is modified to a unit matrix as shown in (4.21) for solving the linear equations.

Gauss Reduction Method

The coefficients (a_{ij}) and the right-hand side constants (b_i) are changed in a systematic way by applying certain operations repeatedly in the Gauss reduction procedure. The iterative process is continued until the square matrix representing the coefficients becomes a unit matrix. For solving a system of n linear equations in n unknowns, n iterations are performed. In the kth iteration, a_{kk} is the *pivot element*, the kth row is the *pivot row*, and the kth column is the *pivot column*. The computational steps follow:

STEP 1. Determine the pivot element (a_{kk}), the pivot row (row k), and the pivot column (column k).

STEP 2. Divide the pivot row numbers by the nonzero pivot element. These values are written in the updated matrix.

STEP 3. Each row values, except the pivot row, are altered as follows: For changing the ith row values, multiply the updated pivot row values by ($-a_{ik}$), add them with the corresponding ith row numbers, and record them in the ith row of the updated matrix. The effect is that the pivot column values are all changed to zero except a_{kk}. The value of a_{kk} is one.

STEP 4. If k is less than n, go to Step 1. If k is equal to n, stop.

The computational procedure is illustrated by the three equations (4.8) through (4.10) of the previous example. The data may be written in the matrix form

x_1	x_2	x_3	b	
2	4	2	20	(4.22a)
3	4	1	18	(4.22b)
5	3	6	48	(4.22c)

Iteration 1:

$$\text{Pivot element} = a_{11} = 2$$

$$\text{Pivot row} = \text{row 1}$$

$$\text{Pivot column} = \text{column 1}$$

The updated matrix is

	x_1	x_2	x_3	b	
(4.22a)/2	1	2	1	10	(4.23a)
(4.23a) × (−3) + (4.22b)	0	−2	−2	−12	(4.23b)
(4.23a) × (−5) + (4.22c)	0	−7	1	−2	(4.23c)

Note that column 1 elements are in the required form for the unit matrix.

Iteration 2:

$$\text{Pivot element} = a_{22} = -2$$

$$\text{Pivot row} = \text{row 2}$$

$$\text{Pivot column} = \text{column 2}$$

The updated matrix is

	x_1	x_2	x_3	b	
(4.24b) × (−2) + (4.23a)	1	0	−1	−2	(4.24a)
(4.23b)/(−2)	0	1	1	6	(4.24b)
(4.24b) × (7) + (4.23c)	0	0	8	40	(4.24c)

The elements of column 1 and column 2 are in the required form for the unit matrix.

Iteration 3: (last iteration)

$$\text{Pivot element} = a_{33} = 8$$

$$\text{Pivot row} = \text{row } 3$$

$$\text{Pivot column} = \text{column } 3$$

The updated matrix is

	x_1	x_2	x_3	b	
(4.25c) + (4.25c)	1	0	0	3	(4.25a)
(4.25c) × (−1) + (4.24b)	0	1	0	1	(4.25b)
(4.24c)/8	0	0	1	5	(4.25c)

The updated matrix is a unit matrix and the values of b, namely 3, 1, and 5, denote the values of x_1, x_2, and x_3, respectively.

A Difficult Case   Sometimes the pivot element is zero during the iterative process of the Gauss reduction method. In such a case, some modifications are needed to apply the technique. The trick is to interchange the pivot row (row k) and the following row (row $k + 1$); then, the operations are continued designating the new kth row as the pivot row. Consider the three equations

$$4x_1 + 2x_2 + 6x_3 = 4 \tag{4.26}$$

$$3x_1 + 1.5x_2 + 6x_3 = 4.5 \tag{4.27}$$

$$5x_1 + 4.5x_2 + 4x_3 = 5.5 \tag{4.28}$$

The coefficients and the values are written in the tabular form:

x_1	x_2	x_3	b	
4	2	6	4	
3	1.5	6	4.5	(4.29)
5	4.5	4	5.5	

The first iteration yields the updated matrix:

x_1	x_2	x_3	b
1	0.5	1.5	1
0	0	1.5	1.5
0	2	-3.5	0.5

(4.30)

The pivot element for the second iteration is $a_{22}(=0)$. The second and the third row values are interchanged and the new pivot element equals 2.

x_1	x_2	x_3	b
1	0.5	1.5	1
0	2	-3.5	0.5
0	0	1.5	1.5

(4.31)

The second and the third iterations yield the following updated tableaus:

x_1	x_2	x_3	b
1	0	2.375	0.875
0	1	-1.75	0.25
0	0	1.5	1.5

(4.32)

x_1	x_2	x_3	b
1	0	0	-1.5
0	1	0	2
0	0	1	1

(4.33)

The values of x_1, x_2, and x_3 are -1.5, 2, and 1, respectively.

A Computer Program for Solving a System of Linear Equations

In this section we formulate a computer model for solving a system of linear equations utilizing the Gauss reduction procedure. The following variable names will be used in the program:

NVARI   = number of variables in the given system of linear equations

NCOLS   = number of columns in the matrix including the constant (b_i) column
                  = NVARI + 1

IROW   = pivot row

C(IROW, IROW)   = pivot element

FACTR   = pivot element value
                  = C(IROW, IROW)

C(I, J)   = coefficient in row *i* for variable *j*. For $j = 1, 2, \ldots,$ NVARI

C(I, NCOLS)   = constant in row $i = b_i$

CHANGE   = this subprogram interchanges the pivot row values and (IROW + 1)th row values in the complex situation.

The variable $C(,)$ is a two-dimensional array of size NVARI by NCOLS. We might employ the following statements to read and print out the input data.

```
      READ 150, NVARI
      NCOLS = NVARI + 1
      DO 153 L = 1, NVARI
      READ 155, (C(L, J), J = 1, NCOLS)
      PRINT 160, (C(L, J), J = 1, NCOLS)
  153 CONTINUE
  150 FORMAT ( )
  155 FORMAT ( )
  160 FORMAT ( )
```

For a given IROW value, the row values would be interchanged when required and the pivot value is established by the instructions

```
      IF (C(IROW, IROW) .NE. O) GO TO 14
      CALL CHANGE (C, IROW, NCOLS)
   14 FACTR = C (IROW, IROW)
```

Here is the subprogram CHANGE that interchanges the numbers of the pivot row (IROW) and the following row (IROW + 1):

```
SUBROUTINE CHANGE(C, IROW, NCOLS)
DIMENSION C(3, 4)
M = IROW + 1
DO 7 K = 1, NCOLS
VAL = C (IROW, K)
C(IROW, K) = C(M, K)
C(M, K) = VAL
7   CONTINUE
RETURN
END
```

Next, the pivot row values are modified so that the pivot element becomes one.

```
DO 20 JCOL = 1, NCOLS
C(IROW, JCOL) = C(IROW, JCOL)/FACTR
20   CONTINUE
```

The following statements would cause the computer to alter the other (NVARI-1) rows so that the pivot column elements become zero:

```
DO 30 I = 1, NVARI
IF (I.EQ.IROW) GO TO 30
X = − C(I, IROW)
DO 30 JCOL = 1, NCOLS
C(I, JCOL) = (C(IROW, JCOL) * X) + C(I, JCOL)
30   CONTINUE
```

By placing the main program instructions that are stated above within the DO loop, the iterative operations are executed NVARI times.

```
DO 10 IROW = 1, NVARI
───────────
10   CONTINUE
```

The updated matrix that would be obtained after executing the last iteration may be printed out by the instructions:

```
      DO 40 K = 1, NVARI
      PRINT 160, (C(K, JCOL), JCOL = 1, NCOLS)
 40   CONTINUE
160   FORMAT ( )
```

We might print out the solution of the linear equations by the following statements:

```
      DO 50 L = 1, NVARI
      PRINT 185, L, C(L, NCOLS)
 50   CONTINUE
185   FORMAT ( )
```

The complete computer program is shown in Figure 4.4.

An Illustration of the Use of the Computer Model   The three-variable problem given in (4.29) can be solved employing the computer program as follows:

The values of NVARI and the matrix elements might be transmitted according to the FORMAT instructions:

```
150   FORMAT (I10)
155   FORMAT (4F4.2)
```

The input data deck has the following appearance:

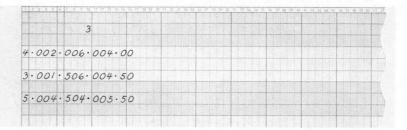

We might employ the following FORMAT statement for printing the initial values and the final values of the matrix:

```
160   FORMAT (4F10.2)
```

```
                DIMENSION C (3,4)
        C
        C   READ AND PRINTOUT THE DATA
                READ 150, NVARI
                PRINT 800
                NCOLS= NVARI+1
                DO 153 L=1, NVARI
                READ 155, (C(L,J), J=1,NCOLS)
                PRINT 160, (C(L,J),J=1,NCOLS)
            153 CONTINUE
        C   MODIFY THE MATRIX SUCH THAT
        C   C(K,K) = 1 FOR K = 1,2,...,NVARI
        C   AND ALL OTHER C(I,J) = 0
                DO 10 IROW=1, NVARI
                IF(C(IROW,IROW).NE.0)GO TO 14
        C   THE PIVOT ELEMENT IS ZERO
                CALL CHANGE(C,IROW,NCOLS)
            14 FACTR = C (IROW,IROW)
        C   MODIFY THE ROW COEFFICIENTS
                DO 20 JCOL=1,NCOLS
                C(IROW,JCOL)= C(IROW,JCOL)/FACTR
            20 CONTINUE
        C   MODIFY THE COLUMN COEFFICIENTS
                DO 30 I = 1,NVARI
                IF(I.EQ.IROW)GO TO 30
                X=-C(I,IROW)
                DO 30 JCOL = 1,NCOLS
                C(I,JCOL)=(C(IROW,JCOL)*X)+C(I,JCOL)
            30 CONTINUE
            10 CONTINUE
        C   PRINTOUT THE LAST REDUCED MATRIX
                PRINT 170
                DO 40 K = 1,NVARI
                PRINT 160, (C(K,JCOL),JCOL= 1,NCOLS)
            40 CONTINUE
                PRINT 170
        C   PRINTOUT THE VALUES OF THE VARIABLES
                PRINT 171

                DO 50 L=1,NVARI
                PRINT 185 , L, C(L,NCOLS)
            50 CONTINUE
           150 FORMAT(I10)
           155 FORMAT(4F4.2)
           160 FORMAT(4F10.2)
           170 FORMAT (3(/))
           171 FORMAT(5X,9H VARIABLE ,6H VALUE )
           185 FORMAT ( I11,F10.3)
           800 FORMAT (1H1)
                STOP
                END

                SUBROUTINE CHANGE(C,IROW,NCOLS)
                DIMENSION C(3,4)
        C   THE ROWS ARE INTERCHANGED
                M = IROW+ 1
                DO 7 K = 1, NCOLS
                VAL = C (IROW,K)
                C(IROW,K) = C(M,K)
                C(M,K) = VAL
             7 CONTINUE
                RETURN
                END
```

FIGURE 4.4 Computer program for solving a system of linear equations [illustrated for the three-variable problem (4.29)].

The solution of the system of equations may be printed out by employing the FORMAT statement:

185   FORMAT (I11, F10.3)

The resulting computer printout is shown in Figure 4.5.

```
4.00        2.00        6.00        4.00
3.00        1.50        6.00        4.50
5.00        4.50        4.00        5.50


1.00           0           0       -1.50
   0        1.00           0        2.00
   0           0        1.00        1.00


VARIABLE  VALUE
     1      -1.500
     2       2.000
     3       1.000
```

FIGURE 4.5   Computer printout for the three-variable problem (4.29).

The matrices can be employed to solve a wide variety of problems. We shall demonstrate the use of matrices by these two examples:

1. Input-output analysis
2. Parts requirements problem

INPUT-OUTPUT ANALYSIS

The input-output model was developed by Professor Leontief in the 1930s; it requires the solution of a system of n linear equations in n unknowns. This model may be employed to determine the levels of activity in the various sectors of a national economy that is conformable with the desired pattern of exogenous demand, such as the export. The activity of a particular sector, say the automobile industry, may be operated at any positive level so long as the required inputs are available. It may be assumed that the outputs produced by each sector, such as automobiles, are just enough to meet the exogenous demand and the demand of the other sectors. We can illustrate the input-output analysis better by a numerical example.

Suppose that the economy of a particular country is grouped into the three sectors: agriculture, industry, and services. The flow of goods in millions of dollars for the economy is shown in Table 4.1. We note in

TABLE 4.1   Flow of Goods in Millions of Dollars

Producer	User			Exogenous Demand	Total Production
	Agriculture	Industry	Services		
Agriculture	40	80	20	60	200
Industry	60	160	60	520	800
Services	10	60	30	400	500
Total usage	110	300	110	980	

Table 4.1 that the total output of the agriculture sector was $200 million. Of this amount, $40 million was used by itself, $80 million went to the industry sector, $20 million to the services sector, and the remaining $60 million went to the exogenous demand. We can obtain the inputs to the sectors from the column values. For example, the inputs of industry are $80 million from agriculture, $160 million from industry, and $60 million from services. Table 4.1 indicates all the internal flow of commodities between sectors as well as each sector's supply to the exogenous demand.

The value added by each sector, such as profit, is determined by applying the expression

$$\begin{pmatrix} \text{Value added} \\ \text{by a sector} \end{pmatrix} = \begin{pmatrix} \text{Total production} \\ \text{by that sector} \end{pmatrix} - \begin{pmatrix} \text{Total input} \\ \text{to that sector} \end{pmatrix} \quad (4.34)$$

Thus,

$$\begin{aligned} \text{Value added by agriculture} &= 200 - 110 = 90 \\ \text{Value added by industry} &= 800 - 300 = 500 \\ \text{Value added by services} &= 500 - 110 = 390 \end{aligned}$$

Table 4.2 shows the flow of goods as well as the value added by each sector for the national economy.

TABLE 4.2 Flow of Goods in Millions of Dollars

Producer	User			Exogenous Demand	Total Production
	Agriculture	Industry	Services		
Agriculture	40	80	20	60	200
Industry	60	160	60	520	800
Services	10	60	30	400	500
Value added by the sector	90	500	390		
Column total	200	800	500	980	1500

The dollar value of the commodity that must be procured by each sector (user) from each of the producing sectors for producing $1 worth of its own goods is subsequently calculated. These values may be obtained by dividing each one of the nine elements denoting the internal flow amounts by the corresponding column total value.

From Producer	To User		
	Agriculture	Industry	Services
Agriculture	40/200	80/800	20/500
Industry	60/200	160/800	60/500
Services	10/200	60/800	30/500

For example, in order to achieve a total production of $800 million by the industry sector, $80 million worth of goods from the agriculture sector is required. Therefore, the dollar value of commodities needed from the agriculture sector to produce $1 worth of goods in the industry sector would be $80/800. In the same way, we can show that $160/800 worth of commodities from industry sector would be needed to produce $1 worth of goods in the same sector. The inputs required in dollar value for yielding $1 worth of commodity in each sector are shown below:

From Producer	To User		
	Agriculture	Industry	Services
Agriculture	0.2	0.1	0.04
Industry	0.3	0.2	0.12
Services	0.05	0.075	0.06

Let

$$V_1 = \text{value of output of the agricultural sector}$$

$$V_2 = \text{value of output of the industry sector}$$

$$V_3 = \text{value of output of the services sector}$$

$$E_1 = \text{exogenous demand of agricultural output}$$

$$E_2 = \text{exogenous demand of industry output}$$

$$E_3 = \text{exogenous demand of services output}$$

When the total outputs of the agricultural, industry, and services sectors are V_1, V_2, and V_3, they would need $0.2V_1$, $0.1V_2$, and $0.04V_3$ amounts of agricultural output, respectively. Hence, the total quantity of agricultural output required, including the exogenous demand, amounts to

$$0.2V_1 + 0.1V_2 + 0.04V_3 + E_1$$

The production of the economy is controlled so that the production level of any sector meets the needs of the sectors as well as the exogenous demand. Thus, for the agricultural sector

$$0.2V_1 + 0.1V_2 + 0.04V_3 + E_1 = V_1$$

Likewise, the equations may be formulated for the industry and the services sectors. The following equations are obtained for the three sectors:

$$0.2V_1 + 0.1V_2 + 0.04V_3 + E_1 = V_1$$

$$0.3V_1 + 0.2V_2 + 0.12V_3 + E_2 = V_2$$

$$0.05V_1 + 0.075V_2 + 0.06V_3 + E_3 = V_3 \qquad (4.35)$$

These equations may be rewritten as follows:

$$-0.8V_1 + 0.1V_2 + 0.04V_3 = -E_1$$

$$0.3V_1 - 0.8V_2 + 0.12V_3 = -E_2$$

$$0.05V_1 + 0.075V_2 - 0.94V_3 = -E_3 \qquad (4.36)$$

These equations can be employed to determine the amount of goods to be produced by each sector in terms of the exogenous demands. We can find out the effects of different sets of exogenous demands on the outputs of the sectors. Suppose the exogenous demands of agriculture, industry, and services were \$60, \$520, and \$400 million, respectively. If the exogenous demands of the agricultural sector and the services sector are kept at the same level, while the output of industry sector is increased to \$550 million, we have

$$E_1 = 60$$
$$E_2 = 550$$
$$E_3 = 400$$

For this set of demands, the outputs can be determined by solving the equations

$$-0.8V_1 + 0.1V_2 + 0.04V_3 = -60$$
$$0.3V_1 - 0.8V_2 + 0.12V_3 = -550$$
$$0.05V_1 + 0.075V_2 - 0.94V_3 = -400 \qquad (4.37)$$

The solution is

$$V_1 = 205.168$$
$$V_2 = 839.957$$
$$V_3 = 503.463$$

It indicates that the output of agriculture must be stepped up by \$5.168 million ($= 205.168$ million $- 200$ million), the output of industry must be stepped up by \$39.957 million ($= 839.957$ million $- 800$ million), and the output of services must be increased by \$3.463 million ($= 503.463$ million $- 500$ million).

The activities of an economy are to be classified into a large number of categories for most detailed planning purposes of the economy. In such cases, a large system of linear equations are designed, and the equations may be solved efficiently by employing the matrices.

PARTS REQUIREMENTS PROBLEM

A manufacturer may be required to produce final assemblies, subassemblies, and/or parts for future sales or to fulfill the customer's orders. When the manufacturer wishes to produce a specified number of final assemblies, subassemblies and/or parts, the requirement of parts that are needed can be determined by employing either of the following two methods:

1. Graphical method.
2. Matrix multiplication procedure.

Assume that the Jones Cabinet Company manufactures coffee tables and side tables. The materials required to produce these tables are shown in the Figures 4.6 and 4.7.

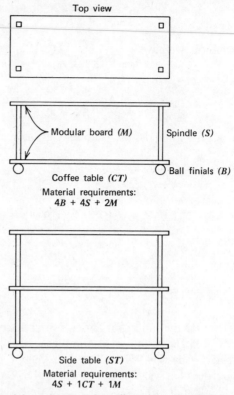

Top view

Modular board (M)   Spindle (S)

Ball finials (B)

Coffee table (CT)

Material requirements:
4B + 4S + 2M

Side table (ST):

Material requirements:
4S + 1CT + 1M

FIGURE 4.6 **Material requirements for coffee table and side table. (Note. Screws and connectors are not considered since they are minor items and are available in large quantity in stock.)**

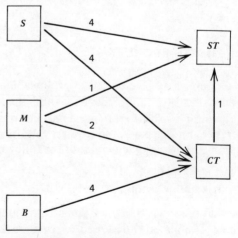

FIGURE 4.7 **Assembly requirements diagram.**

To make one coffee table, the requirements are

Four ball finials
Four spindles
Two modular boards

To make one side table, the requirements are

Four spindles
One coffee table
One modular board

The coffee table may be considered as the subassembly in the manufacturing of side tables. Note in Figure 4.7 that the three arrows entering
side table indicate that four spindles, one modular board, and one coffee
table are required to produce one side table.

Assume that an order is received for the following quantities of these
products:

Four spindles
Four coffee tables
Three modular boards
Two side tables

The company wants to compute the requirement of parts and subassemblies. Determining the component requirements (subassemblies, parts,
and raw materials) for a given order or sales forecast is called *exploding*.

Graphical Method

The graphical representation of the explosion process for an order of one
unit of each product is shown in Figure 4.8. Level 0 indicates the requirements of the order or the desired output from the final assembly. When we
move from one level to the next level, the assembly or the subassembly is
exploded into its subassembly and parts. For example, in order to obtain
one coffee table in level 0, 4 of B, 4 of S, and 2 of M are to be assembled in
level 1. Since the parts cannot be exploded in the consecutive levels, they
are taken as such to the raw material supply level. When 1 CT would be
needed in level 1, to achieve this 4 of B, 4 of S, and 2 of M would be
assembled in level 2. The total requirements of any level can be determined
by adding the parts in the particular row. The requirements of the raw
material supply level is 9 of B, 13 of S, and 6 of M. This implies that if we

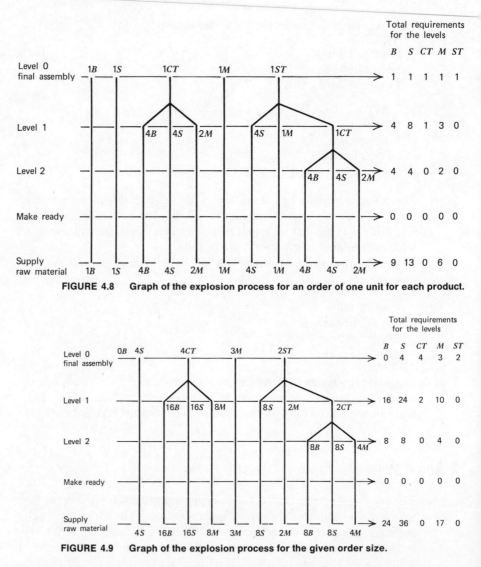

FIGURE 4.8   Graph of the explosion process for an order of one unit for each product.

FIGURE 4.9   Graph of the explosion process for the given order size.

want to obtain one unit of each product in the final assembly, we would need 9 of B, 13 of S, and 6 of M in the assembly process.

Figure 4.9 shows the explosion process for the given order of four spindles, four coffee tables, three modular boards, and two side tables. The units shown in Figure 5.8 are multiplied by the corresponding required final assemblies and the total requirements for each level are calculated. Note that the raw material requirements are the following:

24 units of ball finials
36 units of spindles
17 units of modular boards

When a large number of levels exists for any assembly process, the graphical approach becomes cumbersome and in such cases the matrix multiplication procedure discussed in the following section would be very useful.

Matrix Multiplication Procedure

The requirements for each level as well as the total raw material requirements can be determined by employing the matrices. The assembly requirements diagram shown in Figure 4.7 is represented by the assembly matrix in Table 4.3.

TABLE 4.3   The Assembly Matrix Indicating the Input Requirements to Assemble One Unit of Each Product as Output

		Output				
		B	S	CT	M	ST
	B	0	0	4	0	0
Input requirements	S	0	0	4	0	4
(parts and assemblies)	CT	0	0	0	0	1
	M	0	0	2	0	1
	ST	0	0	0	0	0

B is a part and, thus, it requires no assembly. CT requires the assembly of 4 of B, 4 of S, and 2 of M. The raw materials do not need the assembly of any products, and they can be identified by the columns in the assembly matrix that have zero elements in all the positions. We note in Table 4.3 that the products B, S, and M are the raw materials. The row elements indicate the requirements of the particular material by various products. For example, the row M implies that 2 of M and 1 of M are required to assemble 1 unit of CT and 1 unit of ST, respectively. The final assemblies do not become the input of any product and, hence, they have zero elements in the particular row. Note that ST is the final assembly and it has zero elements in the row ST. Observe that the assembly matrix is a square matrix.

A diagonal matrix is then designed that represents the order quantity of the products in the diagonal elements. The diagonal matrix for the given order of four spindles, four coffee tables, three modular boards, and two side tables is

	B	S	CT	M	ST
B	0	0	0	0	0
S	0	4	0	0	0
CT	0	0	4	0	0
M	0	0	0	3	0
ST	0	0	0	0	2

The diagonal matrix represents the requirement matrix for level 0. The requirements of any level can be established by adding the row values of the corresponding requirements matrix.

For level 0, the requirements are

$$
\begin{bmatrix}
0 & 0 & 0 & 0 & 0 \\
0 & 4 & 0 & 0 & 0 \\
0 & 0 & 4 & 0 & 0 \\
0 & 0 & 0 & 3 & 0 \\
0 & 0 & 0 & 0 & 2
\end{bmatrix}
\quad
\begin{matrix}
\text{Row} \\
\text{Total} \\
0 \text{ of } B \\
4 \text{ of } S \\
4 \text{ of } CT \\
3 \text{ of } M \\
2 \text{ of } ST
\end{matrix}
$$

The requirements of any level, K, can be obtained by employing the expression

$$
[\text{Assembly matrix}] \cdot \begin{bmatrix} \text{Requirements matrix} \\ \text{at level } (K - 1) \end{bmatrix} = \begin{matrix} \text{Requirements} \\ \text{matrix at level } K \end{matrix} \quad (4.38)
$$

Expression 4.38 can be used repeatedly until the resulting requirements matrix becomes a zero matrix in order to determine the requirements at various levels.

For level 1, the requirements are

$$
\begin{bmatrix}
0 & 0 & 4 & 0 & 0 \\
0 & 0 & 4 & 0 & 4 \\
0 & 0 & 0 & 0 & 1 \\
0 & 0 & 2 & 0 & 1 \\
0 & 0 & 0 & 0 & 0
\end{bmatrix}
\times
\begin{bmatrix}
0 & 0 & 0 & 0 & 0 \\
0 & 4 & 0 & 0 & 0 \\
0 & 0 & 4 & 0 & 0 \\
0 & 0 & 0 & 3 & 0 \\
0 & 0 & 0 & 0 & 2
\end{bmatrix}
=
\begin{bmatrix}
0 & 0 & 16 & 0 & 0 \\
0 & 0 & 16 & 0 & 8 \\
0 & 0 & 0 & 0 & 2 \\
0 & 0 & 8 & 0 & 2 \\
0 & 0 & 0 & 0 & 0
\end{bmatrix}
\quad
\begin{matrix}
\text{Row} \\
\text{Total} \\
16 \text{ of } B \\
24 \text{ of } S \\
2 \text{ of } CT \\
10 \text{ of } M \\
0 \text{ of } ST
\end{matrix}
$$

For level 2, the requirements are

Row Total

$$
\begin{bmatrix} 0 & 0 & 4 & 0 & 0 \\ 0 & 0 & 4 & 0 & 4 \\ 0 & 0 & 0 & 0 & 1 \\ 0 & 0 & 2 & 0 & 1 \\ 0 & 0 & 0 & 0 & 0 \end{bmatrix} \times \begin{bmatrix} 0 & 0 & 16 & 0 & 0 \\ 0 & 0 & 16 & 0 & 8 \\ 0 & 0 & 0 & 0 & 2 \\ 0 & 0 & 8 & 0 & 2 \\ 0 & 0 & 0 & 0 & 0 \end{bmatrix} = \begin{bmatrix} 0 & 0 & 0 & 0 & 8 \\ 0 & 0 & 0 & 0 & 8 \\ 0 & 0 & 0 & 0 & 0 \\ 0 & 0 & 0 & 0 & 4 \\ 0 & 0 & 0 & 0 & 0 \end{bmatrix}
\begin{matrix} 8 \text{ of } B \\ 8 \text{ of } S \\ 0 \text{ of } CT \\ 4 \text{ of } M \\ 0 \text{ of } ST \end{matrix}
$$

For level 3, the requirements are

Row Total

$$
\begin{bmatrix} 0 & 0 & 4 & 0 & 0 \\ 0 & 0 & 4 & 0 & 4 \\ 0 & 0 & 0 & 0 & 1 \\ 0 & 0 & 2 & 0 & 1 \\ 0 & 0 & 0 & 0 & 0 \end{bmatrix} \times \begin{bmatrix} 0 & 0 & 0 & 0 & 8 \\ 0 & 0 & 0 & 0 & 8 \\ 0 & 0 & 0 & 0 & 0 \\ 0 & 0 & 0 & 0 & 4 \\ 0 & 0 & 0 & 0 & 0 \end{bmatrix} = \begin{bmatrix} 0 & 0 & 0 & 0 & 0 \\ 0 & 0 & 0 & 0 & 0 \\ 0 & 0 & 0 & 0 & 0 \\ 0 & 0 & 0 & 0 & 0 \\ 0 & 0 & 0 & 0 & 0 \end{bmatrix}
\begin{matrix} 0 \text{ of } B \\ 0 \text{ of } S \\ 0 \text{ of } CT \\ 0 \text{ of } M \\ 0 \text{ of } ST \end{matrix}
$$

Recall that the level requirements are the same as those presented in Figure 4.9. Level 3 denotes the "make ready" stage in Figure 4.9. It signifies that we cannot explode the assembly process further. The total requirement matrix for the given order is determined by adding the requirement matrices at each level, including the level 0:

$$
\begin{bmatrix} 0 & 0 & 0 & 0 & 0 \\ 0 & 4 & 0 & 0 & 0 \\ 0 & 0 & 4 & 0 & 0 \\ 0 & 0 & 0 & 3 & 0 \\ 0 & 0 & 0 & 0 & 2 \end{bmatrix} + \begin{bmatrix} 0 & 0 & 16 & 0 & 0 \\ 0 & 0 & 16 & 0 & 8 \\ 0 & 0 & 0 & 0 & 2 \\ 0 & 0 & 8 & 0 & 2 \\ 0 & 0 & 0 & 0 & 0 \end{bmatrix} + \begin{bmatrix} 0 & 0 & 0 & 0 & 8 \\ 0 & 0 & 0 & 0 & 8 \\ 0 & 0 & 0 & 0 & 0 \\ 0 & 0 & 0 & 0 & 4 \\ 0 & 0 & 0 & 0 & 0 \end{bmatrix}
$$

Row Total

$$
= \begin{bmatrix} 0 & 0 & 16 & 0 & 8 \\ 0 & 4 & 16 & 0 & 16 \\ 0 & 0 & 4 & 0 & 2 \\ 0 & 0 & 8 & 3 & 6 \\ 0 & 0 & 0 & 0 & 2 \end{bmatrix}
\begin{matrix} 24 \text{ of } B \\ 36 \text{ of } S \\ 6 \text{ of } CT \\ 17 \text{ of } M \\ 2 \text{ of } ST \end{matrix}
$$

The row totals of the total requirement matrix indicate the total number of parts, subassemblies, and assemblies that will be supplied or assembled at one time or another in the assembly process. The total requirement of raw materials is

24 units of ball finials
36 units of spindles
17 units of modular boards

During the assembly process, a total of six coffee tables and two side tables would be assembled to meet the order requirements. Note that the level requirements and the total requirements check with the values of the graphical method shown in Figure 4.9. The information of the total requirements of products can be used for various purposes such as procuring the parts and subassemblies in time, planning the production schedules, and forecasting the manpower requirements. We shall formulate a computer model for the matrix multiplication method in the next section.

A Computer Program for the Matrix Multiplication Procedure

In developing a computer model for the parts requirements problem employing the matrix multiplication method, we use the following notations in the program:

NPRODS	= number of items (products) in the problem
LEVEL	= level number
NRSUM	= row total value
KUNITS(J)	= number of units of item J required in the final assembly
IPREQ(I, J)	= element in row I and column J of the assembly matrix
NRESO(I, J)	= element in row I and column J of the requirements matrix for level $(k - 1)$
NRES(I, J)	= element in row I and column J of the requirements matrix for level K
KUMPTS(I, J)	= element in row I and column J of the total requirements matrix
RECORD	= this subprogram prints out the required matrix and the NRSUM values for each level
FIND	= this subprogram calculates the requirements matrix for each level

The variable KUNITS( ) is a one-dimensional array of size NPRODS elements. The variables IPREQ(,), NRESO(,), NRES(,), and KUMPTS(,) are square matrices of size NPRODS by NPRODS.

We might employ the following instructions to read in and print out the input data:

```
      READ 705, NPRODS
      PRINT 201, NPRODS
      DO 10 I = 1, NPRODS
      READ 705, KUNITS(I)
      PRINT 203, I, KUNITS(I)
  10  CONTINUE
      DO 20 J = 1, NPRODS
      READ 200, (IPREQ(J, K), K = 1, NPRODS)
      PRINT 222, (IPREQ(J, K), K = 1, NPRODS)
 705  FORMAT ( )
 201  FORMAT ( )
 203  FORMAT ( )
 200  FORMAT ( )
 222  FORMAT ( )
```

We can formulate the diagonal matrix, initialize the total requirement matrix, and initialize the value of LEVEL by the statements:

```
      DO 25 I = 1, NPRODS
      DO 25 J = 1, NPRODS
      NRES(I, J) = 0
      IF (I.EQ.J) NRES(I, J) = KUNITS(I)
      KUMPTS (I, J) = 0
  25  CONTINUE
      LEVEL = 0
```

Now, for each level the requirement matrix and the row totals can be computed employing expression 4.38. The following statements would cause the computer to calculate the desired values and print them out on the printer:

```
  80  PRINT 800, LEVEL
      CALL    RECORD (NPRODS, NRSUM, NRES)
      DO 40 I = 1, NPRODS
      DO 40 J = 1, NPRODS
      KUMPTS(I, J) = KUMPTS(I, J) + NRES(I, J)
      NRESO(I, J) = NRES(I, J)
  40  CONTINUE
      CALL    FIND (NPRODS, IPREQ, NRESO, NRES)
      LEVEL = LEVEL + 1
      DO 60 I = 1, NPRODS
      DO 60 J = 1, NPRODS
      IF (NRES(I, J) .NE. 0.0) GO TO 80
  60  CONTINUE
```

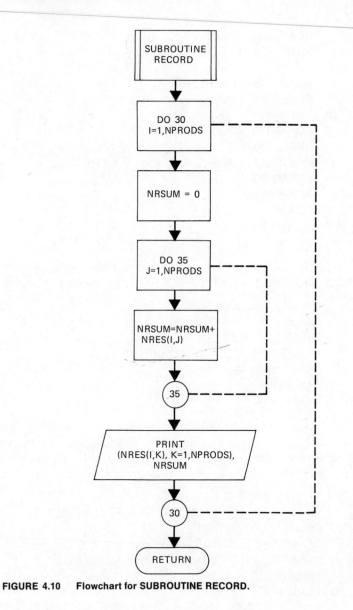

FIGURE 4.10   Flowchart for SUBROUTINE RECORD.

Figure 4.10 depicts the flowchart for the subroutine RECORD. The following instructions form the subprogram:

```
      SUBROUTINE RECORD (NPRODS, NRSUM, NRES)
      DIMENSION NRES ( , )
      DO 30 I = 1, NPRODS
      NRSUM = 0
      DO 35 J = 1, NPRODS
   35 NRSUM = NRSUM + NRES(I, J)
      PRINT 700, (NRES(I, K), K = 1, NPRODS), NRSUM
   30 CONTINUE
  700 FORMAT ( )
      RETURN
      END
```

The subroutine FIND calculates the requirements matrix for each level and Figure 4.11 displays the flowchart for it. The subroutine is comprised of the following statements:

```
      SUBROUTINE FIND (NPRODS, IPREQ, NRESO, NRES)
      DIMENSION IPREQ( , )
      DIMENSION NRESO( , )
      DIMENSION NRES( , )
      DO 100 I = 1, NPRODS
      DO 100 J = 1, NPRODS
      NRES (I, J) = 0
      DO 100 K = 1, NPRODS
      L = IPREQ(I, K) * NRESO(K, J)
      NRES(I, J) = NRES(I, J) + L
  100 CONTINUE
      RETURN
      END
```

The total requirement matrix and the corresponding row totals can be printed out by employing the statement:

```
CALL RECORD (NPRODS, NRSUM, KUMPTS)
```

A complete program composed of the previously formulated program segments is shown in Figure 4.12. The program can be used to solve the parts requirements problem in which the number of products does not exceed 10.

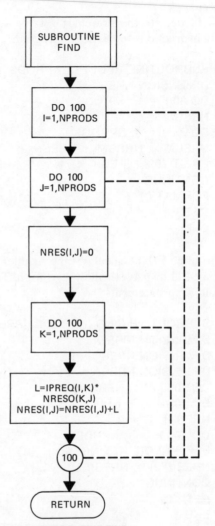

FIGURE 4.11   Flowchart for SUBROUTINE FIND.

An Illustration of the Computer Program   Let us illustrate the use of the computer model by the Jones Cabinet Company example. We can input the value of NPRODS, the requirements at the final assembly, and the assembly matrix by the following specification instructions:

```
705  FORMAT (I1)
200  FORMAT (5I1)
```

```
        PROGRAM  PARTS
        DIMENSION KUNITS(10), IPREQ(10,10)
        DIMENSION NRESO(10,10),  NRES(10,10)
        DIMENSION  KUMPTS (10,10)
C
C  READ AND PRINT THE INPUT DATA
        PRINT 802
        READ 705,NPRODS
        PRINT 201,NPRODS
        PRINT 400
        PRINT 405
C
        DO 10 I=1,NPRODS
        READ  705, KUNITS (I)
        PRINT 203,  I,  KUNITS(I)
     10 CONTINUE
C
        PRINT 410
C
        DO 20 J=1,NPRODS
        READ 200, ( IPREQ(J,K),  K= 1, NPRODS )
        PRINT 222, ( IPREQ(J,K),  K= 1, NPRODS )
     20 CONTINUE
C
C  INITIALIZE NRES AND KUMPTS VALUES
        DO 25 I=1,NPRODS
        DO 25 J=1,NPRODS
        NRES (I,J) = 0
        IF ( I .EQ. J ) NRES(I,J) = KUNITS(I)
        KUMPTS (I,J) = 0
     25 CONTINUE
C
C  PRINT LEVEL
        LEVEL = 0
     80 PRINT  800, LEVEL
        PRINT 415
        PRINT 420
C
C  PRINT THE VALUES FOR THE PARTICULAR LEVEL
        CALL       RECORD ( NPRODS, NRSUM, NRES )
C
C  FIND OUT THE REVISED KUMPTS AND NRESO
        DO 40 I=1,NPRODS
        DO 40 J=1,NPRODS
        KUMPTS (I,J) = KUMPTS(I,J) + NRES(I,J)
        NRESO (I,J) = NRES (I,J)
     40 CONTINUE
C
C  CALCULATE THE NEW NRES VALUES
        CALL       FIND ( NPRODS, IPREQ, NRESO, NRES )
        LEVEL = LEVEL + 1
C
C  TEST IF THIS IS THE LAST LEVEL
        DO 60 I=1,NPRODS
```

FIGURE 4.12 Computer program for solving the parts requirements problem (illustrated for the Jones Cabinet Company Problem).

```
              DO 60 J=1,NPRODS
              IF ( NRES (I,J) .NE. 0.0 ) GO TO 80
          60 CONTINUE
      C
      C   PRINT THE TOTAL REQUIREMENTS
              PRINT 850
              PRINT 415
              PRINT 420
              CALL    RECORD ( NPRODS, NRSUM, KUMPTS )
      C
         200 FORMAT ( 5I1 )
         201 FORMAT ( 5X, 17HNUMBER OF ITEMS = , I2)
         203 FORMAT ( 11X, I2, 3X, I2)
         222 FORMAT ( 10X, 5I3)
         400 FORMAT ( 5X, 30HREQUIREMENTS AT FINAL ASSEMBLY )
         405 FORMAT ( 10X, 10HITEM UNITS   )
         410 FORMAT ( /, 5X, 15HASSEMBLY MATRIX , / )
         415 FORMAT ( 27X, 3HROW )
         420 FORMAT ( 26X, 5HTOTAL )
         705 FORMAT (I1)
         800 FORMAT (/,5X,12HLEVEL NUMBER , I2)
         802 FORMAT ( 1H1 )
         850 FORMAT (/,5X,18HTOTAL REQUIREMENTS )
              STOP
              END


              SUBROUTINE  FIND ( NPRODS, IPREQ, NRESO, NRES )
      C
      C   THIS SUBROUTINE CALCULATES THE REQUIREMENTS OF THE LEVEL
              DIMENSION   IPREQ(10,10)
              DIMENSION   NRESO (10,10)
              DIMENSION   NRES  ( 10,10)
      C
              DO 100 I=1,NPRODS
              DO 100 J=1,NPRODS
              NRES (I,J) = 0
      C
              DO 100 K=1,NPRODS
              L = IPREQ (I,K) * NRESO (K,J)
              NRES (I,J) = NRES (I,J) + L
         100 CONTINUE
      C
              RETURN
              END


              SUBROUTINE RECORD ( NPRODS, NRSUM, NRES )
      C
      C   THIS SUBROUTINE PRINTS THE VALUES OF THE LEVEL
      C
              DO 30 I = 1, NPRODS
              NRSUM = 0
      C
              DO 35 J = 1, NPRODS
          35 NRSUM = NRSUM + NRES ( I,J)
              PRINT  700, ( NRES(I,K), K = 1,NPRODS ), NRSUM
          30 CONTINUE
      C
         700 FORMAT ( 10X, 5I3, I4 )
              RETURN
              END
```

FIGURE 4.12 (*Continued*)

```
NUMBER OF ITEMS = 5
REQUIREMENTS AT FINAL ASSEMBLY
     ITEM UNITS
       1    0
       2    4
       3    4
       4    3
       5    2

ASSEMBLY MATRIX

       0   0   4   0   0
       0   0   4   0   4
       0   0   0   0   1
       0   0   2   0   1
       0   0   0   0   0

LEVEL NUMBER 0
                         ROW
                         TOTAL
       0   0   0   0   0    0
       0   4   0   0   0    4
       0   0   4   0   0    4
       0   0   0   3   0    3
       0   0   0   0   2    2

LEVEL NUMBER 1
                         ROW
                         TOTAL
       0   0  16   0   0   16
       0   0  16   0   8   24
       0   0   0   0   2    2
       0   0   8   0   2   10
       0   0   0   0   0    0

LEVEL NUMBER 2
                         ROW
                         TOTAL
       0   0   0   0   8    8
       0   0   0   0   8    8
       0   0   0   0   0    0
       0   0   0   0   4    4
       0   0   0   0   0    0

TOTAL REQUIREMENTS
                         ROW
                         TOTAL
       0   0  16   0   8   24
       0   4  16   0  16   36
       0   0   4   0   2    6
       0   0   8   3   6   17
       0   0   0   0   2    2
```

FIGURE 4.13   Computer printout for the Jones Cabinet Company problem.

The data deck has the following appearance:

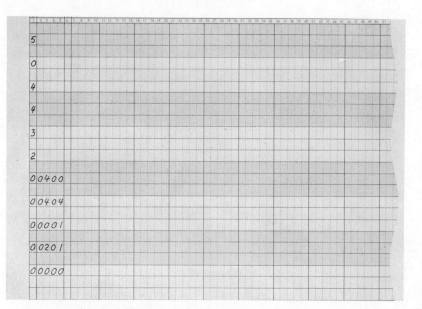

The specification instructions with the statement numbers 201, 203, 222, 700, and 800 command the computer to print out the results as shown in Figure 4.13. Matrices are also highly useful in solving large-size linear programming problems. Linear programming will be discussed in Chapter 5.

PROBLEMS

4.1 A company manufactures and markets washing machines. The number of units produced for the past four months are:

	Month		
1	2	3	4
6000	7000	6500	7200

The requirements of direct labor, materials, and indirect labor per unit for the past months are:

Month	Direct Labor	Materials	Indirect Labor
1	2.5	50.5	1
2	2.8	47.8	1
3	2.7	46.5	1.5
4	2.6	51.2	1

Determine the total requirements of direct labor, materials, and indirect labor for the four months using matrix multiplication method.

4.2 A corporation produces aluminum in three plants situated in different regions. The chemicals, A, B, C, and D are supplied to the plants from a central warehouse. The requirements of the plants for next week are:

	Chemicals in Tons			
Plant	A	B	C	D
P1	100	150	200	250
P2	80	100	60	100
P3	600	400	520	500

The volume and price per ton of the chemicals are:

Chemical	Cubic Feet	Price
A	8	$500
B	9	600
C	6	700
D	10	300

Determine the total volume and the total price of the chemicals needed by each plant. Use the matrix multiplication procedure.

4.3 Determine the values of X_1, X_2, and X_3 for the system of linear equations:

$$3X_1 + 4X_2 + 2X_3 = 6$$
$$10X_1 + 7X_2 - 5X_3 = 21$$
$$4X_1 + 6X_2 + 5X_3 = 9$$

Employ the Gauss reduction method.

4.4 Find the values of X_1, X_2, and X_3 for the system of linear equations:

$$3X_1 - 5X_2 + 4X_3 = -47.5$$

$$4X_1 + 2X_2 + 5X_3 = 7$$

$$2X_1 + 3X_2 + 6X_3 = 9$$

Use the Gauss reduction method.

4.5 The economy of a particular country is classified into four sectors. The commodity flow in millions of dollars is given in the following table:

	User				
Producer	A	B	C	D	Exogenous Demand
A	50	70	120	180	90
B	70	130	160	120	210
C	30	20	40	10	460
D	150	140	80	50	300

(a) Calculate the value added by each sector.
(b) Formulate a system of equations to determine the amounts of commodities to be produced by each sector in terms of the exogenous demands.
(c) If the exogenous demands for the planning periods are $100, $225, $485, and $303 millions for A, B, C, and D, respectively, determine the output of each sector. Design a FORTRAN program to establish the desired values.

4.6 A small company manufactures the end products E and F that contain several common components. The assembly requirements diagram shown below indicates the parts required to fabricate E and F.

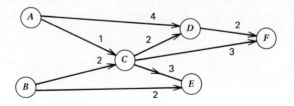

The company has received an order for the supply of two units of A, three units of C, two units of D, three units of E, and one unit of F.

(a) Determine the total requirements of each level by using the graphical approach, as shown in Figure 4.9.
(b) Compose a FORTRAN program to establish the total requirements for each level.

SELECTED REFERENCES

Baumol, W. J., *Economic Theory and Operations Analysis*, Prentice-Hall, Englewood Cliffs, N.J., Third Edition, 1972.

Dorfman, R., P. A. Samuelson, and R. M. Solow, *Linear Programming and Economic Analysis*, McGraw-Hill, New York, 1958.

Hadley, G., *Linear Algebra*, Addison-Wesley, Reading, Mass., 1961.

McMillan, C., and R. F. Gonzalez, *Systems Analysis: A Computer Approach to Decision Models*, Richard D. Irwin, Homewood, Ill., Third Edition, 1973.

Schick, W., and C. J. Merz, Jr., *Fortran for Engineering*, McGraw-Hill, New York, 1972.

Sokolnikoff, S., and R. M. Redheffer, *Mathematics for Physics and Modern Engineering*, McGraw-Hill, New York, Second Edition, 1966.

Stern, M. E., *Mathematics for Management*, Prentice-Hall, Englewood Cliffs, N.J., 1963.

Stone, D., "Input-Output Analysis and the Multi-Product Firm," *Financial Analysts Journal*, Vol. 25, No. 4, July-August, 1969, pp. 96–102.

Teichroew, D., *An Introduction to Management Science—Deterministic Models*, Wiley, New York, 1964.

Vazsonyi, A., *Scientific Programming in Business and Industry*, Wiley, New York, 1958.

allocation of resources—lp

MANAGEMENT IS most often faced with the problem of allocating various limited resources such as men, money, machines, and materials to various competing activities in the most efficient way. Linear programming (LP) is an important mathematical technique most extensively used by the systems analyst to optimize the allocation problem of a given system. The use of linear programming can be found in the government, agriculture, business, and industrial sectors. We shall discuss in this chapter the formulation and solution procedure of linear programming by means of sample problems together with the circumstances under which this tool can be used. A computer program is also presented with computer printout solutions of the sample problems.

MATHEMATICAL FORMULATION

The general form of the linear programming problem is to find the values of the variables, $x_1, x_2, \ldots, x_n$, which maximize

$$Z = c_1 x_1 + c_2 x_2 + \cdots + c_n x_n$$

subject to

$$a_{11} x_1 + a_{12} x_2 + \cdots + a_{1n} x_n \leqq b_1$$
$$a_{21} x_1 + a_{22} x_2 + \cdots + a_{2n} x_n \leqq b_2$$
$$\cdots\cdots\cdots\cdots\cdots\cdots\cdots\cdots\cdots\cdots\cdots\cdots$$
$$a_{m1} x_1 + a_{m2} x_2 + \cdots + a_{mn} x_n \leqq b_m \qquad (5.1)$$

and
$$x_1 \geq 0, x_2 \geq 0, \ldots, x_n \geq 0$$
where
$$x_j = \text{decision variables } (j = 1, 2, \ldots, n)$$
$$m = \text{number of restrictions}$$
$$n = \text{number of decision variables}$$
$$a_{ij}, b_i, c_j = \text{given constant values}$$
$$(i = 1, 2, \ldots, m; j = 1, 2, \ldots, n)$$

The sign "$\leq$" is read "less than or equal to" and the sign "$\geq$" is read "greater than or equal to." The formulation can be best illustrated by a sample problem.

Example 5.1

A manufacturing company produces two kinds of products: air conditioners and refrigerators. Each unit of the products yields a contribution of $100 and $80, respectively. The contribution to profit is selling price minus variable cost. Each product should go through the three departments: machining, stamping, and assembly. The hours of departmental times required by each product per unit as well as the total hours available for each department are as follows:

Department	Air Conditioner	Refrigerator	Total Hours Available
Machining	6	3	120
Stamping	8	12	240
Air conditioner assembly	10	—	170
Refrigerator assembly	—	12	180

The manager wishes to determine how many of each product should be made so that the total contribution is maximized.

For producing the two products, the needed resources that are available in limited quantities are the machining, stamping, and the related assembly line facilities. The competing activities are the manufacturing of air conditioners and refrigerators. We need to establish the number of units of the two products to be produced so that the demanded resources do not exceed the available amounts and, at the same time, the resulting total contribution is maximized.

Let x_1 and x_2 be the number of units of air conditioners and refrigerators produced, respectively. The given data in terms of x_1 and x_2 can be summarized as follows:

	Air Conditioners		Refrigerators		Limitation
Number of units manufactured	x_1		x_2		
Total profit $(=Z)$	$100x_1$	$+$	$80x_2$		
Hours of machining department	$6x_1$	$+$	$3x_2$	$\leq$	120
Hours of stamping department	$8x_1$	$+$	$12x_2$	$\leq$	240
Hours of air conditioner assembly	$10x_1$	$+$	0	$\leq$	170
Hours of refrigerator assembly	0	$+$	$12x_2$	$\leq$	180

When one air conditioner is produced, the profit is \$100 and, hence, the contribution from x_1 air conditioners equals \$$100x_1$. Similarly, the contribution from x_2 refrigerators equals \$$80x_2$. Thus, the total contribution from the two products $(=Z)$ amounts to $(100x_1 + 80x_2)$. The objective is to maximize the total contribution. The objective function then can be expressed in the form:

$$\text{Maximize } Z = 100x_1 + 80x_2$$

To produce one air conditioner, 6 hours of machining department time are required. Hence, to produce x_1 units of air conditioners, $6x_1$ hours of machining department time are needed. Likewise, $3x_2$ hours of machining department time are needed to produce x_2 refrigerators. The total machining department time required then is $(6x_1 + 3x_2)$. Available machining department time is 120 hours. The required machining department time must be less than or equal to 120 hours.

$$6x_1 + 3x_2 \leq 120 \quad \text{for the machining department}$$

In a similar manner, we can show that the total stamping department time required is $(8x_1 + 12x_2)$ hours. The available stamping time is 240 hours. The restriction for the stamping department then can be expressed in this form:

$$8x_1 + 12x_2 \leq 240 \quad \text{for the stamping department}$$

The air conditioner assembly line is used only in the production of air conditioners. The total time needed to produce x_1 air conditioners is $10x_1$ hours of assembly time. Since the total available time is 170 hours, we have the restriction:

$$10x_1 \leq 170 \quad \text{for air conditioner assembly}$$

The refrigerator assembly line is used only in manufacturing refrigerators. The total assembly time required to manufacture x_2 refrigerators amounts to $12x_2$ hours. The available amount of refrigerator assembly time is limited to 180 hours. Therefore, this limited refrigerator assembly time also introduces a similar restriction:

$$12x_2 \leq 180 \qquad \text{for refrigerator assembly}$$

The number of products manufactured, x_1 and x_2, must be equal to or greater than zero. It cannot be negative. We now have the restrictions:

$$x_1 \geq 0, \qquad x_2 \geq 0$$

These restrictions are called the *nonnegativity condition*. The expressions can now be represented in the linear programming form as follows:

$$\text{Maximize } Z = 100x_1 + 80x_2$$

$$\text{Subject to} \qquad 6x_1 + 3x_2 \leq 120$$

$$8x_1 + 12x_2 \leq 240$$

$$10\dot{x}_1 \qquad \leq 170$$

$$12x_2 \leq 180$$

and

$$x_1 \geq 0 \qquad X_2 \geq 0 \qquad\qquad (5.2)$$

The variables denoting the number of air conditioners and refrigerators, x_1 and x_2, are called the *decision variables* in linear programming. Taha (1971) states the general linear programming problem as follows: "The general linear programming problem calls for optimizing (maximizing or minimizing) a linear function of variables, called the 'objective function,' subject to a set of linear equalities and/or inequalities called 'constraints' or 'restrictions' . . . ." The general linear programming problem may have the objective function of maximizing or minimizing. The objective function and the restrictions must be in linear form, implying that the exponents of all variables must be one; this means that the terms such as x_1^3 and $(x_1)(x_7)$ are not permitted. Each restriction may be any one of the three types: "less than or equal to" ($\leq$), "equal to" ($=$), or "greater than or equal to" ($\geq$). These variations are illustrated in the following examples.

Example 5.2

The Norman Furniture Company manufactures two products: desks and bookcases. Each desk yields a contribution of $12 while each bookcase yields a contribution of $9. The production planning division wishes to find out the production program that maximizes the total contribution.

Each desk requires eight man-hours of production shop time, six square feet of storage space, and 45 square feet of timber, whereas each bookcase requires four man-hours, nine square feet of space, and 15 square feet of timber. The maximum available production shop time and storage space are 240 man-hours and 360 square feet per week, respectively. The company has contracted for the supply of at least 450 square feet of timber each week. The company has promised to supply its distributor with at least a total of 20 units of desks, bookcases, or a combination of the two.

Let

$$x_1 = \text{number of desks produced weekly}$$

$$x_2 = \text{number of bookcases produced weekly}$$

The total contribution to profit is the sum of contributions obtained from the two products. Hence, the objective function can be expressed as

$$\text{Maximize } Z = 12x_1 + 9x_2$$

The total man-hours and the total storage space needed for manufacturing x_1 desks and x_2 bookcases cannot exceed the total available quantities. Thus we have the two constraints:

$$8x_1 + 4x_2 \leq 240 \qquad \text{for production shop time}$$

$$6x_1 + 9x_2 \leq 360 \qquad \text{for storage space}$$

Any amount of timber but not less than 450 square feet can be bought each week. The required amount of timber per week is $(45x_1 + 15x_2)$. This means that $45x_1 + 15x_2 \geq 450$ for the timber requirement. The total units produced per week is $(x_1 + x_2)$. Since the company has promised to deliver each week at least 20 units of the products, we have

$$x_1 + x_2 \geq 20 \qquad \text{for market supply}$$

In addition to these constraints, the decision variables cannot be negative. The complete mathematical formulation of the linear programming problem thus becomes:

$$\text{Maximize } Z = 12x_1 + 9x_2$$
$$\text{Subject to} \qquad 8x_1 + 4x_2 \leq 240$$
$$6x_1 + 9x_2 \leq 360$$
$$45x_1 + 15x_2 \geq 450$$
$$x_1 + x_2 \geq 20$$

and

$$x_1 \geq 0 \qquad x_2 \geq 0 \qquad\qquad (5.3)$$

Example 5.3

A small oil refinery must produce 5000 barrels of gasoline *A* and 4000 barrels of gasoline *B*. These two gasolines are produced by blending crudes 1 and 2. The specifications for blending the two crudes are as follows:

	Crude 1	Crude 2
Gasoline *A*	No more than 30%	At least 60%
Gasoline *B*	At least 50%	At least 20%, but no more than 40%

A barrel of crude 1 costs $6 while a barrel of crude 2 costs $9. No more than 4800 barrels of crude 1 can be procured. Any amount of crude 2 is available in the market. The manager wants to set up this problem as a linear programming problem to obtain the least total cost solution.

The most difficult task in formulating the linear programming problems in various situations is perhaps choosing the appropriate decision variables. This example is of such a type. Let us explain first the specifications. For gasoline *A*, the specification denotes that crude 1 must be "no more than 30%." This indicates that the total quantity of crude 1 to be used to produce gasoline *A* cannot exceed the 30% limit of the total amount of gasoline *A*. That is, no more than 1500 barrels ($5000 \times 0.30 = 1500$) of crude 1 can be used in producing 5000 barrels of gasoline *A*.

The following four decision variables are needed to formulate the linear programming problem.
Let

x_1 = barrels of crude 1 to be used to produce gasoline *A*

x_2 = barrels of crude 2 to be used to produce gasoline *A*

x_3 = barrels of crude 1 to be used to produce gasoline *B*

x_4 = barrels of crude 2 to be used to produce gasoline *B*

Decision variables used:

	Crude 1	Crude 2
Gasoline *A*	x_1	x_2
Gasoline *B*	x_3	x_4

Thus, we obtain

Total amount of gasoline *A* to be produced $= x_1 + x_2$

Total amount of gasoline *B* to be produced $= x_3 + x_4$

$$\text{Total amount of crude 1 required} = x_1 + x_3$$

$$\text{Total amount of crude 2 required} = x_2 + x_4$$

The objective is to minimize the total cost of the crudes. Therefore, the objective function is in the form:

$$\text{Minimize } Z = 6(x_1 + x_3) + 9(x_2 + x_4)$$

$$= 6x_1 + 9x_2 + 6x_3 + 9x_4$$

The production quantity of the two gasolines must be exactly equal to the demanded amounts.

$$x_1 + x_2 = 5000 \text{ for gasoline } A$$

$$x_3 + x_4 = 4000 \text{ for gasoline } B$$

The proportion of crude 1 in gasoline A cannot be more than 30%. The constraint is

$$x_1 \leq 0.3(x_1 + x_2)$$

This restriction does not conform to the general form because the variables appear on the right-hand side of the "$\leq$" sign. This is modified as

$$x_1 - 0.3(x_1 + x_2) \leq 0$$

This can be written as

$$0.7x_1 - 0.3x_2 \leq 0 \qquad \text{for crude 1 in gasoline } A$$

The portion of crude 2 in gasoline A must be at least 60%. The constraint is

$$x_2 \geq 0.6(x_1 + x_2)$$

This again does not comply with the general form. This restriction can be written as

$$-0.6x_1 + 0.4x_2 \geq 0 \qquad \text{for crude 2 in gasoline } A$$

To insure that the share of crude 1 in gasoline B is at least 50%, we can use the restriction

$$x_3 \geq 0.5(x_3 + x_4)$$

The equivalent restriction complying with the general form is

$$0.5x_3 - 0.5x_4 \geq 0 \qquad \text{for crude 1 in gasoline } B$$

The amount of crude 2 in gasoline B may be any portion from 20% (lower level) through 40% (upper level). This limitation can be stated by the two restrictions:

$$x_4 \geq 0.20(x_3 + x_4)$$

$$x_4 \leq 0.40(x_3 + x_4)$$

These may be rewritten as follows:

$-0.20x_3 + 0.80x_4 \geq 0$ for the lower level of crude 2 in gasoline B

$-0.40x_3 + 0.60x_4 \leq 0$ for the upper level of crude 2 in gasoline B

The amount of crude 1 that can be bought is up to 4800 barrels. Thus we have

$$x_1 + x_3 \leq 4800 \qquad \text{for crude 1 availability}$$

Finally, the decision variables cannot be negative.

$$x_1 \geq 0, x_2 \geq 0, x_3 \geq 0, x_4 \geq 0$$

To recapitulate, the linear programming problem has the following appearance:

$$\text{Minimize } Z = 6x_1 + 9x_2 + 6x_3 + 9x_4$$

$$
\begin{aligned}
\text{Subject to} \quad x_1 + x_2 &= 5000 \\
x_3 + x_4 &= 4000 \\
0.7x_1 - 0.3x_2 &\leq 0 \\
-0.6x_1 + 0.4x_2 &\geq 0 \\
0.5x_3 - 0.5x_4 &\geq 0 \\
-0.20x_3 + 0.80x_4 &\geq 0 \\
-0.4x_3 + 0.6x_4 &\leq 0 \\
x_1 + x_3 &\leq 4800
\end{aligned}
$$

and

$$x_1 \geq 0 \qquad x_2 \geq 0 \qquad x_3 \geq 0 \qquad x_4 \geq 0 \qquad (5.4)$$

The linear programming problems can be solved by an algebraic procedure called the simplex method. However, the graphical method may be used to solve the linear programming problems with two decision variables such as in Examples 5.1 and 5.2. The graphical approach discussed in the following section should facilitate understanding of the more efficient simplex method presented in the subsequent section.

THE GRAPHICAL METHOD

We shall illustrate the graphical approach by Example 5.1. The mathematical form of the linear programming problem is restated for easy reference:

Maximize $Z = 100x_1 + 80x_2$

Subject to

$$6x_1 + 3x_2 \leqq 120 \quad \text{for machining department}$$
$$8x_1 + 12x_2 \leqq 240 \quad \text{for stamping department}$$
$$10x_1 \qquad \leqq 170 \quad \text{for air conditioners assembly}$$
$$12x_2 \leqq 180 \quad \text{for refrigerators assembly}$$

and

$$x_1 \geqq 0 \qquad x_2 \geqq 0$$

The number of units of air conditioners (x_1) is plotted on the horizontal axis and the number of units of refrigerators is plotted on the vertical axis, as shown in Figure 5.1. The nonnegativity condition insures that x_1 and x_2 can only take either zero or positive values. The points in the shaded quadrant in Figure 5.1 have nonnegative values for both x_1 and x_2. The restrictions are then represented on the graph to determine the *solution space*. The solution space refers to the set of points in the graph that satisfies all of the restrictions including the nonnegativity condition. Therefore, we are basically interested in determining the set of points in the shaded quadrant that does not violate any of the restrictions. The restriction related to the machining department is

$$6x_1 + 3x_2 \leqq 120$$

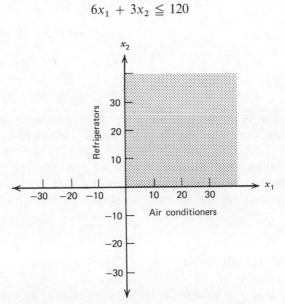

FIGURE 5.1 **Graphical illustration of solution sets that fulfill the nonnegativity condition.**

Suppose that this has the equal sign for representing the constraint on the graph,

$$6x_1 + 3x_2 = 120$$

In order to represent the equation on the graph, we need any two points that fulfill the equation. One way of determining the points is to find those points through which the line should pass on the two axes. This is calculated as follows:

$$\text{Set } x_1 = 0, \quad \text{then } 6(0) + 3x_2 = 120$$
$$3x_2 = 120$$
$$x_2 = 40$$

Thus, the point on the vertical axis is 40 units.

$$\text{Set } x_2 = 0, \quad \text{then } 6x_1 + 3(0) = 120$$
$$6x_1 = 120$$
$$x_1 = 20$$

Thus, the point on the horizontal axis is 20 units. The line (AB) connecting these two points shown in the Figure 5.2 represents the equation. Any point on this line denotes a set of x_1 and x_2 values that will require

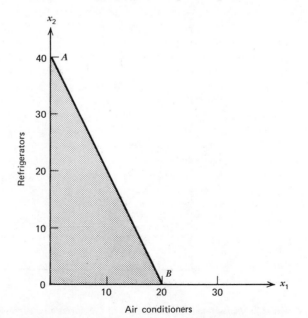

FIGURE 5.2 Graphical representation of the constraint imposed by the machining department.

all of the available machining departmental time. However, the restriction has the inequality sign " $\leq$ ". The line AB divides the quadrant into two parts. The points in one part satisfy the restriction, while the points in the other part do not. Since the machining department does not have to be operated for all of the available hours, as specified by the " $\leq$ " sign, the part that has the origin in it fulfills the restriction. The triangle OAB is the feasible set for this restriction.

Similarly, the restriction for the stamping department is represented on the graph, as shown by the triangle OCD in Figure 5.3. The set of points that satisfies the machining department restriction as well as the stamping department restriction is the area $OCEB$. Likewise, the restrictions for the two assembly lines are also plotted on the graph in Figure 5.4. The *solution space* ($OJFEGH$) that fulfills all the restrictions including the nonnegativity condition is shown by the shaded area in this figure. Any point that does not lie in the solution space violates at least one of the constraints and, thus, it is not a practically possible production program. The solution space is also called the *feasible region*.

Next, the points in the solution space that maximize (or minimize for minimization problems) the objective function is determined. A series of total contribution lines are plotted on the graph to find out the production

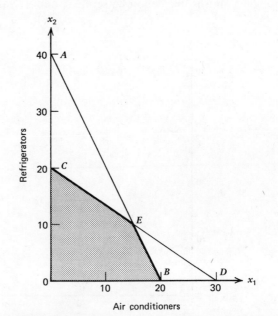

FIGURE 5.3   **Graphical representation of the constraints imposed by the machining department and the stamping department.**

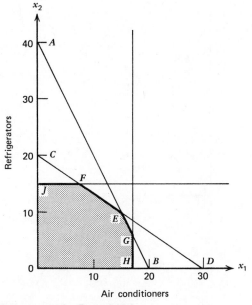

FIGURE 5.4 **Solution space for Example 5.1.**

programs that give the largest contribution. Suppose that the total contribution is $800; we then obtain the equation from the objective function

$$800 = 100x_1 + 80x_2$$

This equation is represented, as explained earlier, by a line on the graph shown in Figure 5.5. Each point on the line that lies in the feasible region refers to a production program, and its total contribution to profit is $800. The line is called the *equal profit line* or *isoprofit line* for $800. For various total profit values, isoprofit lines are drawn on the graph. Figure 5.5 displays isoprofit lines for $400, $800, $1200, $1600, $2000 and $2300. The series of lines are parallel to each other and, the larger the profit, the farther the line is from the origin. The profit line for $2300 touches the feasible region at the corner point E, indicating that this is the only production program that yields the total contribution of $2300. The profit contribution line for any higher contribution would not lie in the solution space. Therefore, the largest total contribution is $2300 and the optimum plan is specified by the corner point E.

Optimum Solution:
Total contribution (Z) = $2300
Number of air conditioners to be produced (x_1) = 15 units
Number of refrigerators to be produced (x_2) = 10 units

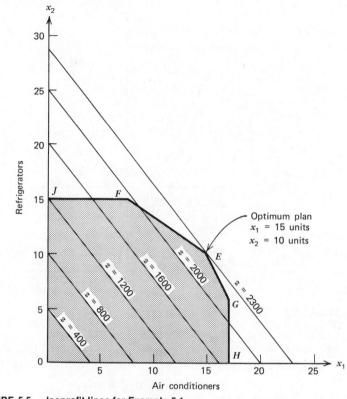

FIGURE 5.5   Isoprofit lines for Example 5.1.

Note that the optimum point E is situated on the lines FE and EG, representing the stamping department and the machining department restrictions. It indicates that the machining department and the stamping department are busy. The point E lies away from the air conditioners assembly and the refrigerators assembly restrictions. It denotes that the two assembly lines are operated below their full capacity for the optimum plan. The linear programming solution not only gives the optimum plan, but also points to the actions to be taken to improve the system. For example, to increase the profit, additional stamping and machining department times may be allotted by operating overtime or procuring additional machines. If the calculated optimum plan were implemented, the excess resources available in the two assembly lines could be removed and allotted to somewhere else where they are needed. We note in Figure 5.5 that one of the corner points represents the optimum solution. In fact, this is one of the important characteristics of the linear programming problem. Although there are innumerable points in the feasible region, the optimum plan can be established by evaluating only the finite number of corner points. We need not test the feasible points that do not lie at the corners. For our illustration, the evaluation of the corner points is given below:

Corner Point	x_1	x_2	Calculation of Z Value
O	0	0	100(0) + 80(0) = 0
J	0	15	100(0) + 80(15) = 1200
F	7.5	15	100(7.5) + 80(15) = 1950
E	15	10	100(15) + 80(10) = 2300 Maximum (*m*)
G	17	6	100(17) + 80(6) = 2180
H	17	0	100(17) + 80(0) = 1700

Figure 5.6 shows the solution space and the optimum solution for Example 5.2. The graphical approach can be used to solve the linear

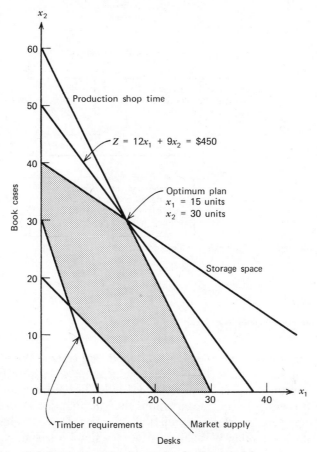

FIGURE 5.6 Solution space and optimum plan for Example 5.2.

programming problems with at most three decision variables. The three-variable problems can be frequently solved using the three-dimensional graph with some difficulty. A summary of the graphical method is shown in Figure 5.7. The simplex method discussed in the next section is a very powerful algorithm that can be applied to solve linear programming problems with any number of variables.

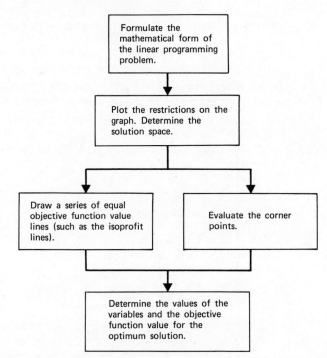

FIGURE 5.7   Summary of steps for the graphical method.

THE SIMPLEX METHOD

The simplex method was developed by Dantzig, Wood, and their associates in 1947. A set of operations is applied repetitively; the solution is improved in each iteration; and, finally, the optimum solution is obtained. It is easier to operate equations than inequalities. Hence, the inequalities in the mathematical formulation are changed to equations first by applying slack variables. We shall illustrate the simplex technique for Example 5.1.

Slack Variables

The first restriction specifies the limitation on the availability of machining departmental time and it is

$$6x_1 + 3x_2 \leq 120$$

It indicates that the total requirement cannot exceed 120 hours. The inequality can be converted to an equation by including a unique slack variable on the left-hand side.

$$6x_1 + 3x_2 + x_3 = 120$$

The slack variable in this equation refers to the unused time in the machining department. Similarly, it may be considered that x_4 is the slack variable referring to the unused time in the stamping department. Likewise, x_5 is the slack variable referring to the unused time in the air conditioners assembly line, and x_6 is the slack variable referring to the unused time in the refrigerators assembly line. The problem is modified as follows:

Maximize $Z = 100x_1 + 80x_2$

Subject to

$$
\begin{aligned}
6x_1 + 3x_2 + x_3 &= 120 \\
8x_1 + 12x_2 + x_4 &= 240 \\
10x_1 + x_5 &= 170 \\
12x_2 + x_6 &= 180 \quad (5.5)
\end{aligned}
$$

and,

$x_j \geq 0$, where $j = 1, 2, \ldots, 6$.

The slack variables are permitted to assume only nonnegative values in order that the original restrictions can be fulfilled. When a slack variable is zero, it specifies that there is no unused time in the related department. If a slack variable takes any negative value, it denotes that the total requirement is larger than the available amount. This is not allowed. Thus, the slack variables cannot take negative values. In the nonnegativity restriction, we include the slack variables. Therefore,

$$x_3 \geq 0 \qquad x_4 \geq 0 \qquad x_5 \geq 0 \qquad x_6 \geq 0$$

Computational Procedure

An initial trial solution is determined in the simplex method. When we obtain the initial trial solution, the method tells how to find the next improved solution, if there is one. For the new trial solution, an improved solution is found and this process is continued repetitively. It also shows when an optimum solution has been established; the calculations are then terminated.

If M equations are given in M variables, we can calculate the values of the variables by the simultaneous solution of the equations (Chapter 4). However, if M equations are given in N variables, where N is larger than M, we set $(N - M)$ variables at zero level and solve the remaining M variables. The resulting solution is called a *basic solution*. To put it in other words, a basic solution for a set of M equations in N variables, where N is larger than M, will have $(N - M)$ variables equal to zero, and the values of the remaining M variables are found from the given set of equations. In the computational procedure, progressively improved basic feasible solutions are determined starting with the initial basic feasible solution. We shall show that the basic feasible solutions represent the corner points in the solution space. In a basic feasible solution, $(N - M)$ variables are set at zero level (called the *nonbasic variables*) and the remaining M variables have nonnegative values (called the *basic variables*). The basic variables, in most situations, take positive values.

An improved solution for a given basic solution is obtained as follows: A nonbasic variable enters into the basis and thus becomes a basic variable. In turn, a basic variable leaves the basis and becomes a nonbasic variable. The resulting solution is also a basic feasible solution. Figure 5.8 summarizes the steps for the simplex method.

The objective function is

$$\text{Maximize } Z = 100x_1 + 80x_2$$

The slack variables do not contribute anything to the objective function. Hence, we may rewrite the objective function as

$$\text{Maximize } Z = 100x_1 + 80x_2 + 0x_3 + 0x_4 + 0x_5 + 0x_6$$

This can be modified as

$$Z - 100x_1 - 80x_2 - 0x_3 - 0x_4 - 0x_5 - 0x_6 = 0 \qquad (5.6)$$

We shall now discuss the six steps shown in Figure 5.8.

STEP 1. **Determine an Initial Solution.** The initial basic solution is obtained easily by selecting the slack variables to be the basic variables. Thus, x_3, x_4, x_5, and x_6 are in basis and x_1 as well as x_2 are the nonbasic variables. Therefore,

$$x_1 = 0$$
$$x_2 = 0$$

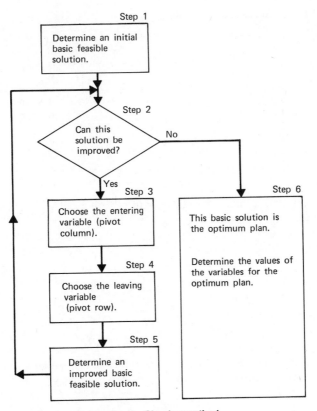

FIGURE 5.8   Summary of steps for the Simplex method.

This indicates that no units are produced and the resulting total contribution is zero. Equations 5.6 and 5.5 represent the initial basic solution and are shown in tabular form in Table 5.1.

In Table 5.1, the coefficients of the variables and the constants are displayed. The first row represents the objective function, and the following four rows represent the four restrictions. Each restriction must have a unique basic variable. The basic variables for the four restrictions are x_3, x_4, x_5, and x_6 in that order. For the sake of uniformity, Z is assumed to be the basic variable for the objective function. The basic variable for any constraint must be shown by the coefficient "1" in that particular row for the variable. The particular column referring to the variable must have zero coefficients in all *other* elements, including the objective function row. For example, x_3 is the basic variable for the first restriction. Thus, the coefficient in that row for the variable x_3 is 1. The column related to

TABLE 5.1   Initial Tableau of Example 5.1

	Basic				Coefficients				Constant
i	Variable	Z	x_1	x_2	x_3	x_4	x_5	x_6	(b_i)
0	Z	1	-100	-80	0	0	0	0	0
1	x_3	0	6	3	1	0	0	0	120
2	x_4	0	8	12	0	1	0	0	240
3	x_5	0	10	0	0	0	1	0	170
4	x_6	0	0	12	0	0	0	1	180

variable x_3 has 0 coefficients in all other cells. The basic variables have the values as shown in Table 5.1.

$$Z = 0$$

$$x_3 = 120$$

$$x_4 = 240$$

$$x_5 = 170$$

$$x_6 = 180$$

The variables x_1 and x_2 are not in the basis and, therefore, they are non-basic variables:

$$x_1 = 0$$

$$x_2 = 0$$

The initial solution is represented by the origin O in the solution space depicted in Figure 5.5.

STEP 2. **Test for Possible Improvement.**   In this step, the basic solution is examined to see if it can be further improved. A negative coefficient in the objective function row indicates the increase in contribution with a unit increase of the related variable. A positive coefficient in the objective function row represents the decrease in contribution with a unit increase of the related variable. The presence of any negative coefficients in the objective function row means that the basic solution can be improved and, thus, that the objective function value can be increased further. If there are any negative coefficients, go to step 3. On the other hand, if each of the coefficients are nonnegative, the solution cannot be improved any more. If so, go to step 6. We note in Table 5.1 that the coefficients of x_1 and x_2 are -100 and -80, respectively. They denote that an improved solution exists and an improved plan can be established.

STEP 3. **Choose the Entering Variable.** In this step, a product (variable) is selected for inclusion in the basis that would be manufactured at some level. The criterion is to maximize the total contribution. The variables x_1 and x_2 have negative coefficients. Hence, these two variables will increase the total contribution by \$100 and \$80, respectively, per unit increase of their value. Since the objective function is of the maximizing type, we select the variable with the largest unit contribution (largest negative coefficient). Therefore, we choose x_1 for increasing its production level. The selected variable, namely x_1, is the entering variable, and the related column in the tableau is called the *pivot column* (see Table 5.2). If ties exist in selecting the entering variable, one of them is chosen arbitrarily.

TABLE 5.2   Initial Tableau with the Pivot Column, Pivot Row, and Pivot Cell

i	Basic Variable	Z	x_1	x_2	x_3	x_4	x_5	x_6	Constant (b_i)	
						Coefficients				
0	Z	1	-100	-80	0	0	0	0	0	
1	x_3	0	6	3	1	0	0	0	120	Pivot
2	x_4	0	8	12	0	1	0	0	240	Row
3	x_5	0	10	0	0	0	1	0	170	
4	x_6	0	0	12	0	0	0	1	180	

Pivot Cell
Pivot Column

STEP 4. **Choose the Leaving Variable.** The entering variable is a nonbasic variable in the current solution and, hence, it has a zero value. The total contribution is increased by \$100 per unit increase of the entering variable x_1. Up to what level can the value of x_1 be increased? When the value of x_1 is increased, the allocation of resources is shifted. The value of x_1 is increased to the maximum possible amount such that the restrictions are not violated. The first restriction from (5.5) is

$$6x_1 + 3x_2 + x_3 = 120$$

This can be rewritten

$$6x_1 = 120 - 3x_2 - x_3$$

When x_2 and x_3 are zero, the maximum value of x_1 is

$$x_1 = \frac{120 - (3 \times 0) - 0}{6} = 20$$

The second restriction from (5.5) is modified and the maximum value of x_1 is obtained, when x_2 and x_4 are zero, as follows:

$$x_1 = \frac{240 - (12 \times 0) - 0}{8} = 30$$

Likewise, we get the maximum values of x_1 from the third and fourth restrictions of (5.5),

$$x_1 = \frac{170 - 0}{10} = 17$$

$$x_1 = \frac{180 - (12 \times 0) - 0}{0} = \text{no limit}$$

Considering the four restrictions, the largest feasible value of x_1 is 17. If x_1 is increased above 17, the third constraint is violated even though the other three constraints may be fulfilled. These calculations can be performed conveniently using Table 5.2.

The maximum value of the entering variable for the restriction i

$$= \frac{\text{Constant} (= b_i)}{\begin{bmatrix} \text{Coefficient of the entering variable} \\ \text{(in pivot column) in the related row} \end{bmatrix}} \qquad (5.7)$$

Therefore,

$$\text{The largest value of } x_1 \text{ for restriction 1} = \frac{120}{6}$$

$$= 20$$

$$\text{The largest value of } x_1 \text{ for restriction 2} = \frac{240}{8}$$

$$= 30$$

$$\text{The largest value of } x_1 \text{ for restriction 3} = \frac{170}{10}$$

$$= 17 \text{ (pivot row)}$$

$$\text{The largest value of } x_1 \text{ for restriction 4} = \frac{180}{0}$$

$$= \text{no limit}$$

The *smallest positive quotient* represents the largest feasible value of the entering variable. The associated row is the *pivot row*. The basic variable in the pivot row leaves the basis in the next solution, and it becomes a nonbasic variable. The entering variable becomes the basic variable for the pivot row in the next solution. Thus, the basic variable x_5 of the pivot row leaves the basis and, in turn, x_1 is recorded as the new basic variable for the pivot row. The leaving variable for the initial solution is x_5. The element that lies at the intersection of the pivot row and the pivot column is called the *pivot cell*. The pivot row and the pivot cell for the initial solution are shown in Table 5.2.

STEP 5. **Determine an Improved Solution.** A new feasible solution is developed in this step. The resulting total contribution is larger than the previous trial solution. The entering variable is a basic variable, and the leaving variable is a nonbasic variable in the improved solution. The improved solution is obtained so that the coefficients of the new basic variable are zero in each row. The only exception is that the coefficient in the pivot row is one, as shown in Table 5.3. The values are calculated using the previous tableau. The row that has the new basic variable is often called the *main row*. The main row for our illustration is the fourth row in Table 5.3. The values for the main row are obtained:

TABLE 5.3   Requirements for Developing the Second Tableau

i	Basic Variable	Z	x_1	x_2	x_3	x_4	x_5	x_6	Constant (b_i)	
				Coefficients						
0	Z	0								
1	x_3	0								
2	x_4	0								
3	x_1	1								←Main row
4	x_6	0								

by dividing each corresponding value in the row of the preceding tableau by the pivot cell value. We divide, therefore, the values given in the fourth row of Table 5.2 by 10 and get the resulting values as follows:

Z	x_1	x_2	x_3	x_4	x_5	x_6	b_i
$\frac{0}{10}$	$\frac{10}{10}$	$\frac{0}{10}$	$\frac{0}{10}$	$\frac{0}{10}$	$\frac{1}{10}$	$\frac{0}{10}$	$\frac{170}{10}$
$= 0$	$= 1$	$= 0$	$= 0$	$= 0$	$= 0.1$	$= 0$	$= 17$

These values are written in the fourth row as shown in Table 5.4. The other rows of the improved solution are calculated by the expression:

$$
\begin{bmatrix} \text{New element} \\ \text{in row } i \text{ and} \\ \text{column } j \end{bmatrix}
$$

$$
= \begin{bmatrix} \text{Corresponding} \\ \text{value in the} \\ \text{preceding} \\ \text{tableau} \end{bmatrix} - \begin{bmatrix} \text{Pivot column} \\ \text{element in} \\ \text{row } i \text{ of the} \\ \text{preceding} \\ \text{tableau} \end{bmatrix} \times \begin{bmatrix} \text{New element} \\ \text{in main row} \\ \text{and column } j \end{bmatrix} \quad (5.8)
$$

The formula (5.8) can be applied for the objective function and restrictions, together with the coefficients and the constant values. The values of the objective function row are calculated as follows:

Variable (Column)	$\left(\begin{array}{c}\text{Value in}\\\text{Table 6.2}\end{array}\right)$	$-$	$\left(\begin{array}{c}\text{Pivot Column}\\\text{Element in}\\\text{Table 6.2}\end{array}\right)$	$\times$	$\left(\begin{array}{c}\text{New Element}\\\text{in Main Row}\\\text{and column } j\end{array}\right)$	$=$	$\left(\begin{array}{c}\text{New Element}\\\text{of the Improved}\\\text{Solution}\end{array}\right)$
Z	1	$-$	(-100)	$\times$	0	$=$	1
x_1	-100	$-$	(-100)	$\times$	1	$=$	0
x_2	-80	$-$	(-100)	$\times$	0	$=$	-80
x_3	0	$-$	(-100)	$\times$	0	$=$	0
x_4	0	$-$	(-100)	$\times$	0	$=$	0
x_5	0	$-$	(-100)	$\times$	.1	$=$	10
x_6	0	$-$	(-100)	$\times$	0	$=$	0
b_i	0	$-$	(-100)	$\times$	17	$=$	1700

Similarly, the values of the other restrictions are also calculated. The values of the improved solution are shown in Table 5.4. We note in Table 5.4 that x_3, x_4, x_1, and x_6 are basic variables and the other two variables, x_2 and x_5, are nonbasic variables. We also obtain the following information:

$$Z = \$1700$$
$$x_3 = 18 \text{ hours}$$
$$x_4 = 104 \text{ hours}$$
$$x_1 = 17 \text{ air conditioners}$$
$$x_6 = 180 \text{ hours}$$
$$\left.\begin{array}{l}x_2 = 0\\x_5 = 0\end{array}\right\} \text{ Nonbasic variables}$$

TABLE 5.4   Second Tableau of Example 5.1

	Basic								Constant
i	Variable	Z	x_1	x_2	x_3	x_4	x_5	x_6	(b_i)
0	Z	1	0	−80	0	0	10	0	1700	
1	x_3	0	0	3	1	0	−0.6	0	18	
2	x_4	0	0	12	0	1	−0.8	0	104	
3	x_1	0	1	0	0	0	0.1	0	17	←Main row
4	x_6	0	0	12	0	0	0	1	180	

(Coefficients spans the Z through x_6 columns.)

It specifies that 17 air conditioners are produced resulting in a total contribution of $1700. The amount of unused times available in the machining department, stamping department, and the refrigerators assembly line are 18, 104, 180 hours, respectively. Note that no refrigerators are produced in this production plan. This plan is represented by the corner point H in the solution space shown in Figure 5.5.

The calculations given in the steps 2 to 5 form *one iteration* in the simplex method. The iterations are performed several times until no improvements can be made for the updated solution.

We go to step 2 after developing an improved basic feasible solution in step 5 and continue the iterative process. Table 5.5 depicts the second tableau with the pivot values designating the required calculations. The third improved tableau is displayed in Table 5.6. The next iteration is continued and the resulting pivot values, and the fourth tableau are shown in Tables 5.7 and 5.8, respectively.

TABLE 5.5   Second Tableau with the Pivot Column, Pivot Row, and Pivot Cell

	Basic								Constant	Step 4
i	Variable	Z	x_1	x_2	x_3	x_4	x_5	x_6	(b_i)	Calculations
0	Z	1	0	−80	0	0	10	0	1700	⟋Pivot Row
1	x_3	0	0	3	1	0	−0.6	0	18	18/3 = 6
2	x_4	0	0	12	0	1	−0.8	0	104	104/12 = 8.67
3	x_1	0	1	0	0	0	0.1	0	17	17/0 = No limit
4	x_6	0	0	12	0	0	0	1	180	180/12 = 15

Pivot Cell — Pivot Column — (pointing to x_2 column, pivot cell at row 1)

TABLE 5.6 Third Tableau of Example 5.1

	Basic				Coefficients				Constant
i	Variable	Z	x_1	x_2	x_3	x_4	x_5	x_6	(b_i)
0	Z	1	0	0	26.67	0	-6	0	2180
1	x_2	0	0	1	0.33	0	-0.2	0	6
2	x_4	0	0	0	-4	1	1.6	0	32
3	x_1	0	1	0	0	0	.1	0	17
4	x_6	0	0	0	-4	0	2.4	1	108

TABLE 5.7 Third Tableau with the Pivot Column, Pivot Row, and Pivot Cell

	Basic				Coefficients				Constant	Step 4
i	Variable	Z	x_1	x_2	x_3	x_4	x_5	x_6	(b_i)	Calculations
0	Z	1	0	0	26.67	0	-6	0	2180	Pivot Row
1	x_2	0	0	1	0.33	0	-0.2	0	6	$6/(-0.2) = -30$
2	x_4	0	0	0	-4	1	1.6	0	32	$32/1.6 = 20$
3	x_1	0	1	0	0	0	0.1	0	17	$17/0.1 = 170$
4	x_6	0	0	0	-4	0	2.4	1	108	$108/2.4 = 45$

Pivot Cell
Pivot Column

TABLE 5.8 Fourth Tableau of Example 5.1 (Optimum Solution)

	Basic				Coefficients				Constant
i	Variable	Z	x_1	x_2	x_3	x_4	x_5	x_6	(b_i)
0	Z	1	0	0	11.67	3.75	0	0	2300
1	x_2	0	0	1	-0.17	0.13	0	0	10
2	x_5	0	0	0	-2.5	0.63	1	0	20
3	x_1	0	1	0	0.25	-0.06	0	0	15
4	x_6	0	0	0	2	-1.5	0	1	60

Note that the third and the fourth improved solutions represent the corner points G and E, respectively, in Figure 5.5. We have shown in the last section that the corner point E denotes the optimum production plan. How can we recognize this fact in the fourth tableau? We observe that no negative coefficients appear in the objective function row. It specifies that the current trial solution cannot be improved and, thus, it represents the optimum solution.

STEP 6. **Determine the Values of Variables at Optimum.** We can obtain the values of the variables for the optimum solution from the last tableau. For our example, the values are

$Z = \$2300$

$x_2 = 10$ refrigerators

$x_5 = 20$ hours of unused time in the air conditioners assembly line

$x_1 = 15$ air conditioners

$x_6 = 60$ hours of unused time in the refrigerators assembly line

$\left.\begin{array}{l} x_3 = 0 \\ x_4 = 0 \end{array}\right\}$ Nonbasic variables

Recall that we have established the same production program as the optimum solution by the graphical method.

We note that the total contribution is increased as we move from one tableau to the other and also that each solution represents a corner point of the solution space. The total contributions for the four solutions are:

$$\$0 \qquad \text{(First tableau, Corner } O)$$
$$\$1700 \qquad \text{(Second tableau, Corner } H)$$
$$\$2180 \qquad \text{(Third tableau, Corner } G)$$
$$\$2300 \qquad \text{(Fourth tableau, Corner } E)$$

These are the important characteristics of the simplex technique.

LINEAR PROGRAMMING VARIATIONS

The linear programming problem with maximizing objective function and "$\leq$" type of restrictions can be solved by the simplex method described in the last section. For solving a linear problem with minimizing objective function and/or "$\geq$" or "$=$" types of restrictions, some modifications are necessary to apply the simplex technique.

Restrictions with "$\geq$" Sign

A constraint with the "greater than or equal to" form is changed to an "equal to" form by subtracting a unique *surplus variable* from the left side of the inequality sign.

For example,

$$10x_1 + 2x_2 \geq 40$$

is modified to

$$10x_1 + 2x_2 - x_3 = 40$$

The surplus variable x_3 indicates the amount that the left-side expression is in excess of 40. When the left-side expression is equal to 40, x_3 is zero. In order for the restriction to be satisfied, the surplus variable must assume only nonnegative values. A fictitious nonnegative variable is added to the left side of the equality sign to have a basic variable in this row. It is designated as the basic variable in the initial tableau. However, it violates the original restriction. The fictitious variable must be zero in the final solution to fulfill the original restriction. This is accomplished by assigning a large penalty value (negative quantity) to the related coefficient in the objective function for a maximization problem. For a minimization problem, a large positive quantity is assigned to the related coefficient in the objective function. The fictitious variable is often called the *artificial variable*.

The mathematical formulation of Example 5.2 is rewritten for easy reference. (See equation 5.3.)

$$\text{Maximize } Z = 12x_1 + 9x_2$$
$$\text{Subject to} \quad 8x_1 + 4x_2 \leq 240$$
$$6x_1 + 9x_2 \leq 360$$
$$45x_1 + 15x_2 \geq 450$$
$$x_1 + x_2 \geq 20$$

and

$$x_1 \geq 0 \quad x_2 \geq 0$$

The modified form of the problem follows:

Maximize

$$Z = 12x_1 + 9x_2 + 0x_3 + 0x_4 + 0x_5 - 1000x_6 + 0x_7 - 1000x_8$$

Subject to

$$8x_1 + 4x_2 + x_3 \qquad\qquad\qquad\qquad = 240$$

$$6x_1 + 9x_2 \qquad + x_4 \qquad\qquad\qquad = 360$$

$$45x_1 + 15x_2 \qquad\qquad - x_5 + \qquad x_6 \qquad\qquad = 450$$

$$x_1 + x_2 \qquad\qquad\qquad\qquad - x_7 + \qquad x_8 = 20$$

and

$$x_j \geq 0 \qquad \text{for } j = 1, 2, \ldots, 8 \qquad\qquad (5.9)$$

Among the eight variables,

x_1 and x_2 are the decision variables.

x_3 and x_4 are the slack variables.

x_5 and x_7 are the surplus variables.

x_6 and x_8 are the artificial variables.

The slack variables and the artificial variables are set in basis for the initial solution. We have, therefore, x_3, x_4, x_6, and x_8 in basis. Table 5.9 shows the initial tableau in the tabular form. The variable x_6 is the basic variable for the fourth row. Thus it must have a coefficient 1 in the fourth row, and its coefficient must be a 0 in all other rows. But x_6 has the coefficient 1000 in the first row, and this is not permitted.

TABLE 5.9   Initial Tableau of Example 5.2*

	Basic	Coefficients									Constant
i	Variable	Z	x_1	x_2	x_3	x_4	x_5	x_6	x_7	x_8	(b_i)
0	Z	1	-12	-9	0	0	0	1000	0	1000	0
1	x_3	0	8	4	1	0	0	0	0	0	240
2	x_4	0	6	9	0	1	0	0	0	0	360
3	x_6	0	45	15	0	0	-1	1	0	0	450
4	x_8	0	1	1	0	0	0	0	-1	1	20

* Note that the tableau is not in required form for the basic variables x_6 and x_8.

The coefficients of the basic variable x_8 is also not in the required form, because it also has the coefficient 1000 in the first row. The tableau must be in standard form for applying the computational steps shown in

Figure 5.8. By multiplying each value in the fourth row and the fifth row by -1000 and adding them up with the related value in the first row, we can obtain the initial basic solution in the required form.

Variable (Column)	$\begin{pmatrix}\text{Fourth Row}\\\text{Value}\\x(-1000)\end{pmatrix}$	$+$	$\begin{pmatrix}\text{Fifth Row}\\\text{Value}\\x-(1000)\end{pmatrix}$	$+$	$\begin{pmatrix}\text{First Row}\\\text{Value}\end{pmatrix}$	$=$	$\begin{pmatrix}\text{New First}\\\text{Row Value}\end{pmatrix}$
Z	0	$+$	0	$+$	1	$=$	1
x_1	$-45,000$	$+$	$(-1,000)$	$+$	-12	$=$	$-46,012$
x_2	$-15,000$	$+$	$(-1,000)$	$+$	-9	$=$	$-16,009$
x_3	0	$+$	0	$+$	0	$=$	0
x_4	0	$+$	0	$+$	0	$=$	0
x_5	$\cdot\,1,000$	$+$	0	$+$	0	$=$	1,000
x_6	$-1,000$	$+$	0	$+$	(1,000)	$=$	0
x_7	0	$+$	1,000	$+$	0	$=$	1,000
x_8	0	$+$	$(-1,000)$	$+$	(1,000)	$=$	0
b_i	$-450,000$	$+$	$(-20,000)$	$+$	0	$=$	$-470,000$

The standard form of the initial tableau is displayed in Table 5.10. The simplex method can be applied to the initial tableau shown in Table 5.10 to establish the optimum solution.

TABLE 5.10   Initial Tableau of Example 5.2

	Basic					Coefficients					Constant
i	Variable	Z	x_1	x_2	x_3	x_4	x_5	x_6	x_7	x_8	(b_i)
0	Z	1	$-46,012$	$-16,009$	0	0	1000	0	1000	0	$-470,000$
1	x_3	0	8	4	1	0	0	0	0	0	240
2	x_4	0	6	9	0	1	0	0	0	0	360
3	x_6	0	45	15	0	0	-1	1	0	0	450
4	x_8	0	1	1	0	0	0	0	-1	1	20

Restrictions with "$=$" Sign

In a constraint with the "equal to" sign, there is no basic variable. A unique artificial variable is added to the left side of the equal sign of each "equal to" form of constraint. They are designated to be the basic variables in the initial solution. As we have discussed earlier, an appropriate large positive or negative coefficient for the artificial variables is assigned in the objective function. This guarantees that the fictitious variables will not be in the basis in the optimum solution.

Minimization Problems

When a LP problem with the minimizing objective function is encountered as in Example 5.3, it is to be converted to a maximization problem. Consider the following problem:

$$\text{Minimize } Y = 30x_1 + 60x_2$$

$$\text{Subject to} \quad x_1 + x_2 \geq 10$$

$$x_2 \leq 15$$

$$6x_1 + 4x_2 \leq 120$$

and

$$x_1 \geq 0 \quad x_2 \geq 0 \tag{5.10}$$

On introducing slack variables, surplus variables, and artificial variables, the problem has the following appearance:

$$\text{Minimize } Y = 30x_1 + 60x_2 + 10{,}000x_4$$

$$\text{Subject to} \quad x_1 + x_2 - x_3 + x_4 = 10$$

$$x_2 + x_5 = 15$$

$$6x_1 + 4x_2 + x_6 = 120$$

and

$$x_j \geq 0 \quad \text{for } j = 1, 2, \ldots, 6 \tag{5.11}$$

The objective function may be modified to the maximizing form by multiplying both sides by -1.

$$\text{Maximize} -Y = -30x_1 - 60x_2 - 10{,}000x_4$$

Let

$$Z = -Y$$

Then

$$\text{Maximize } Z = -30x_1 - 60x_2 - 10{,}000x_4$$

$$\text{Subject to} \quad x_1 + x_2 - x_3 + x_4 = 10$$

$$x_2 + x_5 = 15$$

$$6x_1 + 4x_2 + x_6 = 120$$

and

$$x_j \geq 0 \quad \text{for } j = 1, 2, \ldots, 6 \tag{5.12}$$

Now, the simplex method may be employed to determine the optimum solution. On establishing the optimum plan, we can determine the objective function value for the original LP problem (5.10) by changing the sign of the optimum Z value.

A COMPUTER MODEL FOR THE LP PROBLEMS

In this section we present a general computer program to solve the linear programming problems by the simplex algorithm. We also illustrate how we can use the program for solving LP problems. The program can be utilized to solve problems with up to 10 restrictions and 20 variables. The following symbols are used in the computer program.

MROWS	= number of restrictions
NCOLS	= number of variables (includes the slack variables, artificial variables, and surplus variables)
MPROB	= type of problem When MPROB = 0, it is a minimization problem When MPROB = 1, it is a maximization problem
INDEX	= tableau number
Z	= objective function value
BIGVAL	= the smallest coefficient in the objective function row
IPROW	= number of the restriction that is selected as the pivot row
JPCOL	= variable number that is related to the pivot column
P	= pivot cell value
FIND1	= this subprogram determines the column that has the smallest coefficient in the objective function row
IPROWX	= this subprogram establishes the pivot row
MODIFY	= this subprogram calculates the values for the next improved tableau
A(K, J)	= coefficient in restriction K for variable J
B(K)	= constant value for the restriction K $(= b_k)$
C(J)	= coefficient of variable J in the objective function row
IBASIC(K)	= variable number designating the basic variable for restriction K

The variables B( ) and IBASIC( ) are one-dimensional arrays with MROWS elements. The one-dimensional variable C( ) has NCOLS elements. The variable A(,) is a two-dimensional array of size MROWS by NCOLS.

We might initialize the program by the instructions:

```
      READ 100, MROWS, NCOLS, MPROB
      READ 110, (C(J), J = 1, NCOLS)
      DO 500 I = 1, MROWS
      READ 120, (A(I, J), J = 1, NCOLS)
  500 CONTINUE
      READ 130, (B(I), I = 1, MROWS)
      READ 140, (IBASIC(I), I = 1, MROWS)
  100 FORMAT ( )
  110 FORMAT ( )
  120 FORMAT ( )
  130 FORMAT ( )
  140 FORMAT ( )
```

The input values may be printed out by the following statements.

```
      PRINT 801, MROWS
      PRINT 805, NCOLS
      PRINT 810, MPROB
      PRINT 210, (C(J), J = 1, NCOLS)
      DO 510 I = 1, MROWS
      PRINT 220, IBASIC(I), (A(I, J), J = 1, NCOLS), B(I)
  510 CONTINUE
  801 FORMAT ( )
  805 FORMAT ( )
  810 FORMAT ( )
  210 FORMAT ( )
  220 FORMAT ( )
```

The following statements will cause the computer to change the signs of the objective function coefficients for the maximizing problems.

```
      IF (MPROB.LT.1) GO TO 515
      DO 520 J = 1, NCOLS
  520 C(J) = - C(J)
```

The values of Z and INDEX are initialized by the assignment statements.

```
  515 Z = 0.0
      INDEX = 1
```

When artificial variables are included in the basis of the initial solution, the tableau is not in the standard form. The following set of instructions will change the initial tableau in those situations and bring the tableau to the standard form.

```
      DO 530 K = 1, MROWS
      L = IBASIC(K)
      IF (C(L).EQ.O.O) GO TO 530
      X = −C(L)
      DO 535 J = 1, NCOLS
535   C(J) = C(J) + A(K, J) * X
      Z = Z + B(K) * X
530   CONTINUE
```

The iterative operations are executed from this point on. The updated solution is printed out by the statements.

```
550   CONTINUE
      PRINT 225, INDEX
      PRINT 240, (C(J), J = 1, NCOLS), Z
      DO 545 I = 1, MROWS
545   PRINT 220, IBASIC(I), (A(I, J), J = 1, NCOLS), B(I)
225   FORMAT ( )
240   FORMAT ( )
220   FORMAT ( )
```

The subroutine FIND1 is structured to determine the variable that has the smallest coefficient in the objective function row. The flowchart for SUBROUTINE FIND1 is found in Figure 5.9. The subprogram is given below.

```
      SUBROUTINE FIND1 (C, BIGVAL, JPCOL, NCOLS)
      DIMENSION C(20)
      BIGVAL = 999999.99
      DO 560 J = 1, NCOLS
      IF (BIGVAL.LE.C(J)) GO TO 560
      BIGVAL = C(J)
      JPCOL = J
560   CONTINUE
      RETURN
      END
```

The above subroutine might be executed by including the following

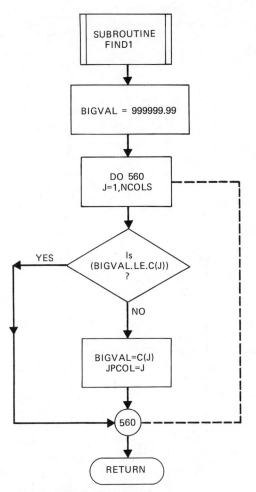

FIGURE 5.9   Flowchart for SUBROUTINE FIND1.

instruction in the main program:

CALL FIND1 (C, BIGVAL, JPCOL, NCOLS)

When the smallest coefficient (BIGVAL) in the objective function row is negative, the pivot row is determined. The subprogram IPROWX is utilized to find out the pivot row. If BIGVAL is a nonnegative value, the current solution is the optimum solution. In such a case, the iterative operations are skipped and the optimum solution is printed out by transferring control to statement 700. The following statements would do this for us.

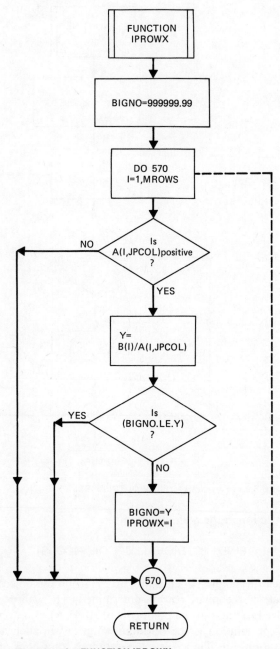

FIGURE 5.10 Flowchart for FUNCTION IPROWX.

```
      IF (BIGVAL) 565, 700, 700
565   CONTINUE
      IPROW = IPROWX (MROWS, JPCOL, A, B, IPROW)
```

Figure 5.10 shows the flowchart for the function IPROWX and the function subprogram is given below.

```
      FUNCTION IPROWX(MROWS, JPCOL, A, B, IPROW)
      DIMENSION A(10, 20), B(10)
      BIGNO = 999999.99
      DO 570 I = 1, MROWS
      IF (A(I, JPCOL) ) 570, 570, 575
575   Y = B (I)/A(I, JPCOL)
      IF (BIGNO.LE.Y) GO TO 570
      BIGNO = Y
      IPROWX = I
570   CONTINUE
      RETURN
      END
```

The improved solution is established by using subprogram MODIFY. Then, the tableau number is increased by one and the control is transferred to the beginning of the iterative operations.

```
      CALL MODIFY(JPCOL, IPROW, MROWS, NCOLS, IBASIC,
7 A, B, C, Z)
      INDEX = INDEX + 1
      GO TO 550
```

Figure 5.11 displays the flowchart for subroutine MODIFY; and the subroutine is given below.

```
      SUBROUTINE MODIFY(JPCOL, IPROW, MROWS, NCOLS,
7 IBASIC, A, B, C, Z)
      DIMENSION A(10, 20), B(10), C(20), IBASIC(10)
      IBASIC(IPROW) = JPCOL
      P = A(IPROW, JPCOL)
      DO 580 J = 1, NCOLS
580   A(IPROW, J) = A(IPROW, J)/P
      B(IPROW) = B(IPROW)/P
      R = C(JPCOL)
      RNEG = −R
      DO 585 J = 1, NCOLS
```

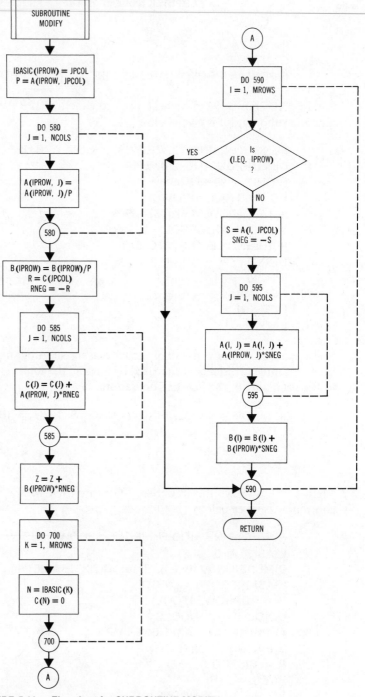

FIGURE 5.11   Flowchart for SUBROUTINE MODIFY.

```
585   C(J) = C(J) + A(IPROW, J) * RNEG
      Z = Z + B(IPROW) * RNEG
      DO 700 K = 1, MROWS
      N = IBASIC(K)
700   C(N) = 0
      DO 590 I = 1 MROWS
      IF (I.EQ.IPROW) GO TO 590
      S = A(I, JPCOL)
      SNEG = -S
      DO 595 J = 1, NCOLS
595   A(I, J) = A(I, J) + A(IPROW, J) * SNEG
      B(I) = B(I) + B(IPROW) * SNEG
590   CONTINUE
      RETURN
      END
```

The following instructions would enable the computer to change the sign of the objective function value for the minimization problems and print out the values of the variables for the optimum plan.

```
700   PRINT 245
      IF (MPROB.EQ.O) Z = -Z
      PRINT 250, Z
      DO 610 I = 1, MROWS
610   PRINT 255, IBASIC(I), B(I)
      X = 0.0
      DO 620 I = 1, NCOLS
      DO 625 J = 1, MROWS
      K = IBASIC(J)
      IF (I.EQ.K) GO TO 620
625   CONTINUE
      PRINT 255, I, X
620   CONTINUE
250   FORMAT ( )
255   FORMAT ( )
      STOP
      END
```

In Figure 5.12, a complete program for the simplex method is presented, it is composed of our previously designed program segments. Many comment statements are included to enhance the readability of the computer program.

```
              DIMENSION A(10,20),B(10),C(20),IBASIC(10)
C
C  READ THE INPUT DATA
        READ 100, MROWS,NCOLS,MPROB
        READ 110, (C(J),J=1,NCOLS)
        DO 500 I=1,MROWS
        READ 120, (A(I,J),J=1,NCOLS)
    500 CONTINUE
        READ 130, (B(I),I=1,MROWS)
        READ 140, (IBASIC(I),I=1,MROWS)
C
C  PRINTOUT THE INPUT DATA
        PRINT 800
        PRINT 801,MROWS
        PRINT 805,NCOLS
        PRINT 810,MPROB
        PRINT 815
        PRINT 820
        PRINT 210, (C(J),J=1,NCOLS)
        DO 510 I=1,MROWS
        PRINT 220, IBASIC(I),(A(I,J),J=1,NCOLS),B(I)
    510 CONTINUE
C
C  CHANGE C VALUES FOR THE MAXIMIZING PROBLEM.
        IF ( MPROB .LT. 1 ) GO TO 515
        DO 520 J=1,NCOLS
    520 C(J)=-C(J)
C
C  INITIALIZE Z AND INDEX
    515 Z=0.0
        INDEX=1
C
        DO 530 K=1,MROWS
        L=IBASIC(K)
        IF(C(L).EQ.0.0) GO TO 530
        X=-C(L)
        DO 535 J=1,NCOLS
    535 C(J)=C(J)+A(K,J)*X
        Z=Z+B(K)*X
    530 CONTINUE
C
    550 CONTINUE
C
C  PRINT THE TABLEAU
        PRINT 225,INDEX
        PRINT 240,           (C(J),J=1,NCOLS),Z
        DO 545 I=1,MROWS
    545 PRINT 220, IBASIC(I),(A(I,J),J=1,NCOLS),B(I)


C
C  FINDOUT THE PIVOT COLUMN
        CALL           FIND1 ( C, BIGVAL, JPCOL, NCOLS )
        IF (BIGVAL)565,700,700
    565 CONTINUE
C
C  DETERMINE THE PIVOT ROW
        IPROW =   IPROWX ( MROWS, JPCOL, A, B, IPROW )
```

FIGURE 5.12 Computer program for the Simplex method to solve the LP problems (illustrated for Example 5.1).

```
C
C   CALCULATE AND FINDOUT THE NEXT TABLEAU
      CALL        MODIFY(JPCOL, IPROW, MROWS, NCOLS, IBASIC,
     7A,  B,  C,  Z  )
      INDEX=INDEX+1
C
      GO TO 550
C
C   PRINT THE OPTIMUM SOLUTION
  700 PRINT 245
C
C   CHANGE Z VALUE FOR THE MINIMIZATION PROBLEM
      IF ( MPROB .EQ. 0 ) Z=-Z
      PRINT 250, Z
      DO 610 I=1,MROWS
  610 PRINT 255, IBASIC(I),B(I)
      X=0.0
      DO 620 I=1,NCOLS
      DO 625 J=1,MROWS
      K=IBASIC(J)
      IF (I.EQ.K) GO TO 620
  625 CONTINUE
      PRINT 255, I,X
  620 CONTINUE
C
  100 FORMAT (3I2)
  110 FORMAT ( 6F6.0 )
  120 FORMAT ( 6F3.0 )
  130 FORMAT ( 4F5.0 )
  140 FORMAT (4I2)
  210 FORMAT ( 4X, 1HZ, 6F9.2 )
  220 FORMAT ( I5, 7F9.2  )
  240 FORMAT ( 4X, 1HZ, 7F9.2  )
  225 FORMAT (  / ,20X,7HTABLEAU,I3,/)
  245 FORMAT ( 2(/), 3X, 8HVARIABLE , 4X, 5HVALUE  )
  250 FORMAT (     9X,1HZ,F10.2)
  255 FORMAT (I10,F10.2)
  800 FORMAT (1H1)
  801 FORMAT (5(/),5X,26HNUMBER OF RESTRICTIONS   =,I3  )
  805 FORMAT (5X,26HNUMBER OF VARIABLES       =,I3   )
  810 FORMAT (5X,26HTYPE OF ORIGINAL PROBLEM =,I3   )
  815 FORMAT (7X,33HZERO DENOTES MINIMIZATION PROBLEM  )
  820 FORMAT (7X,33HONE  DENOTES MAXIMIZATION PROBLEM,/ )
      STOP
      END


      SUBROUTINE  FIND1 ( C, BIGVAL, JPCOL, NCOLS )
C
C   THIS SUBPROGRAM DETERMINES THE PIVOT COLUMN
      DIMENSION C(20)
      BIGVAL=999999.99
C
      DO 560 J=1,NCOLS
      IF ( BIGVAL .LE. C(J) ) GO TO 560
      BIGVAL = C(J)
      JPCOL=J
  560 CONTINUE
C
      RETURN
      END
```

Figure 5.12 (*Continued*)

```
      FUNCTION  IPROWX ( MROWS, JPCOL, A, B, IPROW )
C
C   THIS SUBPROGRAM DETERMINES THE PIVOT ROW
      DIMENSION  A(10,20), B(10)
      BIGNO=999999.99
C
      DO 570 I=1,MROWS
      IF (A(I,JPCOL))570,570,575
  575 Y=B(I)/A(I,JPCOL)
      IF ( BIGNO .LE. Y ) GO TO 570
      BIGNO = Y
      IPROWX = I
  570 CONTINUE
C
      RETURN
      END




      SUBROUTINE MODIFY(JPCOL, IPROW, MROWS, NCOLS, IBASIC,
     7A, B, C, Z )
C
C   THIS SUBROUTINE CALCULATES THE CELL VALUES
C   FOR NEXT TABLEAU
      DIMENSION A(10,20), B(10), C(20), IBASIC(10)
      IBASIC(IPROW)=JPCOL
      P=A(IPROW,JPCOL)
C
      DO 580 J=1,NCOLS
  580 A(IPROW,J)=A(IPROW,J)/P
C
      B(IPROW)=B(IPROW)/P
      R=C(JPCOL)
      RNEG=-R
C
      DO 585 J=1,NCOLS
  585 C(J)=C(J)+A(IPROW,J)*RNEG
      Z=Z+B(IPROW)*RNEG
C
      DO 700 K = 1, MROWS
      N = IBASIC ( K )
  700 C(N) = 0
C
      DO 590 I=1,MROWS
      IF(I.EQ.IPROW) GO TO 590
      S=A(I,JPCOL)
      SNEG=-S
C
      DO 595 J=1,NCOLS
  595 A(I,J)=A(I,J)+A(IPROW,J)*SNEG
C
      B(I)=B(I)+B(IPROW)*SNEG
  590 CONTINUE
C
      RETURN
      END
```

Figure 5.12 (*Continued*)

ILLUSTRATIONS ON USING THE PROGRAM

Illustration of the Computer Program for Example 5.1

The modified formulation for Example 5.1 is depicted in expression 5.5. There are four restrictions and six variables for this maximization problem. This information is transmitted to the computer by a READ statement. The related FORMAT statement is

100   FORMAT (3I2)

The coefficients of the objective function might be read according to the following FORMAT statement:

110   FORMAT (6F6.0)

The data card pertaining to the coefficients of the restrictions may be prepared according to the FORMAT instruction:

120   FORMAT (6F3.0)

The data card that gives the constant values of the constraints may be written according to this FORMAT statement:

130   FORMAT (4F5.0)

The data card that specifies the variable numbers designating the basic variables of the initial solution might be designed according to the FORMAT statement:

140   FORMAT (4I2)

The resulting data deck has the following appearance:

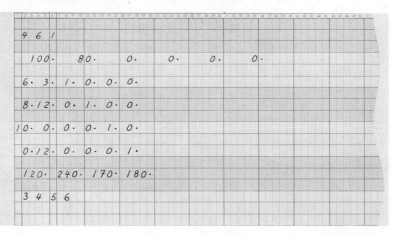

```
NUMBER OF RESTRICTIONS   =  4
NUMBER OF VARIABLES      =  6
TYPE OF ORIGINAL PROBLEM =  1
  ZERO DENOTES MINIMIZATION PROBLEM
  ONE  DENOTES MAXIMIZATION PROBLEM
```

Z	100.00	80.00	0	0	0	0	
3	6.00	3.00	1.00	0	0	0	120.00
4	8.00	12.00	0	1.00	0	0	240.00
5	10.00	0	0	0	1.00	0	170.00
6	0	12.00	0	0	0	1.00	180.00

TABLEAU 1

Z	-100.00	-80.00	-0	-0	-0	-0	0
3	6.00	3.00	1.00	0	0	0	120.00
4	8.00	12.00	0	1.00	0	0	240.00
5	10.00	0	0	0	1.00	0	170.00
6	0	12.00	0	0	0	1.00	180.00

TABLEAU 2

Z	0	-80.00	0	0	10.00	0	1700.00
3	0	3.00	1.00	0	-0.60	0	18.00
4	0	12.00	0	1.00	-0.80	0	104.00
1	1.00	0	0	0	.10	0	17.00
6	0	12.00	0	0	0	1.00	180.00

TABLEAU 3

Z	0	0	26.67	0	-6.00	0	2180.00
2	0	1.00	.33	0	-0.20	0	6.00
4	0	0	-4.00	1.00	1.60	0	32.00
1	1.00	0	0	0	.10	0	17.00
6	0	0	-4.00	0	2.40	1.00	108.00

TABLEAU 4

Z	0	0	11.67	3.75	0	0	2300.00
2	0	1.00	-0.17	.13	.00	0	10.00
5	0	0	-2.50	.63	1.00	0	20.00
1	1.00	0	.25	-0.06	-0.00	0	15.00
6	0	0	2.00	-1.50	0	1.00	60.00

VARIABLE	VALUE
Z	2300.00
2	10.00
5	20.00
1	15.00
6	60.00
3	0
4	0

FIGURE 5.13   Computer printout for Example 5.1.

The tableaus might be printed out according to the specification instructions

210   FORMAT (4X, 1HZ, 6F9.2)
220   FORMAT (I5, 7F9.2)
240   FORMAT (4X, 1HZ, 7F9.2)

Figure 5.13 depicts a computer printout of the results.

Illustration of the Computer Program for Example 5.2

The initial tableau for Example 5.2 is shown in expression 5.9. We note that there are four restrictions and eight variables. The data deck might be designed according to these FORMAT instructions:

100   FORMAT (3I2)
110   FORMAT (8F6.0)
120   FORMAT (8F3.0)
130   FORMAT (4F5.0)
140   FORMAT (4I2)

The data deck has eight cards and has the following appearance:

4 8 1						
1 2·	9·	0·	0·	0· -1000·	0· -1000·
8· 4·	1·	0· 0·	0· 0·	0·	
6· 9·	0·	1· 0·	0· 0·	0·	
45·15·	0·	0· -1·	1· 0·	0·	
1· 1·	0·	0· 0·	0· -1·	1·	
240·	360·	450·	20·		
3 4 6 8					

The solution is printed out according to the FORMAT instructions

210   FORMAT (4X, 1HZ, 8F9.2)
220   FORMAT (I5, 8F9.2, F11.2)
240   FORMAT (4X, 1HZ, 8F9.2, F11.2)

```
NUMBER OF RESTRICTIONS   =  4
NUMBER OF VARIABLES      =  8
TYPE OF ORIGINAL PROBLEM =  1
  ZERO DENOTES MINIMIZATION PROBLEM
  ONE  DENOTES MAXIMIZATION PROBLEM
```

Z	12.00	9.00	0	0	0	-1000.00	0	-1000.00	
3	8.00	4.00	1.00	0	0	0	0	0	240.00
4	6.00	9.00	0	1.00	0	0	0	0	360.00
6	45.00	15.00	0	0	-1.00	1.00	0	0	450.00
8	1.00	1.00	0	0	0	0	-1.00	1.00	20.00

```
             TABLEAU  1
```

Z	-46012.00	-16009.00	0	0	1000.00	0	1000.00	0	-470000.00
3	8.00	4.00	1.00	0	0	0	0	0	240.00
4	6.00	9.00	0	1.00	0	0	0	0	360.00
6	45.00	15.00	0	0	-1.00	1.00	0	0	450.00
8	1.00	1.00	0	0	0	0	-1.00	1.00	20.00

```
             TABLEAU  2
```

Z	0	-671.67	0	0	-22.49	1022.49	1000.00	0	-9880.00
3	0	1.33	1.00	0	.18	-0.18	0	0	160.00
4	0	7.00	0	1.00	.13	-0.13	0	0	300.00
1	1.00	.33	0	0	-0.02	.02	0	0	10.00
8	0	.67	0	0	.02	-0.02	-1.00	1.00	10.00

```
             TABLEAU  3
```

Z	0	0	0	0	-0.10	1000.10	-7.50	1007.50	195.00
3	0	0	1.00	0	.13	-0.13	2.00	-2.00	140.00
4	0	0	0	1.00	-0.10	.10	10.50	-10.50	195.00
1	1.00	0	0	0	-0.03	.03	.50	-0.50	5.00
2	0	1.00	0	0	.03	-0.03	-1.50	1.50	15.00

```
             TABLEAU  4
```

Z	15.00	0	0	0	-0.60	1000.60	0	1000.00	270.00
3	-4.00	0	1.00	0	.27	-0.27	0	0	120.00
4	-21.00	0	0	1.00	.60	-0.60	-0.00	.00	90.00
7	2.00	0	0	0	-0.07	.07	1.00	-1.00	10.00
2	3.00	1.00	0	0	-0.07	.07	.00	-0.00	30.00

```
             TABLEAU  5
```

7	-6.00	0	0	1.00	0	1000.00	0	1000.00	360.00
3	5.33	0	1.00	-0.44	0	0	.00	-0.00	80.00
5	-35.00	0	0	1.67	1.00	-1.00	-0.00	.00	150.00
7	-0.33	0	0	.11	.00	-0.00	1.00	-1.00	20.00
2	.67	1.00	0	.11	.00	-0.00	-0.00	-0.00	40.00

```
             TABLEAU  6
```

7	0	0	1.12	.50	0	1000.00	0	1000.00	450.00
1	1.00	0	.19	-0.08	0	0	.00	-0.00	15.00
5	.00	0	6.56	-1.25	1.00	-1.00	.00	-0.00	675.00
7	0	0	.06	.08	.00	-0.00	1.00	-1.00	25.00
2	0	1.00	-0.13	.17	.00	-0.00	-0.00	.00	30.00

VARIABLE	VALUE
Z	450.00
1	15.00
5	675.00
7	25.00
2	30.00
3	0
4	0
6	0
8	0

FIGURE 5.14 Computer printout for Example 5.2.

The resulting computer printout is shown in Figure 5.14 Note that the data of initial solution is transmitted to the computer system after including the necessary slack variables, surplus variables, and artificial variables. It is *not* required that the initial solution values be in the standard form. To solve the minimization problems, we read the information of the initial solution with the objective function in the minimizing form. For example, to establish the optimum plan for the problem (5.10), we transmit the data given in the formulation (5.11).

PROBLEMS

5.1 The Duckworth Company manufactures two types of industrial scales, Model 310 and Model 340. Each of these models flow through four departments: milling, turning, assembly, and testing. Here are the hours of department time required by each of the models per unit.

	Milling	Turning	Assembly	Testing
Model 310	2	2	2	3
Model 340	1	3	2	4
Hours available per week	60	120	90	120

The contribution of Model 310 is $200 per unit and the contribution of Model 340 is $300 per unit. The management wishes to know how many of each of these models are to be produced per week in order to maximize the total contribution. Set up the linear programming model for the problem.

5.2 The Bergmann Industries manufactures two products, A and B, using three raw materials, R_1, R_2, and R_3. The requirement of raw materials for the production of the products per unit are:

	R_1	R_2	R_3	Profit per Unit
A	4	8	2	$600
B	10	7	1	$800
Amounts Available	36000	11200	6000	

At least 1000 units of A are required and no more than 400 units of B can be sold. The vice-president of production wants to plan the manufacturing such that the total profit is maximized. Formulate the linear programming model for this problem.

5.3 The Drake Enterprises has three products, P_1, P_2, and P_3, in its facilities at Honolulu. The management wishes to ship them to Los Angeles using one of its ships so that the total profit of the products loaded in the ship is maximized. The cargo ship has three holds: forward, center and aft. The capacity limits for the three holds are the following:

Hold	Weight (Tons)	Volume (Cubic Feet)
Forward	2500	50,000
Center	4000	70,000
Aft	2000	40,000

The available amounts of the products as well as the required storage space per ton of the products on the ship are as follows:

Product	Tons Available	Cubic Feet per Ton
P_1	5000	30
P_2	7000	20
P_3	2000	40

The profits per ton of P_1, P_2, and P_3 are \$10, \$12, and \$6, respectively. In order to maintain the trim of the ship, the weight in each hold must be the same percentage of its capacity in tons. How much of the products should be loaded in the ship and how should it be allocated to the three holds? Formulate the linear programming model for the problem.

5.4 The Sweetheart Candy Shop plans to produce two candy mixes, deluxe mix and party mix, out of three ingredients A, B, and C. The objective is to determine the requirement of the ingredients that will maximize the toal profit. Below are the cost and availability of the ingredients:

Ingredient	Cost per Pound	Availability in Pounds
A	\$3.50	Not more than 4000
B	\$2.00	Not less than 2000
C	\$3.00	Not more than 5000

In order to maintain the quality of the mixes, these proportions of the ingredients must be adhered to:

Candy Mix	Specification	Selling Price per Pound
Deluxe Mix	Not less than 60% of B	$7
	Not more than 30% of C	
Party Mix	Not more than 50% of A	$6

Assume that the total profit is the total sales amount less the total cost of the constituents. What is the linear programming formulation for this problem?

5.5 The Hilgert Chemical Company manufactures two chemicals and markets under the brand names Alpha and Beta. The contributions from Alpha and Beta are $90 and $10 per ton, respectively. Two raw materials, K_1 and K_2, are to be procured from outside sources for producing Alpha and Beta. The production requirements and raw materials available per day are:

Raw Material	Alpha per Ton	Beta per Ton	Amount Available per Day
K_1	3 tons	2 tons	At most 30 tons
K_2	10 tons	3 tons	Not less than 30 tons

For one ton of Alpha production two man-hours are required, and for one ton of Beta production, three man-hours are required. The availability of man-hours per day is limited to 12. Assume that the company can sell any amount of Alpha. No more than 12 tons of Beta can be sold on any day. The company wishes to determine how many tons of Alpha and Beta are to be produced daily to maximize the total contribution. Set up the linear programming model for the problem.

5.6 Consider Problem 5.1.
(a) Determine the optimum solution by the graphical method.
(b) Determine the optimum solution by the Simplex method.

5.7 Consider Problem 5.2.
(a) Determine the optimum manufacturing plan by the graphical method.
(b) Determine the optimum manufacturing plan by the Simplex method.

5.8 Consider Problem 5.5.
(a) Determine the optimum solution using the graphical method.
(b) Find out the optimum solution by the Simplex method.

5.9　Write a FORTRAN program to determine the optimum loading plan for Problem 5.1.

5.10　Compose a FORTRAN program to find out the optimum requirement of ingredients for Problem 5.2.

5.11　Develop a FORTRAN program to solve the oil refinery problem (Example 5.3).

5.12　Consider Problem 5.5. The production planning department is considering the possibility of manufacturing and marketing a new product, Gamma, together with Alpha and Beta. The production requirements for Gamma are as follows:

	Gamma per Ton
K_1	9 tons
K_2	2 tons
Man-hours	2

The contribution from Gamma per ton is $100. Give your recommendation to the planning department on the production policy to be implemented by the company so that the total contribution is maximized. (*Hint.* Construct the linear programming problem for the product mix of Alpha, Beta, and Gamma. Solve the problem. Compare the total contribution with that obtained for the original Problem 5.5.)

─────────────── **SELECTED REFERENCES** ───────────────

Ackoff, R. L., and M. W. Sasieni, *Fundamentals of Operations Research*, Wiley, New York, 1968.

Bowman, E. H., and R. B. Fetter, *Analysis for Production and Operations Management*, Richard D. Irwin, Homewood, Ill., 1967.

Buffa, E. S., *Operations Management: Problems and Models*, Wiley, New York, 1968.

Dantizig, G. B., *Linear Programming and Extensions*, Princeton University Press, Princeton, N.J., 1963.

Ferguson, R. O., and L. F. Sargent, *Linear Programming*, McGraw-Hill, New York, 1958.

Hillier, F. S., and G. J. Lieberman, *Introduction to Operations Research*, Holden-Day, San Francisco, 1967.

Kwak, N. K., *Mathematical Programming with Business Applications*, McGraw-Hill, New York, 1973.

Levin, R. I., and R. P. Lamone, *Linear Programming for Management Decisions*, Richard D. Irwin, Homewood, Ill., 1969.

Plane, D. R., and G. A. Kochenberger, *Operations Research for Managerial Decisions*, Richard D. Irwin, Homewood, Ill., 1972.

Taha, H. A., *Operations Research: An Introduction*, Macmillan, New York, 1971.

Thierauf, R. J., and R. A. Grosse, *Decision Making Through Operations Research*, Wiley, New York, 1970.

Thompson, G. E., *Linear Programming: An Elementary Introduction*, Macmillan, New York, 1971.

Vajda, S., *Mathematical Programming*, Addison-Wesley, Reading, Mass., 1961.

Wagner, H. M., *Principles of Management Science*, Prentice-Hall, Englewood Cliffs, N.J., 1970.

CHAPTER 6 allocation of resources-assignment method

THE ASSIGNMENT PROBLEM is a type of allocation problem; it arises in the context of associating each of the N requirements to each of the N available means of satisfying them (requirements). For example, when a company has six projects to be performed and six engineers available to whom these projects can be allotted, the company faces the problem of assigning the projects to the engineers in an efficient manner. One of the important characteristics of the assignment problem is to allocate resources on a one-to-one basis. Each requirement must be assigned to one and only one means of satisfying it. Each resource must be assigned to no more than one requirement.

Another characteristic is that the total number of requirements equals the total number of resources. The effectiveness coefficients (such as processing time, expenditure, or profit) indicate what might result from allocating each requirement to each resource. The effectiveness coefficients are usually represented in a table. The objective is to establish the assignment that minimizes (maximizes) the sum of effectiveness coefficients of the selected combinations.

ILLUSTRATIVE EXAMPLES AND SOLUTION PROCEDURES

We shall illustrate the assignment problem and the computational procedure by two numerical examples.

Example 6.1 Machine Assignment Problem

Suppose that a small machine shop has five machines available, and five jobs are to be processed. Any job can be processed on any of the machines. The estimated processing time in hours for the job-machine combinations are shown in Table 6.1. The criterion is to determine the assignment that minimizes the total processing time.

TABLE 6.1   Hours for Processing the Jobs

	Machines				
Jobs	M1	M2	M3	M4	M5
J1	6	2	7	4	8
J2	8	5	4	3	1
J3	10	6	4	6	2
J4	6	8	6	5	3
J5	7	5	5	1	4

In Table 6.1 the number of resources equals the number of requirements. In other words, the effectiveness coefficients are arranged to form a square matrix. This problem has the minimizing objective function. Since each job is assigned to any one machine, only one element is chosen in each row. Likewise, since each machine is allotted to any one job, only one element is selected in each column. The optimum assignment that minimizes the total time can be determined by employing the Hungarian method.

In the first step of the Hungarian method we decide the smallest value in each row. The smallest value is subtracted from each element in that particular row. This is known as *row reduction*. The revised tableau is shown in Table 6.2.

TABLE 6.2   Reduced Matrix of Table 6.1

	Machines					Subtracted
Jobs	M1	M2	M3	M4	M5	Value
J1	4	0	5	2	6	2
J2	7	4	3	2	0	1
J3	8	4	2	4	0	2
J4	3	5	3	2	0	3
J5	6	4	4	0	3	1

TABLE 6.3   Reduced Matrix of Table 6.1

		Machines			
Jobs	M1	M2	M3	M4	M5
J1	1	0	3	2	6
J2	4	4	1	2	0
J3	5	4	0	4	0
J4	0	5	1	2	0
J5	3	4	2	0	3
Subtracted value	3	0	2	0	0

We observe in Table 6.2 that there is at least one zero in each row. In the second step, we determine the smallest value in each column of the revised matrix. The smallest value is subtracted from each revised element of the associated column. This is called *column reduction*. Table 6.3 indicates the values obtained after column reduction. In the reduced matrix there is at least one zero in each row and at least one zero in each column.

The zero elements of the modified matrix are the potential combinations of the optimal assignment. Now we examine the modified matrix to see if an optimum assignment can be determined by employing the zero elements. We draw the *minimum* number of horizontal and vertical lines required to cover all zeros. Each zero must be covered by at least one line.

TABLE 6.4   The Minimum Number of Lines for the Reduced Matrix

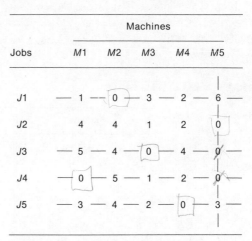

		Machines			
Jobs	M1	M2	M3	M4	M5
J1	1	0	3	2	6
J2	4	4	1	2	0
J3	5	4	0	4	0
J4	0	5	1	2	0
J5	3	4	2	0	3

There may be different ways of drawing the minimum lines. Any one of the several possible sets of lines is adequate to perform the test.

When the number of lines equals the number of rows (columns) of the matrix, it denotes that an optimal assignment can be established immediately. On the other hand, if the number of lines required to cover all zeros is less than the number of rows (columns) of the matrix, an optimal solution cannot be determined right away. In such cases, additional calculations are required to find out the optimal assignment. Table 6.4 displays a minimum set of lines drawn for the reduced matrix. We note in Table 6.4 that five lines are necessary to cover all zeros. Since the number of lines equals the number of rows (columns) of the matrix, we can make an optimal assignment immediately.

The optimal assignment can be determined from the final modified matrix as follows:

STEP 1. (a) Count the number of unmarked zeros in each row.
(b) If any row has exactly one unmarked zero, select that element. Label the element (□) to signify the assignment. All other zeros, if any, in the particular column cannot be considered for assignment. Hence delete these elements (X).

STEP 2. (a) Count the number of unmarked zeros in each column.
(b) If any column has exactly one unmarked zero, select that element. Label the element (□) to signify the assignment. All other zeros, if any, in the particular row cannot be considered for assignment. Therefore delete these elements (X).

Apply the two steps repetitively until either all zeros are marked or at least two unmarked zeros lie in each row and column. The optimum solution is determined when all zeros are marked. When a few unmarked zeros exist, the remaining allocations for the optimum assignment can be established by the trial and error approach. The optimum solution for our example is shown in Table 6.5. The optimum assignment and the associated total processing time are given below:

Optimum Assignment	Time
J1 to M2	2 hours
J2 to M5	1 hour
J3 to M3	4 hours
J4 to M1	6 hours
J5 to M4	1 hour
Total time = 14 hours	

TABLE 6.5   Optimum Solution for the Machine Assignment Problem

	Machines				
Jobs	M1	M2	M3	M4	M5
J1	1	[0]	3	2	6
J2	4	4	1	2	[0]
J3	5	4	[0]	4	⊗
J4	[0]	5	1	2	⊗
J5	3	4	2	[0]	3

Note in Table 6.5 that one and only one element is selected in each row and in each column. We have stated that additional computations are required to determine the optimal assignment whenever the number of lines used to cover all zeros is less than the number of rows. This is illustrated in the following example.

Example 6.2   Car Assignment Problem

Suppose that an auto leasing company has a surplus of one car in each of the cities S1, S2, S3, S4, and S5 and a deficit of one car in each of the cities D1, D2, D3, D4, and D5. Table 6.6 shows the distances in miles between cities with a surplus and cities with a deficit. The company wishes to dispatch the five cars so that the total miles traveled is minimized.

TABLE 6.6   Distances between Cities in Miles

	To				
From	D1	D2	D3	D4	D5
S1	7	4	5	1	4
S2	5	3	6	5	8
S3	8	7	4	2	3
S4	4	1	5	5	8
S5	6	6	6	6	6

The number of rows in this minimization problem equals the number of columns. First, row reduction and column reduction are performed. Then we draw a minimum number of lines to cover all zeros. The calculations are displayed in Table 6.7.

TABLE 6.7   The Minimum Number of Lines Drawn for the Reduced Matrix

		(a) Reduced Matrix				
		To				Subtracted
From	D1	D2	D3	D4	D5	Value
S1	6	3	4	0	3	1
S2	2	0	3	2	5	3
S3	6	5	2	0	1	2
S4	3	0	4	4	7	1
S5	0	0	0	0	0	6
Subtracted value	0	0	0	0	0	

	(b) Minimum Number of Lines				
	To				
From	D1	D2	D3	D4	D5
S1	6	3	4	0	3
S2	2	0	3	2	5
S3	6	5	2	0	1
S4	3	0	4	4	7
S5	—0—	0	—0—	0	—0—

Note that only three lines are used; therefore, there is no optimal assignment among these zero elements. The iterative procedure described below can be employed to solve these problems.

STEP 1. Draw the minimum number of horizontal and vertical lines that will cover all zeros. If the number of lines equals the number of rows, it indicates that an optimum solution can be established and stop. Otherwise, go to Step 2.

STEP 2. Select the smallest element in the matrix that is not covered by any line.

STEP 3. The modified matrix is obtained as follows: Subtract the smallest element from all elements that are not covered by any line. Add the smallest element to all elements that lie at the intersection of *two* lines. The remaining elements of the matrix are not changed, and they are recorded as they are in the modified matrix. Then go to Step 1.

These steps are employed repeatedly until the number of lines equals the number of rows of the matrix. The iterative calculations are applied on the reduced matrix shown in Table 6.7. The smallest value that is not covered by any line is 1 (element $S3-D5$), as can be seen in Table 6.7*b*. The resulting modified matrix is depicted in Table 6.8. Again, the iterative operations are employed, and Table 6.9 shows the subsequent matrix. Table 6.10 indicates that the modified matrix displayed in Table 6.9 can be used to establish the solution. The optimum assignment and the associated total mileage are given below:

Optimum Assignment	Distance
$S1$ to $D4$	1 mile
$S2$ to $D1$	5 miles
$S3$ to $D5$	3 miles
$S4$ to $D2$	1 mile
$S5$ to $D3$	6 miles
Total Distance =	16 miles

A summary of the algorithm is shown in Figure 6.1.

TABLE 6.8   Modified Matrix

	To				
	$D1$	$D2$	$D3$	$D4$	$D5$
$S1$	5	3	3	0	2
$S2$	1	0	2	2	4
$S3$	5	5	1	0	0
$S4$	2	0	3	4	6
$S5$	0	1	0	1	0

TABLE 6.9   Modified Matrix Obtained in the Second Iteration

(a) Minimum Number of Lines

From	D1	D2	D3	D4	D5
S1	5	3	3	0	2
S2	1	0	2	2	4
S3	— 5 —	5 —	1 —	0 —	0 —
S4	2	0	3	4	6
S5	— 0 —	1 —	0 —	1 —	0 —

(b) Modified Matrix

From	D1	D2	D3	D4	D5
S1	4	3	2	0	1
S2	0	0	1	2	3
S3	5	6	1	1	0
S4	1	0	2	4	5
S5	0	2	0	2	0

TABLE 6.10   Optimum Solution for the Car Assignment Problem

(a) Minimum Number of Lines

From	D1	D2	D3	D4	D5
S1	— 4 —	3 —	2 —	0 —	1 —
S2	— 0 —	0 —	1 —	2 —	3 —
S3	5	6	1	1	0
S4	1	0	2	4	5
S5	— 0 —	2 —	0 —	2 —	0 —

TABLE 6.10 (*Continued*)

(*b*) Optimum Solution

From	D1	D2	D3	D4	D5
S1	4	3	2	[0]	1
S2	[0]	⊠	1	2	3
S3	5	6	1	1	[0]
S4	1	[0]	2	4	5
S5	⊠	2	[0]	2	⊠

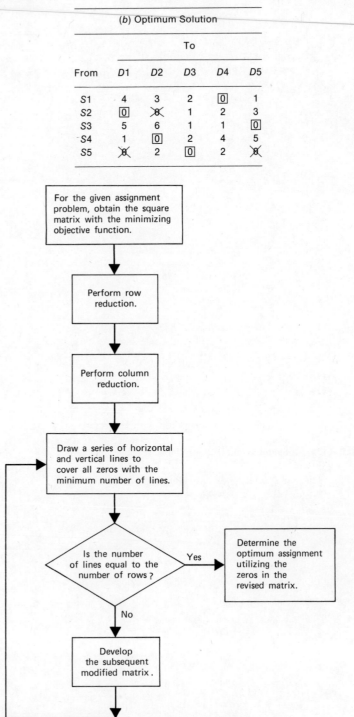

FIGURE 6.1   **Summary of steps for the assignment method.**

SPECIAL CASES OF THE ASSIGNMENT PROBLEMS

For some assignment problems, the matrix must be changed and brought to the standard form before employing the assignment technique. An assignment problem is said to be in the standard form if it fulfills two conditions:

1. The number of rows in the matrix is equal to the number of columns.
2. The objective is to minimize the sum of effectiveness coefficients of the selected combinations.

The original matrix is modified when any of the two restrictions are not satisfied; then, the assignment method is applied. We shall discuss the modifications necessary for these two special situations:

1. The unbalanced case where the number of rows is not equal to the number of columns.
2. The maximizing objective function.

Unbalanced Case

The unbalanced case arises when the number of rows in the effectiveness matrix is not equal to the number of columns. By adding fictitious rows *or* columns (dummy rows or dummy columns), the effectiveness matrix can be modified to a square matrix. The elements in the dummy rows or dummy columns are zeros. Then the assignment algorithm is applied to establish the optimum solution. The computational procedure is illustrated by the contractors selection problem.

Example 6.3   Contractors Selection Problem

The XYZ company is planning to construct three buildings, *B*1, *B*2, and *B*3, in Denver. The bid figures in $10,000 units received for the construction of the buildings are shown in Table 6.11. Any contractor can get at most one construction job. Management wishes to select the contractors for the buildings that will result in a minimum total cost.

TABLE 6.11   Bid Values for Constructing the Buildings ($10,000 Units)

Buildings	Contractors			
	*C*1	*C*2	*C*3	*C*4
*B*1	9	4	6	8
*B*2	7	6	8	5
*B*3	8	7	5	5

TABLE 6.12   Optimum Assignment of Buildings to the Contractors

(a) Modified Matrix

Buildings	Contractors			
	C1	C2	C3	C4
B1	9	4	6	8
B2	7	6	8	5
B3	8	7	5	5
Dummy	0	0	0	0

(b) Reduced Matrix

Buildings	Contractors				Subtracted Value
	C1	C2	C3	C4	
B1	5	0	2	4	4
B2	2	1	3	0	5
B3	3	2	0	0	5
Dummy	0	0	0	0	0
Subtracted value	0	0	0	0	

(c) Minimum Number of Lines

Buildings	Contractors			
	C1	C2	C3	C4
B1	— 5 —	0 —	2 —	4
B2	— 2 —	1 —	3 —	0
B3	— 3 —	2 —	0 —	0
Dummy	— 0 —	0 —	0 —	0

(d) Optimum Selection

Buildings	Contractors			
	C1	C2	C3	C4
B1	5	[0]	2	4
B2	2	1	3	[0]
B3	3	2	[0]	⊗
Dummy	[0]	⊗	⊗	⊗

We note in Table 6.11 that there are three rows and four columns in the matrix. Therefore, we add a dummy row; the resulting square matrix is shown in Table 6.12a. The assignment technique is employed to the square matrix. Table 6.12 displays the operations performed on the matrix as well as the optimum selection. Table 6.12d indicates that contractor C1 would not be offered any of the building projects.

Maximizing Objective Function

The sum of effectiveness coefficients of the selected combinations is maximized in the maximizing assignment problems. The effectiveness of the maximizing problems is usually measured by profit instead of cost. We can demonstrate the modifications to be performed on the effectiveness matrix by a numerical example.

Example 6.4 Salesmen Assignment Problem

The Conway Appliance Company wishes to assign four salesmen, S1, S2, S3, and S4 to the business territories, T1, T2, and T3 so as to maximize the total profit. The profit potential of each salesman in the three territories is shown in Table 6.13 in $1000 units per month. Each territory can be served by one man, and any man can be assigned at most to one territory.

When the number of rows is not equal to the number of columns, dummy rows or dummy columns are added as needed, and a square matrix is obtained. We note in Table 6.13 that one dummy column is required to make it a square matrix. The square matrix that includes the dummy column is shown in Table 6.14a. By assigning any salesman to the fictitious territory, no profit will be realized. The elements are thus zeros in the dummy column. The problem can be changed to a minimizing problem by calculating the *relative costs*. The relative costs are determined by subtracting each profit value from the largest profit.

TABLES 6.13   Profit Potential of Salesmen in Various Territories
(in $10,000 Units)

		Territories	
Salesmen	T1	T2	T3
S1	1	4	7
S2	8	3	1
S3	5	6	2
S4	4	1	7

TABLE 6.14 Relative Cost Matrix Calculations for the Salesmen Assignment Problem

(a) Modified Square Matrix

Salesmen	T1	T2	T3	Dummy
		Territories		
S1	1	4	7	0
S2	8	3	1	0
S3	5	6	2	0
S4	4	1	7	0

(b) Relative Cost Matrix

Salesmen	T1	T2	T3	Dummy
		Territories		
S1	7	4	1	8
S2	0	5	7	8
S3	3	2	6	8
S4	4	7	1	8

As can be seen in Table 6.14a, the largest profit is 8. Table 6.14b displays the relative cost values. Maximizing the total profit is equal to minimizing the total relative cost. The assignment algorithm can be employed to Table 6.14b to find out the optimum assignment. On determining the optimum assignment, the associated total profit can be obtained by using the original profit values given in Table 6.13. The salesman who gets the fictitious territory would not be assigned any territory. In the next section, the assignment method is applied to the relative cost matrix, and the optimum solution is established by employing a computer model.

COMPUTER MODEL FOR THE ASSIGNMENT PROBLEMS

We shall develop in this section a computer program for solving the assignment problems. The effectiveness matrix is assumed to be in the standard form:

(a) Number of rows equals number of columns.
(b) Minimizing objective function.

The following variables are used in the program:

N = number of rows in the square matrix
IA(K, L) = the element that lies in the Kth row and Lth column in the updated revised matrix IA
IB(K, J) = the element that lies in the Kth row and Lth column in the modified matrix IB
LINES = number of lines drawn in the matrix IB to cover all zeros
LARGNO = a large number compared to the elements of the effectiveness matrix IA
KBIG = the smallest element in the particular row or column
ILNO = the smallest element in the revised matrix that is not covered by any line
REDUC 1 = this subprogram is used to perform the row reduction
REDUC 2 = this subprogram is used to perform the column reduction
CHANGE = this subprogram is used to copy the elements of matrix IA into matrix IB
MODIFY = this subprogram is employed to draw the minimum number of lines to cover all zeros and also to decide the value of LINES
SMALL = this subprogram is employed to determine the value of ILNO
ADD = this subprogram is employed to establish the next revised matrix

The variables IA(,), and IB(,) are two-dimensional arrays of size N rows by N columns. The flowchart for the main program is shown in Figure 6.2. We may transmit the input data to the computer memory and also print out the values by these statements:

```
      READ 50, N, LARGNO
      PRINT 140, N
      PRINT 150, LARGNO
      DO 60 I = 1, N
      READ 70, (IA(I, J), J = 1, N
      PRINT 80, (IA(I, J), J = 1, N)
   60 CONTINUE
   50 FORMAT ( )
  140 FORMAT ( )
  150 FORMAT ( )
   70 FORMAT ( )
   80 FORMAT ( )
```

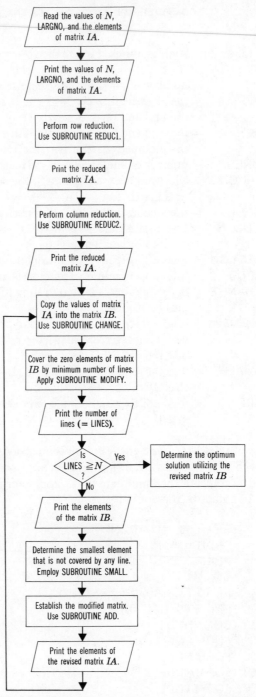

FIGURE 6.2   Summary of steps for the main program of the assignment method.

The row reduction can be accomplished and the reduced matrix can be printed out by the following instructions:

```
      CALL REDUC1(IA, N)
      DO 700 I = 1, N
700   PRINT 80, (IA(I, M), M = 1, N)
```

For executing the column reduction and for printing out the resulting reduced matrix, we might employ these statements:

```
      CALL REDUC2(IA, N)
      DO 701 I = 1, N
701   PRINT 80, (IA(I, M), M = 1, N)
```

Next, matrix IA is copied into matrix IB. The zero elements of IB are covered by the minimum number of lines. Then, the matrix is tested to see if the optimum assignment can be established by using the zero elements. These operations can be performed by utilizing the following statements:

```
90    CALL CHANGE(IA, IB, N)
      CALL MODIFY(LARGNO, IB, N, LINES)
      PRINT 7, LINES
      IF (LINES.GE.N) GO TO 110
7     FORMAT ( )
```

When the optimum solution cannot be established from the current revised matrix, the current matrix is printed out. Subsequently, a modified matrix is developed, and the new matrix is printed out. Control is then transferred to the beginning of the iterative operations. The following set of instructions would cause the computer to execute the calculations:

```
      DO 702 I = 1, N
702   PRINT 80, (IB(I, M), M = 1, N)
      CALL SMALL(N, ILNO, IB)
      CALL ADD(IA, IB, ILNO, LARGNO, N)
      DO 703 I = 1, N
703   PRINT 80, (IA(I, M), M = 1, N)
      GO TO 90
```

When the optimum solution can be determined from the updated matrix, control is transferred to statement 110, and the program is terminated.

```
110   CONTINUE
      STOP
      END
```

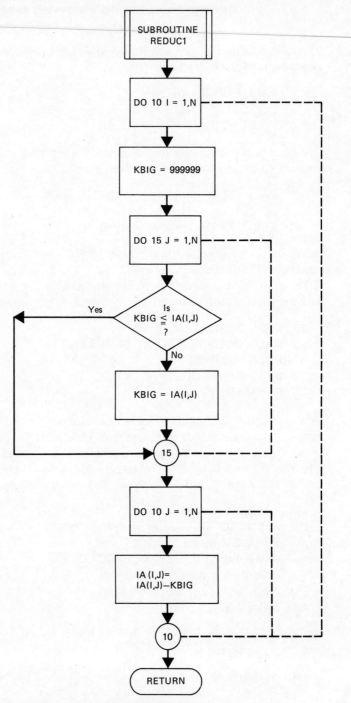

FIGURE 6.3   Flowchart for SUBROUTINE REDUC1.

The computational procedure and the computer programs for the sub-programs are discussed next. The subprogram REDUC1 is employed to determine the smallest element in each row and to subtract the smallest element from each element in the related row. Figure 6.3 shows the flow-chart for SUBROUTINE REDUC1. The subroutine includes the following statements:

```
      SUBROUTINE REDUC1(IA, N)
      DIMENSION IA( , )
      DO 10 I = 1, N
      KBIG = 999999
      DO 15 J = 1, N
      IF (KBIG.LE.IA(I, J)) GO TO 15
      KBIG = IA(I, J)
   15 CONTINUE
      DO 10 J = 1, N
      IA(I, J) = IA(I, J) − KBIG
   10 CONTINUE
      RETURN
      END
```

The subprogram REDUC2 is employed to determine the smallest ele-ment in each column and to subtract the smallest element from each element in the related column. Figure 6.4 shows the flowchart for SUB-ROUTINE REDUC2. The subroutine is composed of the instructions shown below:

```
      SUBROUTINE REDUC2(IA, N)
      DIMENSION IA( , )
      DO 10 J = 1, N
      KBIG = 999999
      DO 15 I = 1, N
      IF (KBIG.LE.IA(I, J)) GO TO 15
      KBIG = IA(I, J)
   15 CONTINUE
      DO 30 I = 1, N
      IA(I, J) = IA(I, J) − KBIG
   30 CONTINUE
   10 CONTINUE
      RETURN
      END
```

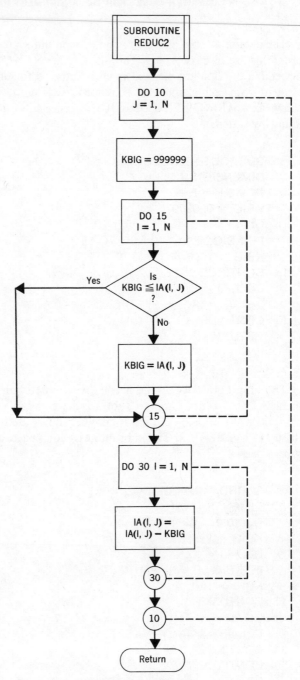

FIGURE 6.4   Flowchart for SUBROUTINE REDUC2.

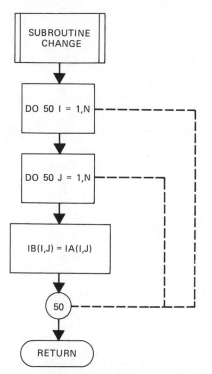

FIGURE 6.5 Flowchart for **SUBROUTINE CHANGE.**

Subprogram CHANGE is used to copy the elements of matrix IA into matrix IB. Figure 6.5 displays the flowchart. Subprogram CHANGE includes the following statements:

```
      SUBROUTINE CHANGE(IA, IB, N)
      DIMENSION IA( , ), IB( , )
      DO 50 I = 1, N
      DO 50 J = 1, N
      IB(I, J) = IA(I, J)
   50 CONTINUE
      RETURN
      END
```

Figure 6.6 displays a summary of computational steps for SUBROU-TINE MODIFY. We might initialize the subprogram by these in-structions:

```
L2 = LARGNO * 2
LINES = 0
```

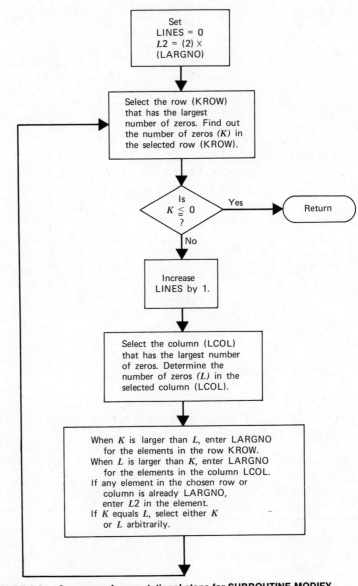

FIGURE 6.6   Summary of computational steps for SUBROUTINE MODIFY.

The following set of instructions would cause the computer to choose the row that has the largest number of zeros:

```
350  K = -999999
     DO 100 I = 1, N
     IZEROS = 0
     DO 110 J = 1, N
     IF (IB(I, J).EQ.O) IZEROS = IZEROS + 1
110  CONTINUE
     IF (IZEROS.LE.K) GO TO 100
     K = IZEROS
     KROW = I
100  CONTINUE
```

The row that has the largest number of zeros is identified by the variable KROW, which has K zeros. If there are no zeros in the modified matrix, the computations are terminated. Otherwise, a line is drawn to cover the zeros. The total number of lines is thus increased by one.

```
IF (K.LE.O) GO TO 360
LINES = LINES + 1
```

The following set of statements enables the computer to select the column with the largest number of zeros.

```
     L = -999999
     DO 200 J = 1, N
     IZEROS = 0
     DO 210 I = 1, N
     IF (IB(I, J).EQ.O) IZEROS = IZEROS + 1
210  CONTINUE
     IF (IZEROS.LE.L) GO TO 200
     L = IZEROS
     LCOL = J
200  CONTINUE
```

The variable LCOL specifies the column that has the largest number of zeros. The column LCOL has L zeros. If K is larger than L, a horizontal line must be drawn through row KROW. A large value LARGNO is recorded for each element of row KROW. If any of these elements are already LARGNO, the particular element is at the intersection of two lines. The element is identified by entering the value L2.

```
          IF (K.LE.L) GO TO 310
          DO 320 J = 1, N
          IF (IB(KROW, J).EQ.LARGNO) GO TO 701
          IB(KROW, J) = LARGNO
          GO TO 320
     701  IB(KROW, J) = L2
     320  CONTINUE
          GO TO 330
```

When L is larger than or equal to K, a vertical line must be drawn through the column LCOL. If any element in the column is LARGNO, the element lies at the intersection of two lines. The value L2 is assigned to the particular element and other elements are given a large value, LARGNO. The iterative operations are applied until no zeros appear in the matrix.

```
     310  DO 340 I = 1, N
          IF (IB(I, LCOL).EQ.LARGNO) GO TO 700
          IB(I, LCOL) = LARGNO
          GO TO 340
     700  IB(I, LCOL) = L2
     340  CONTINUE
     330  GO TO 350
     360  RETURN
          END
```

Subprogram MODIFY is composed of our previously developed program segments. Subprogram SMALL is employed to determine the smallest element ILNO in the revised matrix IB that is not covered by any line. Figure 6.7 shows the flowchart for SUBROUTINE SMALL, and the subprogram is given below:

```
          SUBROUTINE SMALL (N, ILNO, IB)
          DIMENSION IB( , )
          ILNO = 999999
          DO 600 I = 1, N
          DO 600 J = 1, N
          IF (IB(I, J).GT.ILNO) GO TO 600
          ILNO = IB(I, J)
     600  CONTINUE
          RETURN
          END
```

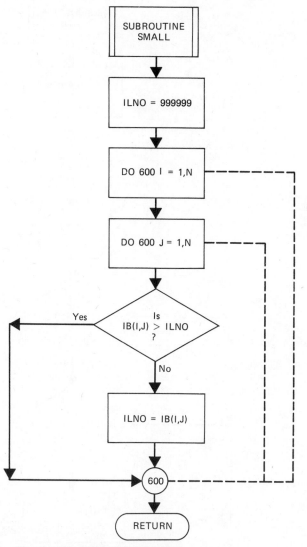

FIGURE 6.7   Flowchart for SUBROUTINE SMALL.

Subprogram ADD is employed to develop the next modified matrix. If any element is L2 in matrix IB, the modified elemental value is obtained by adding ILNO to the corresponding value of matrix IA. When an element is less than LARGNO, it denotes that it is not covered by any line. The revised values for these are obtained by subtracting ILNO from the corresponding values of matrix IA. All other elements are not changed.

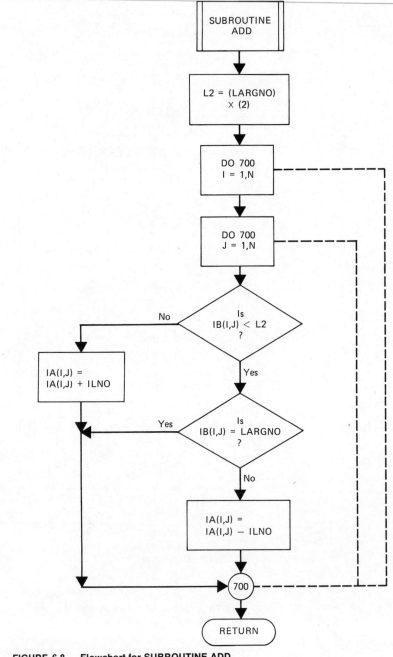

FIGURE 6.8   Flowchart for SUBROUTINE ADD.

Figure 6.8 depicts the flowchart for subprogram ADD. The instructions are given below:

```
     SUBROUTINE ADD (IA, IB, ILNO, LARGNO, N)
     DIMENSION IA( , )
     DIMENSION IB( , )
     L2 = LARGNO * 2
     DO 700 I = 1, N
     DO 700 J = 1, N
     IF (IB(I, J).LT.L2) GO TO 80
     IA(I, J) = IA(I, J) + ILNO
     GO TO 700
  80 IF (IB(I, J).EQ.LARGNO) GO TO 700
     IA(I, J) = IA(I, J) − ILNO
 700 CONTINUE
     RETURN
     END
```

The complete computer program composed of our previously designed program segments for the assignment method is shown in Figure 6.9.

ILLUSTRATION ON USING THE COMPUTER PROGRAM

Let us employ the computer program to solve Example 6.4, Salesmen Assignment Problem. The computer model can be used to solve problems up to the matrix size of 10 rows and 10 columns by writing the DIMEN-SION statements for IA and IB accordingly in the main program as well as in the subprograms. Figure 6.9 shows the DIMENSION statements composed for the main program and the subprograms. The data are read and printed out according to the FORMAT statements.

```
  50  FORMAT (I2, I4)
 140  FORMAT (1H1, 16HNUMBER OF ROWS =, I2)
 150  FORMAT (1X, 20HSELECTED LARGE NO. =, I4)
  70  FORMAT (4I2)
  80  FORMAT (4I5)
```

The number of lines drawn to cover the zeros in the modified matrix is printed out according to the FORMAT statement:

```
  7  FORMAT (1X, 17HNUMBER OF LINES =, I3)
```

```
            DIMENSION IA(10,10), IB(10,10)
            READ 50, N, LARGNO
            PRINT 140, N
            PRINT 120
            PRINT 150, LARGNO
            PRINT 120
C
            DO 60 I = 1,N
C  READ AND PRINTOUT THE DATA
            READ 70,(IA(I,J), J = 1,N)
            PRINT 80,(IA(I,J), J = 1,N)
         60 CONTINUE
            PRINT 120
            CALL REDUC1(IA,N)
C  PRINTOUT MATRIX AFTER ROW REDUCTION
C
            DO 700 I = 1,N
        700 PRINT 80, (IA (I,M), M= 1,N)
            PRINT 120
C
C
            CALL REDUC2(IA,N)
C  PRINTOUT MATRIX AFTER COLUMN REDUCTION
C
            DO 701 I = 1,N
        701 PRINT 80, (IA (I,M), M =1,N )
            PRINT 120
C
C
         90 CALL CHANGE(IA,IB,N)
C
C
            CALL MODIFY(LARGNO,IB,N,LINES)
            PRINT 7, LINES
            PRINT 120
C  PRINTOUT MODIFIED MATRIX WITH LINES
C
            IF (LINES.GE.N) GO TO 110
            DO 702 I = 1,N
        702 PRINT 80, (IB (I,M), M = 1,N )
            PRINT 120
C
C
            CALL SMALL(N,ILNO,IB)
C
C
            CALL ADD(IA,IB,ILNO,LARGNO,N)
C  PRINTOUT THE REVISED MATRIX
C
            DO 703 I = 1,N
        703 PRINT 80, (IA (I,M), M = 1,N )
            PRINT 120
            GO TO 90
```

FIGURE 6.9   Computer program for the assignment problems (illustrated for Example 6.4).

```
C
  110 CONTINUE
C
    7 FORMAT (1X,17HNUMBER OF LINES =, I3)
   50 FORMAT (I2,I4)
   70 FORMAT (4I2)
   80 FORMAT (4I5)
  120 FORMAT (1H0)
  140 FORMAT(1H1,16HNUMBER OF ROWS =, I2)
  150 FORMAT (1X,20HSELECTED LARGE NO. =, I4 )
C
      STOP
      END


      SUBROUTINE REDUC1 (IA,N)
C  THIS PREFORMS THE ROW REDUCTION
C
      DIMENSION IA(10,10)
C
      DO 10 I = 1,N
      KBIG = 999999
C
      DO 15 J = 1,N
      IF (KBIG.LE.IA(I,J)) GO TO 15
      KBIG = IA(I,J)
   15 CONTINUE
C
      DO 10 J = 1,N
      IA(I,J) = IA(I,J) - KBIG
   10 CONTINUE
C
      RETURN
      END


      SUBROUTINE REDUC2 (IA,N)
C  THIS PERFORMS THE COLUMN REDUCTION
C
      DIMENSION IA(10,10)
C
      DO 10 J = 1,N
      KBIG = 999999
C
      DO 15 I = 1,N
      IF (KBIG.LE.IA(I,J)) GO TO 15
      KBIG = IA(I,J)
   15 CONTINUE
C
      DO 30 I = 1,N
      IA(I,J) = IA(I,J) - KBIG
   30 CONTINUE
   10 CONTINUE
C
      RETURN
      END
```

FIGURE 6.9 (Continued)

```
      SUBROUTINE CHANGE (IA,IB,N)
C  THIS COPIES IA TO IB
C
      DIMENSION IA(10,10), IB(10,10)
C
      DO 50 I = 1,N
C
      DO 50 J = 1,N
      IB(I,J) = IA(I,J)
   50 CONTINUE
C
      RETURN
      END



      SUBROUTINE ADD (IA,IB,ILNO,LARGNO,N)
C  THIS DETERMINES THE REVISED MATRIX
C
      DIMENSION IA(10,10)
      DIMENSION IB(10,10)
      L2 = LARGNO * 2
C
      DO 700 I = 1,N
C
      DO 700 J = 1,N
      IF (IB(I,J) .LT. L2) GO TO 80
      IA(I,J) = IA(I,J) + ILNO
      GO TO 700
   80 IF (IB(I,J).EQ.LARGNO) GO TO 700
      IA(I,J) = IA(I,J) - ILNO
  700 CONTINUE
C
      RETURN
      END



      SUBROUTINE SMALL (N,ILNO,IB)
C  THIS FINDS THE SMALLEST NUMBER
C
      DIMENSION IB(10,10)
      ILNO = 999999
C
      DO 600 I = 1,N
C
      DO 600 J = 1,N
      IF (IB(I,J).GT. ILNO) GO TO 600
      ILNO = I (I,J)
  600 CONTINUE
C
      RETURN
      END
```

FIGURE 6.9 (*Continued*)

```
      SUBROUTINE MODIFY (LARGNO,IB,N,LINES)
C  THIS DRAWS THE MINIMUM NUMBER OF LINES
C
      DIMENSION IB(10,10)
      L2 = LARGNO * 2
      LINES = 0
350 K = -999999
C
      DO 100 I = 1,N
      IZEROS = 0
C
      DO 110 J = 1,N
      IF (IB(I,J).EQ.0) IZEROS = IZEROS + 1
110 CONTINUE
      IF (IZEROS.LE.K) GO TO 100
      K = IZEROS
      KROW = I
100 CONTINUE
      IF (K.LE.0) GO TO 360
      LINES = LINES + 1
      L = -999999
C
      DO 200 J = 1,N
      IZEROS = 0
C
      DO 210 I = 1,N
      IF (IB(I,J).EQ.0) IZEROS = IZEROS + 1
210 CONTINUE
      IF (IZEROS.LE.L) GO TO 200
      L = IZEROS
      LCOL = J
200 CONTINUE
      IF (K.LE.L) GO TO 310
C
      DO 320 J = 1,N
      IF (IB (KROW, J) .EQ. LARGNO ) GO TO 701
      IB (KROW, J) = LARGNO
      GO TO 320
701 IB (KROW,J) = L2
320 CONTINUE
      GO TO 330
C
310 DO 340 I = 1,N
      IF (IB(I,LCOL).EQ.LARGNO) GO TO 700
      IB (I, LCOL) = LARGNO
      GO TO 340
700 IB (I,LCOL) = L2
340 CONTINUE
330 GO TO 350
C
360 RETURN
      END
```

FIGURE 6.9 (Continued)

The matrix must be in the standard form for employing our computer model. Hence, Table 6.14*b* is used as the input matrix for the computer program. The data deck has the following appearance:

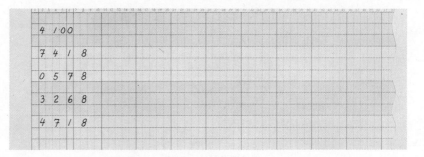

The resulting computer printout is shown in Figure 6.10. There are five matrices in the computer printout. The first matrix denotes the initial matrix in the standard form. The second and third matrices indicate the reduced matrices obtained after row reduction and column reduction, respectively. The fourth matrix specifies how the three lines were drawn. Of the three lines, one horizontal line passes through row 3 and the two vertical lines pass through column 1 and column 3. The fifth matrix is the final modified matrix. The optimum solution can be determined by using the zero elements in the final modified matrix as follows:

	Territories			
Salesmen	T1	T2	T3	Dummy
S1	6	2	0	0
S2	⓪	4	7	1
S3	2	⓪	5	ⓧ
S4	3	5	0	0

There are two zeros in rows *S*1 and *S*4, and there are two zeros in columns *T*3 and dummy. Therefore, the two salesmen, *S*1 and *S*4, may be assigned according to the two alternatives:

Alternative (1): *S*1 to *T*3   *S*4 to dummy
Alternative (2): *S*1 to dummy   *S*4 to *T*3

NUMBER OF ROWS = 4

SELECTED LARGE NO. = 100

```
7    4    1    8
0    5    7    8
3    2    6    8
4    7    1    8


6    3    0    7
0    5    7    8
1    0    4    6
3    6    0    7


6    3    0    1
0    5    7    2
1    0    4    0
3    6    0    1
```

NUMBER OF LINES = 3

```
100    3  100    1
100    5  100    2
200  100  200  100
100    6  100    1


6    2    0    0
0    4    7    1
2    0    5    0
3    5    0    0
```

NUMBER OF LINES = 4

FIGURE 6.10   Computer printout for the salesmen assignment problem.

The two possible optimal assignments and the total profit associated with them are given below:

Alternative (1):

Optimum Schedule	Profit
$S1$ to $T3$	$70,000
$S2$ to $T1$	80,000
$S3$ to $T2$	60,000
$S4$ to dummy	0
Total profit	$210,000

Alternative (2):

Optimum Schedule	Profit
S1 to dummy	$ 0
S2 to T1	80,000
S3 to T2	60,000
S4 to T3	70,000
Total profit	$210,000

Note that when an assignment problem has multiple solutions, each of the solutions yields the same objective function value.

PROBLEMS

6.1 The El Paso Mercantile Company has four departmental managers, A, B, C, and D to be assigned to four product divisions: soap, oil, margarine, and refinery. Each manager can be assigned to one and only one division, and also each division should be supervised by only one manager. The rating indexes for the managers to these product divisions have been computed analyzing the manager's experience, education, capability, and aptitude. A small rating index for any manager-division combination is preferred over a large index value. The index values are shown in the table.
(a) Based on this information, establish the optimum assignment of managers to divisions that minimizes the total index.
(b) Compute the total index at optimality.

		Division		
Manager	Soap	Oil	Margarine	Refinery
A	8	11	5	11
B	13	10	11	6
C	4	10	5	2
D	1	8	2	9

6.2 The Wallowa Equipment Rental Company has a surplus of one horse trailer in cities P, Q, R, and S and a deficit of one horse trailer in cities A, B, C, and D. The distances in miles between the cities with a surplus and the cities with a deficit are shown on the next page:

| | | To | | |
	A	B	C	D
P	8	5	8	2
Q	9	3	6	9
R	4	10	7	4
S	3	5	6	1

From (rows P, Q, R, S)

The management wishes to determine the assignment of surplus trailers to the cities with a deficit to minimize the total mileage traveled. If alternate optimal assignments exist, find out the optimal assignments. Compute the total distance for the optimum assignment.

6.3   In a small machine shop, five machines are available to process five jobs; and their associated processing times, in hours, are presented in the table. Any one of these jobs can be processed in one and only one machine. Also, no machine can be assigned to process more than one job. The machine burden rate is $10.00 per machine-hour.

(a) Determine the assignment(s) that minimizes the total processing time.

(b) What is the total processing time for the optimum assignment? What is the total machine cost at optimality?

| | Machine | | | | |
Job	M_1	M_2	M_3	M_4	M_5
J_1	13	5	3	9	9
J_2	11	10	2	5	9
J_3	8	10	4	9	13
J_4	4	1	4	2	3
J_5	9	4	2	8	10

6.4   The Dreyfuss and Associates Company has received four contracts for the development, design, and manufacture of computer systems. The personnel department has screened the applicants desired to be project leaders for these contracts and came up with a list of five potential candidates. Their ratings are given in the table. The ratings have been computed after analyzing the applicants' education, experience, personality, aptitude, and other data. A small rating index is to be preferred in the assignment of a project. However, any of the project leaders can be assigned at the maximum to only one project.

(a) Based on this information, determine the assignment of projects to the prospective project leaders that will result in a minimum total index for the computer systems development program.

(b) What is the total index for the optimum assignment?

(c) If the candidate that hasn't got any project is not to be selected, find out who that candidate would be.

	Project			
Candidate	1	2	3	4
Alan	3	8	3	5
Bill	1	7	11	4
Clark	2	4	10	6
Don	4	9	5	3
Ed	7	8	4	2

6.5 The Tillman Seafood Industries has five fishing vessels at different locations in the sea. One fishing vessel will be assigned to each of the four fishing areas, A, B, C, and D. The estimated travel time in hours for the vessels to the areas are given below. The objective is to find the assignment of vessels to the areas that will result in a minimum total travel time. For the optimum assignment, what is the total time?

		Fishing Area			
		A	B	C	D
	Eugene	6	4	9	2
	Fresno	3	3	4	7
Fishing	Grand				
vessel	Rapids	4	8	3	3
	Hillsboro	2	1	7	5
	Independence	5	6	5	8

6.6 A company has three managers, S_1, S_2, and S_3, for assigning to the business districts, D_1, D_2, D_3 and D_4. Each sales manager should be allotted a maximum of one district, and any one of the district's sales activities should be supervised at the maximum by one manager. The estimated profits for the sales manager-business district combinations are given in the table in $1000 units.

(a) Determine the optimum assignment that maximizes the total profit.

(b) What is the total profit (in $1000 units) at optimum condition?

(c) If your solution in (a) above would be put into action, which business district will not be assigned with a manager?

	D_1	D_2	D_3	D_4
S_1	4	2	8	6
S_2	6	3	8	10
S_3	5	7	8	3

6.7   The William's Grocery Store wishes to display the three products, *A*, *B*, and *C* in the four available locations: *P*, *Q*, *R*, and *S*. Each product can be displayed at most in one location. In any location, more than one product cannot be displayed. The estimated daily sales in dollars for the location-product combinations are as follows:

	Product		
Location	A	B	C
P	70	80	100
Q	100	60	70
R	90	60	70
S	20	50	60

(a) Determine the optimum assignment that maximizes the total daily sales.

(b) Find out the total daily sales for the solution in (a) above.

6.8   Here are the cost data for an assignment problem:

	A	B	C	D	E	F	G
1	6	4	11	7	8	5	9
2	3	15	17	13	8	7	12
3	2	0	8	19	9	14	11
4	14	1	9	4	10	6	8
5	9	6	10	3	6	7	2
6	12	11	4	1	5	12	13
7	5	8	13	3	9	1	11
8	7	6	12	4	8	10	7

It is desired to establish the optimum assignment that minimizes the total cost.

(a) Write a FORTRAN program to determine the final reduced matrix.

(b) Establish the optimum assignment(s) using the final reduced matrix obtained in (a).

──────────── SELECTED REFERENCES ────────────

Ackoff, R. L., and M. W. Sasieni, *Fundamentals of Operations Research*, Wiley, New York, 1968.

Churchman, C. W., R. L. Ackoff, and E. L. Arnoff (eds.), *Introduction to Operations Research*, Wiley, New York, 1958.

Diroccaferrera, G. F., *Introduction to Linear Programming Processes*, South-Western Publishing, Cincinnati, 1967.

Fabrycky, W. J., P. M. Ghare, and P. E. Torgersen, *Industrial Operations Research*, Prentice-Hall, Englewood Cliffs, N.J., 1972.

Hillier, F. S., and G. J. Lieberman, *Introduction to Operations Research*, Holden-Day, San Francisco, 1967.

Kwak, N. K., *Mathematical Programming with Business Applications*, McGraw-Hill, New York, 1973.

Levin, R. I., and R. P. Lamone, *Linear Programming for Management Decisions*, Richard D. Irwin, Homewood, Ill., 1969.

Llewellyn, R. W., *Linear Programming*, Holt, Rinehart and Winston, New York, 1964.

Miller, D. W., and M. K. Starr, *Executive Decisions and Operations Research*, Prentice-Hall, Englewood Cliffs, N.J., Second Edition, 1969.

Riggs, J. L., *Economic Decision Models for Engineers and Managers*, McGraw-Hill, New York, 1968.

Sasieni, M., A. Yaspan, and L. Friedman, *Operations Research: Methods and Problems*, Wiley, New York, 1959.

Teichroew, D., *An Introduction to Management Science: Deterministic Models*, Wiley, New York, 1964.

analysis for capital investment decisions

THE SURVIVAL AND GROWTH of an enterprise depends mostly on how effectively the managers and engineers administer the personnel and other resources and plan the course of future activities. A considerable amount of capital outlay is required for implementing these activities apart from other resources such as manpower and raw materials. In determining the course of future actions, the set of possible alternatives that yield the desired results is determined, evaluated, and the one that fulfills the economic criteria is selected. The alternatives are also called investment proposals. For example, suppose that the DZK Company is planning to remodel its warehouse. The company wishes to install a material handling system for moving raw materials, finished products, and supplies. The feasible alternatives included are:

A manual system using two-wheel trucks and four-wheel trucks

A fork-truck system

A powered belt or roller conveyor system

An automatic conveyorized system

Each alternative usually includes a series of disbursements and receipts of cash flow at different points in time. The cash flow of each investment proposal is represented on a cash flow diagram. The horizontal line in the diagram denotes time. The vertical line indicates the cash flow. The movement of time is from left to right. The upward arrows represent receipts ⁤ the downward arrows disbursements. For example, the cash receipts of $6000 and $7000 at the end of the second and fourth years', and the disbursements of $2000, $3000, $2000, and $1000 at present and at the end of the first, third, and fourth years as an alternative can be shown as follows:

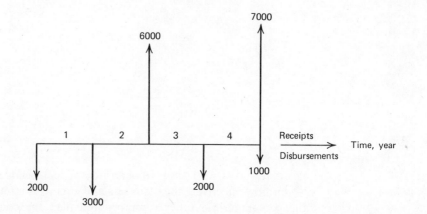

The investment proposals are reduced to a common economic basis for comparison. The interest formulas, the method of calculating the economic equivalence, as well as the techniques for comparing alternatives on an equivalent economic basis are discussed in this chapter.

THE TIME VALUE OF MONEY

A person can invest capital in many ways to produce more income than he could without investing it. Capital is productive, and money (called interest) is paid for the use of capital. The interest rate specifies the ratio between the gain that is received at the end of a period of time, usually one year, and the money invested at the beginning of that period. Consider that an amount of $200 is deposited in a bank account and at the end of that year $220 is received from the account. The gain is $20 (= $220 − $200).

$$\text{Interest rate} = \frac{\text{Gain received at the end of a specified year}}{\text{Amount invested at the beginning of the year}}(100)\%$$

$$= \frac{20}{200}(100)\%$$

$$= 10\%$$

(7.1)

In other words, $200 on hand today is equal to $220 a year from today according to 10% interest rate. A particular amount of money at one point in time is not equal to the same amount of money at some other point in time because of the relationship between interest rate and time span. This is known as the time value of money.

Simple Interest

The money earned in simple interest is proportional to the length of time that the principal is invested. Only the principal amount yields interest. The interest is returned together with the principal amount at the end of the loan period. The interest is calculated employing formula (7.2).

$$I = Pni \qquad (7.2)$$

where

I = total interest earned by simple interest

P = principal or present amount

n = number of interest periods

i = interest rate

Example 7.1

At a simple interest rate of 8% per annum, $1000 is borrowed for a period of four years. Find the total amount that is due at the end of the loan period.

The interest earned for the four-year period would be from (7.2),

$$I = (\$1000)(4)(0.08)$$
$$= \$320.00$$

Thus, the total amount that is due after four-years is $1320.00 (principal plus interest).

Compound Interest

The general practice of the business world is based on compound interest. When money is borrowed for several interest periods, the interest earned in each period is calculated at the end of the period and added to the principal amount. This new total amount is considered the principal for the following time period. In this approach, interest together with the principal has earning power in future time periods. The calculations of compound interest for Example 7.1 are shown in Table 7.1.

TABLE 7.1 Calculation of Compound Interest for Example 7.1

Year k	Principal at Beginning of Year A_k	Interest Earned During Year $B_k = A_k i$	Total Amount At End of Year $A_k + B_k = A(1 + i)^k$
1	$1000.00	$1000.00 × 0.08 = 80.00	$1000(1.08) = 1080.00
2	1080.00	1080.00 × 0.08 = 86.40	$1000(1.08)^2 = 1166.40
3	1166.40	1166.40 × 0.08 = 93.31	$1000(1.08)^3 = 1259.71
4	1259.71	1259.71 × 0.08 = 100.78	$1000(1.08)^4 = 1360.49

The interest formulas useful in converting dollars at a given point in time to an equivalent amount at some other point in time are presented in the next section.

INTEREST FORMULAS

The following symbols are used in the derivation of interest formulas:

i = annual interest rate

n = number of annual interest periods

P = principal sum or present amount at the time point regarded to be the present

A = single payment in a uniform series of n continuous payments made at the end of each annual interest period

F = future sum flowing at the end of the nth annual interest period equal to the compound amount of a present sum P, or the sum of the compound amounts of the payments A at the interest rate i.

If P dollars are invested for a period of n years at the annual interest rate i, the total amount at the end of the nth year (F) is calculated as follows:

Interest earned for the first year $= Pi$

Total amount at the end of the first year $= P + Pi = P(1 + i)$

Interest earned for the second year $= P(1 + i)i$
Total amount at the end of the second year $= P(1 + i) + P(1 + i)i =$
$P(1 + i)^2$
Likewise, total amount at the end of the third year $= P(1 + i)^3$

Therefore, total amount at the end of the nth year

$$F = P(1 + i)^n \qquad (7.3)$$

or

$$F = P(F/P, i, n) \qquad (7.4)$$

where

$$(F/P, i, n) = (1 + i)^n$$

$(F/P, i, n)$ is the factor to find the future-worth of a present sum P. The value of F can be determined, where P, i, and n are known, employing expression 7.4. The values of this factor and other factors developed in this section are presented in Appendix II for different interest rates. For example, the future worth for the present sum $10,000 at the end of 8 years using an interest rate of 5% can be found by applying equation 7.4.

$$F = 10,000(F/P, 5\%, 8)$$
$$= 10,000(1.477) = \$14,770$$

When equation 7.3 is multiplied by $[1/(1 + i)^n]$, the result is

$$P = F\left[\frac{1}{(1 + i)^n}\right] \qquad (7.5)$$

This equation can be employed to find the present-worth of a future sum.

$$P = F(P/F, i, n) \qquad (7.6)$$

where

$$(P/F, i, n) = \left[\frac{1}{(1 + i)^n}\right]$$

$(P/F, i, n)$ is the factor to find the present-worth of a future sum. For example, if $14,770 will be received 8 years from now, its present-worth at 5% annual interest rate is obtained from expression 7.6.

$$P = 14,770(P/F, 5\%, 8)$$
$$= 14,770(0.67684) = \$10,000$$

In many economic studies, a series of payments occurring uniformly at the end of each year is encountered. If A dollars are deposited at the end of each annual period for n years at interest rate i, the future sum that will be received at the end of the nth year can be represented by the cash flow diagram as follows:

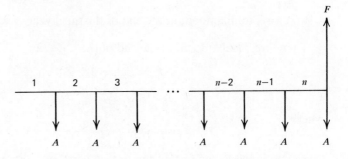

Contribution of the A dollars invested at the end of year n to $F = A$

Contribution of the A dollars invested at the end of year $(n - 1)$ to $F = A(1 + i)^1$

Contribution of the A dollars invested at the end of year $(n - 2)$ to $F = A(1 + i)^2$

Contribution of the A dollars invested at the end of year $(n - 3)$ to $F = A(1 + i)^3$ and,

Contribution of the A dollars invested at the end of year 1 to $F = A(1 + i)^{n-1}$

The value of F is the sum of the series. Hence,

$$F = A + A(1 + i) + A(1 + i)^2 + A(1 + i)^3 + \cdots + A(1 + i)^{n-1}$$
(7.7)

Multiplying equation 7.7 by $(1 + i)$, we obtain

$$F(1 + i) = A(1 + i) + A(1 + i)^2 + A(1 + i)^3 + \cdots + A(1 + i)^n$$
(7.8)

subtracting equation 7.7 from equation 7.8, we obtain

$$F(1 + i) - F = -A + A(1 + i)^n$$

$$F = A\left[\frac{(1 + i)^n - 1}{i}\right]$$
(7.9)

or

$$F = A(F/A, i, n)$$
(7.10)

where

$$(F/A, i, n) = \left[\frac{(1 + i)^n - 1}{i}\right]$$

$(F/A, i, n)$ is the factor to find the future-worth of a uniform series. The value of F can be computed, where A, i, n are known, applying expression 7.10. For example, suppose that $800 payments are made at the end of each year for 7 consecutive years at 9% annual interest. The value of the future amount accumulated at the end of the seventh year is computed using equation 7.10.

$$F = 800(F/A, 9\%, 7)$$
$$= 800(9.200) = \$7360$$

Here is the cash flow diagram.

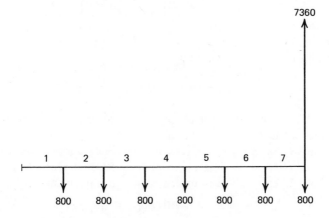

Multiplying equation 7.9 by $[i/(1 + i)^n - 1]$ gives

$$A = F\left[\frac{i}{(1 + i)^n - 1}\right] \qquad (7.11)$$

This equation may be used to find the value of the equal payment series of a future sum.

$$A = F(A/F, i, n) \qquad (7.12)$$

where

$$(A/F, i, n) = \left[\frac{i}{(1 + i)^n - 1}\right]$$

$(A/F, i, n)$ is the factor to find the uniform payment series of a given future sum. The value of A may be calculated, where F, i, and n are known, by employing equation 7.12. For example, consider that a total of $7360 is to be received at the end of the seventh year by depositing a series of seven equal payments at the end of the preceding seven years at an interest rate of 9%. The amount that should be deposited each year is obtained from equation 7.12.

$$A = 7360(A/F, 9\%, 7)$$
$$= 7360(0.10869) = \$800$$

We have shown in equation 7.3 that

$$F = P(1 + i)^n$$

substituting $P(1 + i)^n$ for F in equation 7.11 gives

$$A = P(1 + i)^n \left[\frac{i}{(1 + i)^n - 1} \right]$$
$$= P \left[\frac{i(1 + i)^n}{(1 + i)^n - 1} \right] \quad (7.13)$$

or

$$A = P(A/P, i, n) \quad (7.14)$$

where

$$(A/P, i, n) = \left[\frac{i(1 + i)^n}{(1 + i)^n - 1} \right]$$

$(A/P, i, n)$ is the factor to find the uniform annual payment series for a present-worth. The value of A can be calculated, where P, i, and n are known, by applying expression 7.14. For example, consider that $10,000 is invested at 6% interest to provide 5 equal year-end payments for the next 5 years. The amount of each equal payment will be

$$A = 10,000(A/P, 6\%, 5)$$
$$= 10,000(0.23740) = \$2374$$

Here is the cash flow diagram:

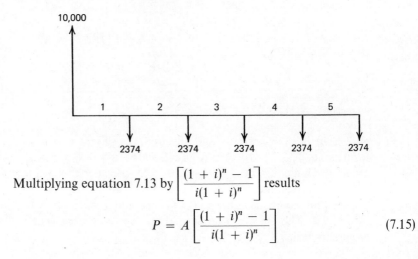

Multiplying equation 7.13 by $\left[\dfrac{(1 + i)^n - 1}{i(1 + i)^n} \right]$ results

$$P = A \left[\frac{(1 + i)^n - 1}{i(1 + i)^n} \right] \quad (7.15)$$

or

$$P = A(P/A, i, n) \tag{7.16}$$

where

$$(P/A, i, n) = \left[\frac{(1 + i)^n - 1}{i(1 + i)^n} \right]$$

$(P/A, i, n)$ is the factor to find the present-worth of a uniform annual payment series. The value of P can be computed, where A, i, and n are known, by using equation 7.16. For example, the present-worth of a series of 5 equal annual payments of \$2374 at an interest rate of 6% will be

$$P = 2374(P/A, 6\%, 5)$$
$$= 2374(4.212) = \$10,000$$

A summary of the interest formulas is shown in Table 7.2.

TABLE 7.2   Summary of Interest Formulas

Formula Name	To Find	Given*	Factor	Equation
Future-worth of a present sum	F	P	$(F/P, i, n)$	$F = P(1 + i)^n$
Present-worth of a future sum	P	F	$(P/F, i, n)$	$P = F\left[\dfrac{1}{(1 + i)^n}\right]$
Future-worth of a uniform series	F	A	$(F/A, i, n)$	$F = A\left[\dfrac{(1 + i)^n - 1}{i}\right]$
Uniform series value of a future sum	A	F	$(A/F, i, n)$	$A = F\left[\dfrac{i}{(1 + i)^n - 1}\right]$
Uniform series value of a present sum	A	P	$(A/P, i, n)$	$A = P\left[\dfrac{i(1 + i)^n}{(1 + i)^n - 1}\right]$
Present-worth of a uniform series	P	A	$(P/A, i, n)$	$P = A\left[\dfrac{(1 + i)^n - 1}{i(1 + i)^n}\right]$

* Values of i and n are also given.

Example 7.2

The Sigma Metal Company is considering the installation of a new testing procedure in the welding section. The proposal may be implemented at the end of 1975. For each of the following five years, \$5000 would be saved. The installation expenses of \$6000 at the end of 1975 and maintenance costs of \$3000 and \$1000 would be incurred at the end of 1977 and 1979, respectively. Calculate the total net savings at the end of 1975 for the five-year period assuming a 6% interest rate.

The cash flow diagram is as follows:

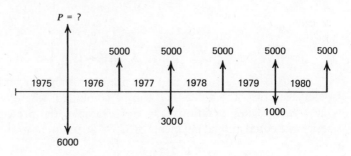

The equivalent value of total savings at the end of 1975 is

$$P = A(P/A, 6\%, 5)$$
$$= 5000(4.212) = \$21,060.00$$

$$\begin{bmatrix} \text{The equivalent} \\ \text{value of total} \\ \text{expenses at the} \\ \text{end of 1975} \end{bmatrix} = \begin{bmatrix} \text{Installations} \\ \text{expenses at} \\ \text{the end of} \\ 1975 \end{bmatrix} + \begin{bmatrix} \text{The equivalent value} \\ \text{of total maintenance} \\ \text{cost at the end of} \\ 1975 \end{bmatrix}$$

$$\begin{bmatrix} \text{Installation expenses} \\ \text{at the end of 1975} \end{bmatrix} = \$6000.00$$

The equivalent value at the end of 1975 of the maintenance cost that is spent at the end of 1977 is

$$P = F(P/F, 6\%, 2)$$
$$= 3000(0.89000) = \$2670.00$$

The equivalent value at the end of 1975 of the maintenance cost at the end of 1979 is

$$P = F(P/F, 6\%, 4)$$
$$= 1000(0.79209) = \$792.09$$

Thus, the equivalent value of total maintenance cost at the beginning of 1975 is

$$= 2670.00 + 792.09$$
$$= \$3462.09$$

$$\begin{bmatrix} \text{The equivalent value of total} \\ \text{expenses at the end of 1975} \end{bmatrix} = 6000.00 + 3462.09$$

$$= \$9462.09$$

$$\begin{bmatrix} \text{Total net savings at} \\ \text{the end of 1975} \end{bmatrix} = \begin{bmatrix} \text{Equivalent value of} \\ \text{total savings at the} \\ \text{end of 1975} \end{bmatrix} - \begin{bmatrix} \text{Equivalent value of} \\ \text{total expenses at the} \\ \text{end of 1975} \end{bmatrix}$$

$$= 21,060.00 - 9462.09$$

$$= \$11,597.91$$

METHODS OF COMPARING INVESTMENT ALTERNATIVES

The typical investment alternative involves receipts and disbursements spread over a span of time. The alternatives are evaluated by reducing them to a common economic basis. The cash receipts and expenditures are converted to a common economic basis using the time value of money. The interest formulas are employed to convert the cash flows to equivalent sums at specified dates. We shall discuss the computational procedures for the two most widely used methods of comparison:

1. Equivalent present amount
2. Equivalent uniform annual amount

The terms used frequently in investment alternatives for plant and equipment are first cost, salvage value, and economic life. The first cost or initial investment of a piece of equipment is the sum of cost of purchase, sales tax, transportation expenditure, installation expense, and other initial expenditures. Salvage value of the equipment is the net cash flow related to the termination of the asset's service. It is the gross receipts from the sale of the asset minus the expenditures, if any, required by its removal and sale. When the equipment is used for a long span of time, the maintenance cost and operating costs increase with age and the salvage value decreases with age. The economic life of an asset is determined and used in the economic studies. The economic life of an asset is established based on the estimates of the cash flow and the associated time of occurrence in the future. The economic life of the piece of equipment specifies the length of time that the equipment can be operated economically.

EQUIVALENT PRESENT AMOUNT

The equivalent present amount method of comparison consists of converting all the disbursements and receipts to a single present sum. The first costs are already at time zero and, hence, no interest formulas are applied to the first cost to find its present value. The single present sum is calculated for each alternative. If net income is measured, the alternative that yields the largest present sum is selected. If net costs are measured, the alternative that yields the smallest present sum is selected. In making the equivalent present amount comparison, the *same* number of years of future service must be provided for each alternative. This method can be employed in comparing alternatives having identical economic lives.

Example 7.3

A mobile home manufacturing company is planning to purchase a truck for material handling purposes. The information for the two proposals, item 364 and item 183, is as follows:

Item number	364	183
Economic life	3 years	3 years
First cost	$5000	$9000
Disbursement at the end of year 1	$ 300	$ 250
Disbursement at the end of year 2	$ 450	$ 300
Disbursement at the end of year 3	$ 700	$ 420
Salvage value	$1000	$3000

Show which truck should be purchased using an interest rate of 8 percent. Note that item 364 and item 183 have the same economic life. Therefore, the equivalent present amounts of the proposals can be compared in selecting the economical proposal. The present amount of net costs for each alternative may be computed as follows:

Here is the cash flow diagram for item 364:

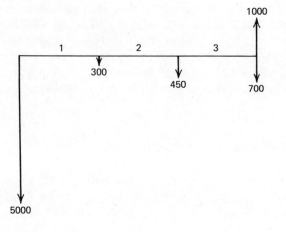

Item 364

First cost	= \$5000.00
Present amount of disbursement at the end of year 1	
= $300(P/F, 8\%, 1) = \$300(0.92593)$	= 277.78
Present amount of disbursement at the end of year 2	
= $450(P/F, 8\%, 2) = \$450(0.85734)$	= 385.80
Present amount of disbursement at the end of year 3	
= $700(P/F, 8\%, 3) = \$700(0.79383)$	= 555.68
Present amount of all monies paid out for 3 years	= \$6219.26
Less present amount of salvage value	
= $1000(P/F, 8\%, 3) = \$1000(0.79383)$	= 793.83
Present amount of net costs for 3 years	\$5425.43

Here is the cash flow diagram for item 183:

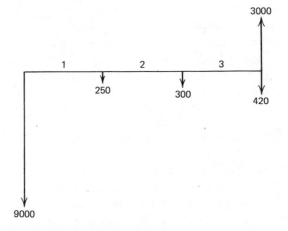

Item 183

First cost	= \$9000.00
Present amount of disbursement at the end of year 1	
= $250(P/F, 8\%, 1) = \$250(0.92593)$	= 231.48
Present amount of disbursement at the end of year 2	
= $300(P/F, 8\%, 2) = \$300(0.85734)$	= 257.20
Present amount of disbursement at the end of year 3	
= $420(P/F, 8\%, 3) = \$420(0.79393)$	= 333.41
Present amount of all monies paid out for 3 years	= \$9822.09
Less present amount of salvage value	
= $3000(P/F, 8\%, 3) = \$3000(0.79383)$	= 2381.49
Present amount of net costs for 3 years	\$7440.60

Of the two proposals, item 364 has the smallest present sum of net costs. Hence, item 364 should be chosen.

A Computer Program

These variables are used in the development of our computer model for the equivalent present amount technique:

IMAX   = number of alternatives (proposals) considered
LIFEY   = economic life in years
RATE   = annual interest rate represented as a fraction (when interest rate is 8%, RATE = 0.08)
PCOST   = first cost
SVAL   = salvage value
ITEM(K) = identification number of the Kth alternative
XMC(N) = disbursement at the end of the Nth year
PWMC   = present amount of XMC(N)
CPWMC   = present amount of all monies paid out for LIFEY periods
PWSV   = present amount of salvage value
CPW(K) = present amount of net costs for the Kth alternative
INDEX   = proposal number that has the smallest present amount of net costs
SMALLV = the smallest value among the computed CPW(K) values

The variables ITEM( ) and CPW( ) are one-dimensional arrays of size IMAX elements. The one-dimensional data array XMC( ) has LIFEY elements. Figure 7.1 exhibits the flowchart for the computer program. The values of IMAX, LIFEY, and RATE might be transmitted to the computer memory and printed out by the statements:

```
       READ 10, IMAX, LIFEY, RATE
       PRINT 700, IMAX
       PRINT 705, LIFEY
       PRINT 710, RATE
  10   FORMAT ( )
 700   FORMAT ( )
 705   FORMAT ( )
 710   FORMAT ( )
```

The input data for the Ith proposal are ITEM(I), PCOST, SVAL, and XMC values. We might transmit the input data to the computer memory

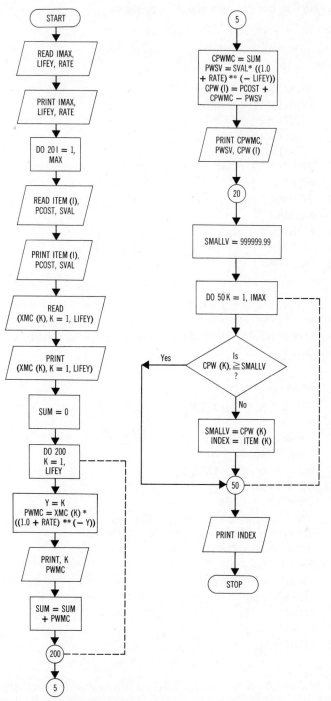

FIGURE 7.1 **Flowchart for selecting the proposal by the equivalent present amount method of comparison.**

and also print out the values by the instructions:

```
      READ 30, ITEM(I), PCOST, SVAL
      PRINT 715, ITEM(I)
      PRINT 720, PCOST
      PRINT 725, SVAL
      READ 40, (XMC(K), K = 1, LIFEY)
      PRINT 735, (XMC(K), K = 1, LIFEY)
  30  FORMAT ( )
  40  FORMAT ( )
 715  FORMAT ( )
 720  FORMAT ( )
 725  FORMAT ( )
 735  FORMAT ( )
```

Next, the present amount of all monies paid out and the present amount of net cost for the Ith alternative are computed. Then, the calculated values are printed out. These operations can be performed by employing these statements:

```
      SUM = 0
      DO 200 K = 1, LIFEY
      Y = K
      PWMC = XMC(K) * ((1.0 + RATE) ** (-Y))
      PRINT 740, K, PWMC
      SUM = SUM + PWMC
 200  CONTINUE
      CPWMC = SUM
      PWSV = SVAL * ((1.0 + RATE) ** (-LIFEY))
      CPW(I) = PCOST + CPWMC - PWSV
      PRINT 745, CPWMC
      PRINT 750, PWSV
      PRINT 755, CPW(I)
 740  FORMAT ( )
 745  FORMAT ( )
 750  FORMAT ( )
 755  FORMAT ( )
```

We could determine the present amount of net cost for each alternative by placing the above two program segments within the following DO loop:

```
     DO 20 I = 1, IMAX
     _____
 20  CONTINUE
```

The following set of instructions would cause the computer to find the smallest present amount of net cost and print out the associated proposal number:

```
      SMALLV = 999999.99
      DO 50 K = 1, IMAX
      IF (CPW(K).GE.SMALLV) GO TO 50
      SMALLV = CPW(K)
      INDEX = ITEM(K)
  50  CONTINUE
      PRINT 770, INDEX
 770  FORMAT ( )
```

The complete computer program composed of our previously designed program segments is shown in Figure 7.2. This computer model can be used to evaluate at most five investment proposals. The economic life must not exceed 10 years.

An Illustration of the Computer Program   Let us employ the computer program to solve Example 7.3. The data deck is prepared according to the FORMAT statements:

```
 10  FORMAT (2I3, F5.2)
 30  FORMAT (I5, 2F9.2)
 40  FORMAT (3F8.2)
```

The first card in the data deck gives the values of IMAX, LIFEY, and RATE, and it has the appearance:

| 2 | 3 | 0.08 |

```
      DIMENSION CPW(5),XMC(10)
      DIMENSION ITEM(5)
C
      READ 10, IMAX, LIFEY, RATE
      PRINT 700, IMAX
      PRINT 705, LIFEY
      PRINT 710, RATE
C
      DO 20 I=1,IMAX
      READ 30, ITEM(I), PCOST, SVAL
      PRINT 715, ITEM(I)
      PRINT 720, PCOST
      PRINT 725, SVAL
      READ 40,(XMC(K),K=1,LIFEY)
      PRINT 730
      PRINT 735,(XMC(K),K=1,LIFEY)
      SUM=0
C
C   CALCULATE THE PWMC FOR EACH XMC VALUE
      DO 200 K=1,LIFEY
      Y=K
      PWMC=XMC(K)*((1.0+RATE)**(-Y))
      PRINT 740, K, PWMC
      SUM=SUM+PWMC
  200 CONTINUE
C
C   CALCULATE CPW, CFW, AND EUAV VALUES
      CPWMC=SUM
      PWSV=SVAL*((1.0+RATE)**(-LIFEY))
      CPW(I)=PCOST+CPWMC-PWSV
C
C   PRINTOUT THE COMPUTED VALUES
      PRINT 745, CPWMC
      PRINT 750, PWSV
      PRINT 755, CPW(I)
      PRINT 765
   20 CONTINUE
C
C   FIND OUT THE MOST PREFERRED ITEM
      SMALLV=999999.99
C
      DO50 K=1,IMAX
      IF(CPW(K).GE.SMALLV) GO TO 50
      SMALLV=CPW(K)
      INDEX=ITEM(K)
   50 CONTINUE
```

FIGURE 7.2   Computer program for the equivalent present amount method of comparison (illustrated for Example 7.3).

```
C
C  PRINTOUT THE SELECTED ITEM
       PRINT 770, INDEX
C
   10 FORMAT(2I3,F5.2)
   30 FORMAT(I5, 2F9.2)
   40 FORMAT(3F8.2)
  700 FORMAT(1H1,2X,12HNO. OF ITEMS,3X,1H=,I4)
  705 FORMAT(3X, 11HLIFE, YEARS, 4X,1H=,I4 )
  710 FORMAT(3X,13HINTEREST RATE,2X,1H=,F5.2,//)
  715 FORMAT(3X,11HITEM NUMBER,6X,1H=,I4)
  720 FORMAT(3X,12HINITIAL COST,5X,1H=,F9.2)
  725 FORMAT(3X,13HSALVAGE VALUE,4X,1H=,F9.2)
  730 FORMAT(3X,18HMAINTENANCE COST =)
  735 FORMAT(3X,3(F8.2,2X))
  740 FORMAT(3X,13HPWMC FOR YEAR,I3,2H =,F9.2)
  745 FORMAT(10X,5HCPWMC,5X,1H=,F9.2)
  750 FORMAT(10X,4HPWSV,6X,1H=,F9.2)
  755 FORMAT(10X,3HCPW,7X,1H=,F9.2)
  765 FORMAT(// )
  770 FORMAT(3X,15HSELECT THE ITEM,I5)
      STOP
      END
```

FIGURE 7.2 (*Continued*)

```
        NO. OF ITEMS     =    2
        LIFE, YEARS      =    3
        INTEREST RATE    =   .08


        ITEM NUMBER          =  364
        INITIAL COST         =  5000.00
        SALVAGE VALUE        =  1000.00
        MAINTENANCE COST =
          300.00     450.00      700.00
        PWMC FOR YEAR   1 =     277.78
        PWMC FOR YEAR   2 =     385.80
        PWMC FOR YEAR   3 =     555.68
                CPWMC     =    1219.26
                PWSV      =     793.83
                CPW       =    5425.43


        ITEM NUMBER          =  183
        INITIAL COST         =  9000.00
        SALVAGE VALUE        =  3000.00
        MAINTENANCE COST =
          250.00     300.00      420.00
        PWMC FOR YEAR   1 =     231.48
        PWMC FOR YEAR   2 =     257.20
        PWMC FOR YEAR   3 =     333.41
                CPWMC     =     822.09
                PWSV      =    2381.50
                CPW       =    7440.60


        SELECT THE ITEM   364
```

FIGURE 7.3   Computer printout for Example 7.3.

273

Then, IMAX set of cards are arranged, each set giving the values of one alternative. The subsequent cards in the data deck have the following appearance:

364	5000·00	1000·00					
300·00	450·00	700·00					
183	9000·00	3000·00					
250·00	300·00	420·00					

The resulting computer printout is shown in Figure 7.3 on page 273.

EQUIVALENT UNIFORM ANNUAL AMOUNT

The equivalent uniform annual amount method of comparison consists of converting all the disbursements and receipts into a series of equal amounts occurring at the end of each year in the life of an asset. This method of analysis is similar to the equivalent present amount method and it yields the same conclusions as the equivalent present amount method. People normally tend to think in terms of annual amounts. Hence, the uniform annual amount method is used more often than the present amount method. The equivalent uniform annual amount method can be utilized in comparing alternatives having identical economic lives as well as in comparing alternatives having different economic lives.

Example 7.4

A construction company is contemplating the purchase of a new air compressor. Two alternatives, type *A* compressor and type *B* compressor, are considered and their estimates are as follows:

	Type *A*	Type *B*
First cost	$12,000	$16,000
Economic life	3 years	4 years
Disbursement at the end of year 1	$ 700	$ 400
Disbursement at the end of year 2	$ 1,200	$ 600

	Type A	Type B
Disbursement at the end of year 3	$ 1,700	$ 800
Disbursement at the end of year 4	—	$ 1,000
Salvage value	$ 6,000	$11,000

If the interest on the capital is 5 percent, determine which compressor should be chosen? The two proposals have different economic lives. Hence, the equivalent uniform annual amount method of comparison can be used in selecting the economical alternative. The uniform annual amounts for each proposal may be calculated as follows:

Here is the cash flow diagram for type A:

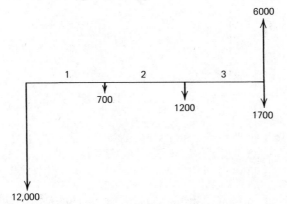

Type A

First cost	= $12,000.00
Present amount of disbursement at the end of year 1	
$= \$700(P/F, 5\%, 1) = \$700(0.95238)$	= 666.67
Present amount of disbursement at the end of year 2	
$= \$1200(P/F, 5\%, 2) = \$1200(0.90703)$	= 1,088.44
Present amount of disbursement at the end of year 3	
$= \$1700(P/F, 5\%, 3) = \$1700(0.86384)$	= 1,468.53
Present amount of all monies paid out for 3 years	$15,223.64
Less	
Present amount of salvage value	
$= \$6000(P/F, 5\%, 3) = \$6000(0.86384)$	= 5,183.04
Present amount of net costs for 3 years	$10,040.60
Uniform annual net cost	
$= \$10,040.60(A/P, 5\%, 3) = \$10,040.60(0.36721)$	= 3,687.00

The cash flow diagram for type B is as follows:

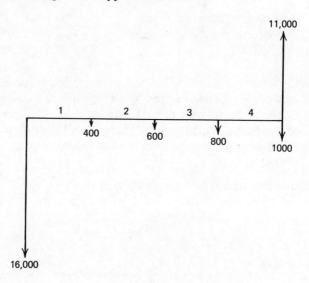

Type B

First cost	=	$16,000.00
Present amount of disbursements at the end of year 1		
= $400(P/F, 5\%, 1) = $400(0.95238)$	=	380.95
Present amount of disbursements at the end of year 2		
= $600(P/F, 5\%, 2) = $600(0.90703)$	=	544.22
Present amount of disbursements at the end of year 3		
= $800(P/F, 5\%, 3) = $800(0.86384)$	=	691.07
Present amount of disbursements at the end of year 4		
= $1000(P/F, 5\%, 4) = $1000(0.82270)$	=	822.70
Present amount of all monies paid out for 4 years		$18,438.94
Less		
Present amount of salvage value		
= $11,000(P/F, 5\%, 4) = $11,000(0.82270)$	=	9,049.70
Present amount of net costs for 4 years		$ 9,389.24
Uniform annual net cost		
= $9389.24(A/P, 5\%, 4) = $9389.24(0.28201)$	=	$ 2,647.86

Note that the calculation of uniform annual net cost involves the two steps:

STEP 1. Calculate the present amount of net costs for the total number of periods. This step is the same as the equivalent present amount method.

STEP 2. Find the uniform annual net cost from the present amount of net costs using the interest factor $(A/P, i, n)$.

Because the uniform annual net cost of type B is the smallest among the two computed values, type B compressor is preferred. The equivalent uniform annual method of comparison may be applied to Example 7.3 as follows:

Item 364
Present amount of net costs for 3 years   = \$5425.43
Uniform annual net cost
   = \$5425.43$(A/P, 8\%, 3)$ = \$5425.43(0.38803)   = \$2105.25

Item 183
Present amount of net costs for 3 years   = \$7440.60
Uniform annual net cost
   = \$7440.60$(A/P, 8\%, 3)$ = \$7440.60(0.38803)   = \$2887.20

Among the uniform annual net costs of item 364 and item 183, item 364 has the smallest cost. Therefore, item 364 is selected. Recall that this is the same decision reached for the equivalent present amount method.

DETERMINATION OF ECONOMIC LIFE

In comparing alternatives by the present amount and the annual amount methods, we have used the economic life periods of the assets. We develop a computational procedure for establishing the economic life of an asset in this section. The following are the steps in determining the economic life of an asset:

STEP 1. Obtain the cost estimates of the first cost, disbursement, and salvage value at the end of each year for different years as well as the annual interest rate.

STEP 2. Calculate the equivalent uniform annual net cost for different replacement time intervals (such as replacing every year and replacing every two years).

STEP 3. Select the replacement time interval that has the smallest annual net cost. The selected time interval denotes the economic life of the asset.

The disbursement of any year includes operating cost and maintenance cost for that particular year. The disbursements increase frequently from one year to the next because of progressive deterioration and obsolescence. The longer we keep an asset, in most cases, the lower the salvage value becomes. The algorithm can be illustrated by a numerical example.

Example 7.5

A new machine, item 780, can be purchased and installed for $4000. The disbursement for any year j occurring at the end of the particular year can be represented by the equation

$$D_j = A + B(j) \tag{7.17}$$

where

D_j = disbursement occurring at the end of the year j

$A = \$100$

$B = \$150$

The salvage value at the end of the year j is expected to conform to the formula

$$S_j = C^j(F) \tag{7.18}$$

where

S_j = salvage value at the end of the year j

$C = 0.8$

F = first cost ($= \$4000$)

What is the economic life of the machine for the annual interest rate of 10 percent?

The disbursement occurring at the end of the year j is

$$D_j = 100 + 150_j \tag{7.19}$$

Thus, the disbursements for different years are the following:

End of Year (j)	Disbursement (D_j)
1	$250.00
2	400.00
3	550.00
4	700.00
5	850.00
6	1000.00

The salvage value at the end of year j is estimated by applying formula (7.18):

$$S_j = (0.8)^j(F) \tag{7.20}$$

Hence, the salvage value at different points in time are

End of Year (j)	Salvage Value (S_j)
1	$3200.00
2	2560.00
3	2048.00
4	1638.40
5	1310.72
6	1048.58

If the machine is replaced at the end of the year K, the cash flow consists of the initial cost, the K disbursements occurring one at the end of each intervening year, and the salvage value realized at the end of the year K. Figure 7.4 shows the cash flow diagrams for various replacement time intervals.

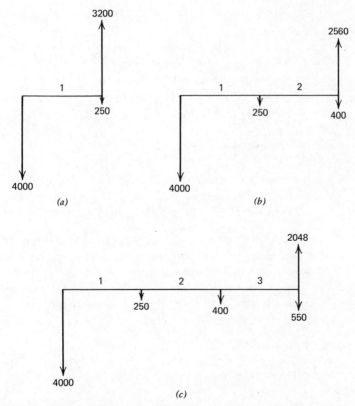

FIGURE 7.4   Cash flow diagrams for different replacement time intervals: (a) one-year, (b) two-year, and (c) three-year replacement intervals.

The equivalent uniform annual amount method of comparison is used to determine the uniform annual net cost, as follows:

Replacement time interval = 1 year (Figure 7.4a)

Initial cost	= \$4000.00
Present amount of disbursement at the end of year 1	
= \$250(P/F, 10%, 1) = \$250(0.90909)	= 227.27
Present amount of all monies paid out for 1 year	4227.27
Less	
Present amount of salvage value	
= \$3200(P/F, 10%, 1) = \$3200(0.90909)	= 2909.09
Present amount of net costs for 1 year	1318.18
Uniform annual net cost	
= \$1318.18(A/P, 10%, 1) = \$1318.18(1.1000)	= 1450.00

Replacement time interval = 2 years (Figure 7.4b)

Initial cost	= \$4000.00
Present amount of disbursements at the end of year 1	
= \$250(P/F, 10%, 1) = \$250(0.90909)	= 227.27
Present amount of disbursements at the end of year 2	
= \$400(P/F, 10%, 2) = \$400(0.82645)	= 330.58
Present amount of all monies paid out for 2 years	4557.85
Less	
Present amount of salvage value	
= \$2560(P/F, 10%, 2) = \$2560(0.82645)	= 2115.71
Present amount of net costs for 2 years	2442.14
Uniform annual net cost	
= \$2442.14(A/P, 10%, 2) = \$2442.14(0.57619)	= 1407.14

Likewise, the uniform annual net costs are calculated for various replacement time intervals. The replacement time interval that has the smallest annual net cost is the economic life of item 780. In the next section we shall design a computer model for the computational procedure and solve this example by employing the computer program.

A Computer Program

The following variables are used in the computer program:

RATE = annual interest rate expressed as a fraction

MAXY = maximum number of years considered

ITEMNO = identification number of the asset
PVAL = initial cost
FACTOR = the parameter C in the formula (8.18) to determine the salvage value
XMC = disbursement at the end of the year K
PWMC = present amount of XMC
CPWMC = cumulative present amount of all disbursements
SVJ = salvage value at the end of the year J
PWSV = present value of SVJ
CPWJ = present amount of net costs
EUAV = uniform annual net cost
LIFE = economic life of the asset
SMALL = the smallest value among the calculated EUAV values
AF = value of the interest factor (A/P, i, n)

Figure 7.5 shows the flowchart for the computer model.

We can employ the following set of statements to transmit the input data to the computer memory, print out the values, and initialize the value of SMALL:

```
        READ 700, RATE, MAXY
        READ 720, ITEMNO, PVAL, A, B, FACTOR
        PRINT 800, RATE
        PRINT 805, MAXY
        PRINT 810, ITEMNO
        PRINT 815, PVAL
        PRINT 820, A
        PRINT 825, B
        PRINT 830, FACTOR
        SMALL = 999999.99
700     FORMAT ( )
720     FORMAT ( )
800     FORMAT ( )
805     FORMAT ( )
810     FORMAT ( )
815     FORMAT ( )
820     FORMAT ( )
825     FORMAT ( )
830     FORMAT ( )
```

For the replacement interval of J years, the cumulative present amount

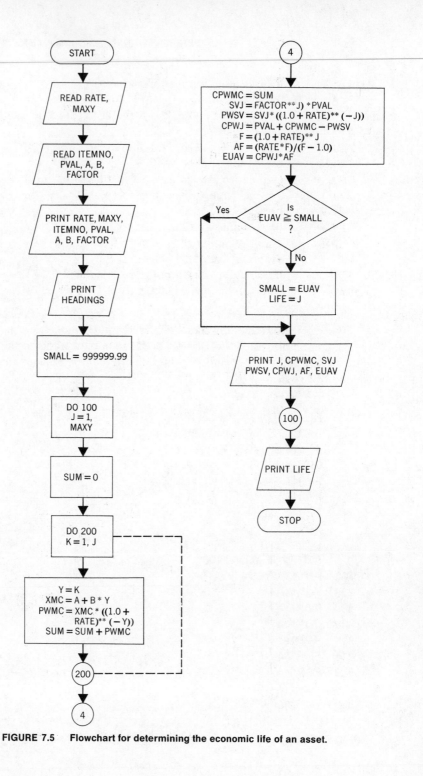

FIGURE 7.5 Flowchart for determining the economic life of an asset.

of J year-end disbursements is computed by including the statements:

```
        SUM = 0
        DO 200 K = 1, J
        Y = K
        XMC = A + B * Y
        PWMC = XMC * ((1.0 + RATE) ** (−Y))
        SUM = SUM + PWMC
200     CONTINUE
        CPWMC = SUM
```

The following set of instructions would cause the computer to calculate the uniform annual net cost for J years, test if it is the smallest amount, and print out the computed values:

```
        SVJ = (FACTOR ** J) * PVAL
        PWSV = SVJ * ((1.0 + RATE) ** (−J))
        CPWJ = PVAL + CPWMC − PWSV
        F = (1.0 + RATE) ** J
        AF = (RATE * F)/(F − 1.0)
        EUAV = CPWJ * AF
        IF (EUAV.GE.SMALL) GO TO 150
        SMALL = EUAV
        LIFE = J
150     PRINT 840, J, CPWMC, SVJ, PWSV, CPWJ, AF, EUAV
840     FORMAT ( )
```

We could determine the uniform annual net cost for different replacement time intervals up to MAXY years and establish the economic life by placing the above two program segments within the following DO loop:

```
        DO 100 J = 1, MAXY
        _____

100     CONTINUE
```

The economic life of the asset is printed out by including the following statements

```
        PRINT 250, LIFE
250     FORMAT ( )
```

The complete computer program composed of our previously designed program segments is shown in Figure 7.6. To solve Example 7.5 using the

```
C   READ THE INPUT DATA
        READ 700, RATE, MAXY
        READ 720, ITEMNO, PVAL, A, B, FACTOR
C
C   PRINTOUT THE INPUT DATA
        PRINT 800, RATE
        PRINT 805, MAXY
        PRINT 810, ITEMNO
        PRINT 815, PVAL
        PRINT 820, A
        PRINT 825, B
        PRINT 830, FACTOR
C
C   PRINT THE TITLE
        PRINT 835
C
        SMALL = 999999.99
        DO 100 J=1,MAXY
        SUM=0
C
C   CALCULATE THE PWMC VALUE
        DO 200 K=1,J
        Y=K
        XMC=A+B*Y
        PWMC=XMC*((1.0+RATE)**(-Y))
        SUM=SUM+PWMC
  200 CONTINUE
C
C   CALCULATE EUAV VALUES
        CPWMC=SUM
        SVJ=(FACTOR**J)*PVAL
        PWSV=SVJ*((1.0+RATE)**(-J))
        CPWJ=PVAL+CPWMC-PWSV
        F=(1.0+RATE)**J
        AF=(RATE*F)/(F-1.0)
        EUAV=CPWJ*AF
        IF ( EUAV .GE. SMALL ) GO TO 150
        SMALL = EUAV
        LIFE = J
C
C   PRINTOUT THE COMPUTED VALUES
  150 PRINT 840,J,CPWMC,SVJ,PWSV,CPWJ,AF,EUAV
  100 CONTINUE
        PRINT 250, LIFE
C
  250 FORMAT (/,10X, 15HECONOMIC LIFE =, I3, 6H YEARS )
  700 FORMAT(F4.2,I2)
  720 FORMAT(I4,F8.2,2F7.2,F4.2)

  800 FORMAT(1H1,2X,13HINTEREST RATE,11X,2H= ,F4.2)
  805 FORMAT(3X,26HNO. OF YEARS CONSIDERED = ,I2)
  810 FORMAT(//,3X,8HITEM NO.,5X,1H=,I4)
  815 FORMAT(3X,14HINITIAL COST =,F8.2)
  820 FORMAT(3X,1HA,12X,2H= ,F7.2)
  825 FORMAT(3X,1HB,12X,2H= ,F7.2)
  830 FORMAT(3X,6HFACTOR,7X,1H=,4X,F4.2)
  835 FORMAT(//,9X,1HJ,4X,5HCPWMC,6X,3HSVJ,7X,
     14HPWSV,5X,4HCPWJ,9X,2HAF,5X,4HEUAV)
  840 FORMAT (/, 5X,I5, 4F10.2, F10.6, F10.2 )
        STOP
        END
```

FIGURE 7.6   Computer program for determining the economic life of an asset (illustrated for Example 7.5).

computer program, the data deck is prepared according to the FORMAT statements 700 and 720:

700 FORMAT (F4.2, I2)
720 FORMAT (I4, F8.2, 2F7.2, F4.2)

The data deck for Example 7.5 has the following appearance:

0.1010

780 4000·00 100·00 150·000·80

The computer printout shown in Figure 7.7 indicates that the five-year replacement interval has the smallest uniform annual net cost. The economic life of item 780, therefore, is five years.

```
INTEREST RATE          =  .10
NO. OF YEARS CONSIDERED = 10


ITEM NO.       = 780
INITIAL COST = 4000.00
A            =  100.00
B            =  150.00
FACTOR       =    .80
```

J	CPWMC	SVJ	PWSV	CPWJ	AF	EUAV
1	227.27	3200.00	2909.09	1318.18	1.100000	1450.00
2	557.85	2560.00	2115.70	2442.15	.576190	1407.14
3	971.07	2048.00	1538.69	3432.38	.402115	1380.21
4	1449.18	1638.40	1119.05	4330.13	.315471	1366.03
5	1976.97	1310.72	813.85	5163.11	.263797	1362.02
6	2541.44	1048.58	591.89	5949.55	.229607	1366.06
7	3131.57	838.86	430.47	6701.10	.205405	1376.44
8	3738.03	671.09	313.07	7424.96	.187444	1391.77
9	4352.97	536.87	227.69	8125.29	.173641	1410.88
10	4969.84	429.50	165.59	8804.25	.162745	1432.85

```
ECONOMIC LIFE =  5 YEARS
```

FIGURE 7.7   Computer printout for Example 7.5.

PROBLEMS

7.1   If a sum of $1200 is loaned for five years at 7% simple interest rate per year, what will be the future amount at the end of the loan period?

7.2   How much will be accumulated at the end of a five-year period, if $1200 is invested now, assuming interest rate at 7% compounded annually? Prepare a table similar to Table 7.1.

7.3   Alan is planning to go for a vacation to Europe five years from now. He can invest $700 one year from now at 6% interest compounded annually to be used for his vacation. How much will be available at the time of his vacation?

7.4   How much must be deposited on January 1, 1975 in order to draw $5000 on December 31, 1981? Assume an interest rate of 9% compounded annually.

7.5   Fredrick's grandmother wants to plan for financing Fred's college education. How much must she deposit on his twelfth birthday to provide $4000 on each birthday from the eighteenth to the twenty-first inclusive? Use interest at 8% compounded annually.

7.6   Three trucks are being considered for purchase by a truck renting company. The company wishes to buy the truck with the smallest (present) amount of net disbursements. Estimates for the three trucks are as follows:

	Truck *A*	Truck *B*	Truck *C*
Life	5 years	5 years	5 years
First cost	$20,000	$30,000	$25,000
Maintenance cost for first year	1,000	700	600
Maintenance cost for second year	1,200	900	1,000
Maintenance cost for third year	1,500	1,300	1,200
Maintenance cost for fourth year	1,800	1,300	1,500
Maintenance cost for fifth year	2,000	1,400	1,600
Terminal salvage value	3,000	5,000	6,000

Compare the equivalent present amounts of the net disbursements using an interest rate of 6%.

7.7   An electronic firm is planning to lease computer system 1000 or buy computer system 750. It is estimated that the computing services will be used for the next six years. The cost estimates for the alternative plans are given below. Assume an i of 8%.

	Leasing system 1000	Buy system 750
First cost	$15,000	$80,000
Annual disbursements for leasing system 1000	17,000	—
Maintenance cost for first year	—	8,000
Annual increase of maintenance cost	—	1,500
Salvage value at the end of sixth year	—	40,000

Write a computer program to select the economic plan using the equivalent present amount method.

7.8 A manufacturer requires a drilling machine. Each of the three machines, $D100$, $D120$, and $D90$ will provide the required services. Estimated cash flows for the alternatives are as follows:

	Machine $D100$	Machine $D120$	Machine $D90$
First cost	$10,000	$12,000	$8000
Estimated life	3 years	4 years	4 years
Maintenance cost for first year	450	350	500
Annual increase of maintenance cost	250	150	300
Salvage value at the end of the life time	6100	7800	3200

Compare the equivalent uniform annual amounts using an i of 9%.

7.9 An oil company is planning to purchase a piece of chemical processing equipment. The following values are estimated:

First cost, including installation (F) $= \$30,000$

Operating and maintenance costs for year j occurring
at the end of the year $= A + B(j)$
where

$$A = \$200$$

$$B = \$500$$

Salvage value at the end of year j $\qquad = C^j(F)$
where

$$C = 0.85$$

Determine the economic life of the equipment. Assume an i of 5%.

7.10 Mr. Green is considering the purchase of a new sports car. He wishes to know whether it is economical to sell it and to buy a similar new car after two years, three years, or four years. The estimated values are as follows:

First cost (F) $\qquad = \$18,000$
Maintenance cost for year j occurring at the end of
  the year $\qquad D_j = A + B(j)$
where

$$A = \$200$$

$$B = \$450$$

Salvage value at the end of year j, $\qquad S_j = C^j(F)$
where

$$C = 0.85$$

Compare the uniform annual net costs for the alternative plans using an i of 5%.

7.11 A company wants to determine whether to install a roller conveyor system (Plan A), or an automatic conveyorized system (Plan B), for moving products in the warehouse. The estimated values are:

	Plan A	Plan B
First cost, including installation (F)	$= \$210,000$	\$290,000
Maintenance cost for year j occurring		
at the end of the year,	$A = \$\ \ 1000$	1000
$D_j = A + B(j)$	$B = \$\ \ 5000$	4000
Salvage value at the end of year j,		
$S_j = C^j(F)$	$C =\ \ \ \ \ \ 0.85$	0.90

Design a FORTRAN program to establish the economic life of the two plans and to select the plan that yields the smallest uniform annual net costs. Use an i of 10%.

──────────────SELECTED REFERENCES──────────────

Barish, N. N., *Economic Analysis for Engineering and Managerial Decision-Making*, McGraw-Hill, New York, 1962.

Bullinger, C. E., *Engineering Economy*, McGraw-Hill, New York, 1958.

DeGarmo, E. P., *Engineering Economy*, Macmillan, New York, Fourth Edition, 1967.

Fabrycky, W. J., P. M. Ghare, and P. E. Torgersen, *Industrial Operations Research*, Prentice-Hall, Englewood Cliffs, N.J., 1972.

Grant, E. L., and W. G. Ireson, *Principles of Engineering Economy*, Ronald Press, New York, Fifth Edition, 1970.

McMillan, C., and R. F. Gonzalez, *Systems Analysis—A Computer Approach to Decision Models*, Richard D. Irwin, Homewood, Ill., Third Edition, 1973.

Manne, A. S., *Economic Analysis for Business Decisions*, McGraw-Hill, New York, 1961.

Morris, W. T., *Engineering, Economy*, Richard D. Irwin, Homewood, Ill., 1960.

Riggs, J. L., *Economic Decision Models for Engineers and Managers*, McGraw-Hill, New York, 1968.

Shubin, J. A., *Managerial and Industrial Economics*, Ronald Press, New York, 1961.

Smith, G. W., *Engineering Economy: Analysis of Capital Expenditures*, The Iowa State University Press, Ames, Iowa, 1968.

Spencer, M. H., and L. Siegelman, *Managerial Economics*, Richard D. Irwin, Homewood, Ill., 1959.

Taylor, G. A., *Managerial and Engineering Economy*, Van Nostrand-Reinhold, New York, 1964.

Thuesen, H. G., W. J. Fabrycky, and G. J. Thuesen, *Engineering Economy*, Prentice-Hall, Englewood Cliffs, N.J., Fourth Edition, 1971.

8 **quantity control**

CHAPTER

THE THEORY of learning curve and line of balance techniques can be utilized in planning and controlling of production systems. These models can be used as valuable aids in planning manpower requirements, forecasting unit cost, estimating delivery schedules, and monitoring the production operations. In this chapter we shall discuss the solution procedures and develop computer models for these methods.

THEORY OF LEARNING CURVE

Most industries manufacture their products in quantities. The workmen perform their operations, in such cases, repeatedly. Whenever these people practice the activities again and again, they learn as they work. Performance of these activities is improved not only the second time but each succeeding time. Because of this progressive improvement, the production time for the second unit is smaller than the production time for the first unit, the time for the third unit is still smaller than the time for the second unit, and so on. As the workers become more efficient, the direct labor time per unit decreases. The theory of learning curve is applicable for such performance improvement situations.

In many production surveys, it has been found that there prevailed a consistent reduction in the accumulated average production time of *doubled quantities*. Consider that the production time for the first unit of a cargo ship is 150,000 man-hours and the ship building industry's learning rate is 80%. The accumulated average time for two units (double the amount of 1 unit) is

(Accumulated average time for 1 unit) × Learning rate
= 150,000 × 0.80 = 120,000 man-hours
Total time to produce 2 units = 120,000 × 2 = 240,000 man-hours

Subsequently, the accumulated average time for 4 units (double the amount of 2 units) is

(Accumulated average time for 2 units) × Learning rate
= 120,000 × 0.80 = 96,000 man-hours
Total time to produce 4 units = 96,000 × 4 = 384,000 man-hours

The method can be used to determine the accumulated average times and the total times to produce 8, 16, 32, 64, 128, 256, 512 units, and so on. The computed values are depicted in Table 8.1.

TABLE 8.1 **Total Production Time Computations for a Learning Rate of 80 Percent**

Number of Units Completed	Accumulated Average Time, Man-Hours	Total Production Time, Man-Hours
1	150,000	150,000 × 1 = 150,000
2	150,000 × 0.8 = 120,000	120,000 × 2 = 240,000
4	120,000 × 0.8 = 96,000	96,000 × 4 = 384,000
8	96,000 × 0.8 = 76,800	76,800 × 8 = 614,400
16	76,800 × 0.8 = 61,440	61,440 × 16 = 983,040
32	61,440 × 0.8 = 49,152	49,152 × 32 = 1,572,864
64	49,152 × 0.8 = 39,322	39,322 × 64 = 2,516,582
128	39,322 × 0.8 = 31,457	31,457 × 128 = 4,026,532
256	31,457 × 0.8 = 25,166	25,166 × 256 = 6,442,496

Production time for the first unit = 150,000 man-hours.

The rate of improvement is smaller in machine operations because most of the work is done by the machines and there is less for workers to learn. On the other hand, man-paced operations are more susceptible to learning and, hence, the learning rate is larger than the machine-paced operations.

DETERMINING THE LEARNING CURVE FORMULA

The relationship between the number of units produced and the accumulated average production time can be expressed mathematically as follows:

$$Y_x = a \cdot x^b \tag{8.1}$$

where

Y_x = accumulated average time for producing x units

x = number of completed units

a = time required to manufacture the first unit

b = value associated with the learning rate

The value of b for our example can be determined by using the information presented in Table 8.1.

For $x = 2$ $Y_2 = 120,000$

For $x = 4$ $Y_4 = 96,000$ and $a = 150,000$

$$Y_2 = 120,000 = 150,000(2)^b \tag{8.2}$$

$$Y_4 = 96,000 = 150,000(4)^b \tag{8.3}$$

Dividing (8.2) by (8.3), we obtain

$$\frac{120,000}{96,000} = \frac{150,000(2)^b}{150,000(4)^b}$$

$$\frac{100}{80} = \left(\frac{1}{2}\right)^b$$

$$\log\left(\frac{100}{80}\right) = b \cdot \log\left(\frac{1}{2}\right)$$

$$b = \frac{\log(100/80)}{\log(1/2)}$$

$$= -0.322$$

Therefore, the mathematical relationship is

$$Y_x = a(x)^b \tag{8.4}$$

where

a = time required to produce the first unit

and

$$b = \frac{\log\left(\dfrac{100}{\text{Learning rate in percent}}\right)}{\log(1/2)}$$

The learning curve formula (8.4) can be used to compute the accumulated average time for any number of completed units.

SOLUTION PROCEDURE FOR THE LEARNING CURVE PROBLEMS

Frequently, it will be necessary to determine the production time for individual units (such as for the first, second, and third unit), the accumulated production time as well as the accumulated average time for a given number of units. By analyzing the past production records of a particular product, we may obtain two types of information:

Case 1. The production time for the first unit and the learning rate.
Case 2. The cumulative production times and the associated total number of units produced for any two quantities.

We shall present calculations for these two cases of input data.

Case 1

The learning rate and the production time for the first unit are known and, therefore, by applying equation 8.4, the learning curve equation is determined. We shall illustrate the procedure for the data given in the previous section.

Production time for the first unit (a) = 150,000 man-hours
Learning rate = 80%

$$b = \frac{\log\left(\dfrac{100}{\text{learning rate in percent}}\right)}{\log(1/2)}$$

$$= \frac{\log(100/80)}{\log(1/2)} = -0.322$$

The learning curve equation is

$$Y = 150{,}000X^{-0.322}$$

The production time for the Xth unit is

$$q = \left[\begin{array}{c}\text{Cumulative time to}\\\text{produce } X \text{ units}\end{array}\right] - \left[\begin{array}{c}\text{Cumulative time to}\\\text{produce } (X - 1) \text{ units}\end{array}\right] \quad (8.5)$$

The accumulated average time for x and $(x - 1)$ units are

$$Y_x = ax^b$$
$$Y_{x-1} = a(x - 1)^b$$

The accumulated time required to produce x units is

$$(Y_x) \cdot (X) \quad (8.6.1)$$

The accumulated time required to produce $(x - 1)$ units is

$$(Y_{x-1}) \cdot (X - 1) \quad (8.6.2)$$

Therefore, from (8.6.1) and (8.6.2) we have

$$\text{Production time for the } X\text{th unit} = (Y_x)(X) - (Y_{x-1})(X - 1) \quad (8.7)$$

The production time for the Xth unit, the accumulated production time, and the accumulated average production time for X completed units can be determined using the equations 8.7, 8.6.1, and 8.4, respectively.

A summary of the solution procedure follows:

1. Obtain the values of TIME1 and RATE where
   TIME1 = production time for the first unit
   RATE = learning rate as a percentage.
2. Compute the value of B using the equation 8.4 where
   B represents the variable b.
3. Calculate the values of AATX using the equation 8.4 where
   AATX = accumulated average time for X units
   AATX = $(\text{TIME1}) \cdot (X)^B$
4. Calculate APTF and PTX applying the equations 8.6 and 8.7 where
   APTF = accumulated production time for X units
   APTF1 = accumulated production time for $(X - 1)$ units
   PTX = production time for Xth unit
   APTF = (AATX)(X)
   APTF1 = $[(\text{TIME1}) \cdot (X - 1)^B](X - 1)$
   PTX = APTF − APTF1

A Computer Model for Case 1

We shall develop a computer program to calculate and print out for 1 through 10 units, the values of AATX, APTF, and PTX for our example problem.

The values of TIME1 and RATE are to be read and printed out. The following statements would do this:

```
      READ 650, TIME1, RATE
650   FORMAT (2F10.0)
      PRINT 650, TIME1, RATE
```

The value of B would be computed by including the statement

```
  B = ALOG(100./RATE)/ALOG(1.0/2.0)
```

When $X = 1$, we have $APTF1 = 0$. It is initialized, therefore, as

```
  APTF1 = 0
```

The following DO loop computes and prints out the values of AATX, APTF, and PTX for the first, second, ..., and tenth unit.

```
      DO 34 J = 1, 10
      F = J
      AATX = (TIME1) * (F ** B)
      APTF = (AATX) * F
      PTX = APTF − APTF1
      PRINT 200, J, AATX, APTF, PTX
200   FORMAT (I10, 3F10.0)
      APTF1 = APTF
 34   CONTINUE
```

The general program is presented in Figure 8.1. The input data for our

```
      READ 650, TIME1, RATE
650   FORMAT (2F10.0)
      PRINT 650, TIME1, RATE
      B=ALOG(100.0/RATE)/ALOG(1.0/2.0)
      AFTF1=0
      DO 34 J=1,10
      F=J
      AATX=(TIME1)*(F**B)
      APTF=(AATX)*F
      PTX=APTF-APTF1
      PRINT 200,J,AATX,APTF,PTX
200   FORMAT (I10,3F10.0)
      APTF1=APTF
 34   CONTINUE
      STOP
      END
```

FIGURE 8.1   Computer program for learning curve problems (Case 1).

numerical example has the following appearance:

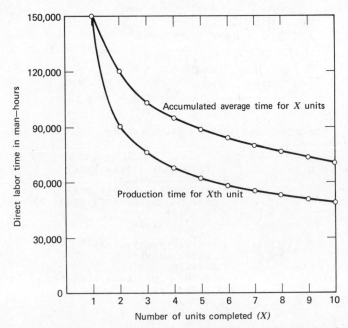

Figure 8.2 shows the resulting computer printout. Figure 8.3 exhibits the accumulated average time for X units as well as the production time for the Xth unit that were obtained for our example.

```
150000.        80.
       1    150000.    150000.    150000.
       2    120000.    240000.     90000.
       3    105316.    315947.     75947.
       4     96000.    384000.     68053.
       5     89346.    446728.     62728.
       6     84252.    505515.     58787.
       7     80173.    561214.     55699.
       8     76800.    614400.     53186.
       9     73942.    665482.     51082.
      10     71476.    714765.     49283.
```

FIGURE 8.2   Computer printout for the cargo ship production problem (Case 1).

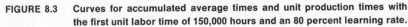

FIGURE 8.3   Curves for accumulated average times and unit production times with the first unit labor time of 150,000 hours and an 80 percent learning rate.

Case 2

For two production quantities, the cumulative production times and the associated total number of units are known. Suppose that the data are:

Cumulative production time		Total number of units
$C1$	..	$X1$
$C2$	..	$X2$

$$Y_{X1} = C1/X1 \quad \text{and} \quad Y_{X2} = C2/X2$$

Substituting the values of Y_{X1}, Y_{X2}, $X1$, and $X2$ in the general learning curve equation 8.4, we get

$$Y_{X1} = a(X1)^b \tag{8.8}$$

$$Y_{X2} = a(X2)^b \tag{8.9}$$

Dividing (8.8) by (8.9),

$$\frac{Y_{X1}}{Y_{X2}} = \frac{a(X1)^b}{a(X2)^b} = \left(\frac{X1}{X2}\right)^b$$

$$\log(Y_{X1}/Y_{X2}) = b \cdot \log(X1/X2)$$

$$b = \frac{\log(Y_{X1}/Y_{X2})}{\log(X1/X2)} \tag{8.10}$$

From (8.8), we obtain

$$a = Y_{X1}/[(X1)^b] \tag{8.11}$$

Since the values of the parameters, a and b, have been computed, the learning curve has been established. The production time for the Xth unit, the accumulated production time, and the accumulated average time for X completed units can be computed applying the relationships (8.7), (8.6.1), and (8.4).

$$Y_2 = Y_1 \text{ (learning rate)}$$

Therefore,

$$\text{Learning rate} = Y_2/Y_1 \tag{8.12}$$

The learning rate, hence, can be determined using the equation 8.12. A summary of the solution procedure follows:

1. Obtain the values of APTX(1), APTX(2), QTY(1), and QTY(2) where
   APTX(1) = Cumulative production time (C1) for X1 units
   APTX(2) = Cumulative production time (C2) for X2 units
   QTY(1) = X1 units
   QTY(2) = X2 units
2. Calculate AATN(1) and AATN(2) where
   AATN(1) = Accumulated average time for X1 units ($= Y_{X1}$)
   AATN(2) = Accumulated average time for X2 units ($= Y_{X2}$)

3. Calculate the value of *b* applying the relationship (8.10).
4. Calculate the value of *a* applying the relationship (8.11).
5. Calculate and print out the values of K, AATX, APTF and PTX applying the equation 8.4, 8.6.1, and 8.7.
6. Determine the learning rate, RATE, according to the equation 8.12:

$$\text{RATE} = \frac{AV(1)}{AV(2)} (100)$$

where

$AV(1) = $ AATX for one unit and
$AV(2) = $ AATX for two units

A Computer Model for Case 2

For the computational steps we shall structure a general computer model. DIMENSION statements are formulated for the four subscripted variables.

```
DIMENSION APTX(2)
DIMENSION QTY(2)
DIMENSION AATN(2)
DIMENSION AV(2)
```

The values of APTX and QTY are read and printed out by including the statements:

```
      DO 10 J = 1, 2
      READ 101, APTX(J), QTY(J)
101   FORMAT (2F10.0)
      PRINT 101, APTX(J), QTY(J)
 10   CONTINUE
```

We might calculate the values of Y_{X1}, Y_{X2}, B, and A with the following instructions:

```
AATN(1) = APTX(1)/QTY(1)
AATN(2) = APTX(2)/QTY(2)
Y = AATN(1)/AATN(2)
X = QTY(1)/QTY(2)
B = ALOG(Y)/ALOG(X)
A = AATN(1)/(QTY(1) ** B)
```

When X = 1, the accumulated average time for zero units is zero and therefore, we initialize,

    APTF1 = 0.0

We could calculate and print out the values of AATX, APTF, and PTX for the first through the tenth unit by the following DO loop:

```
        APTF1 = 0.0
        DO 77 K = 1, 10
        F = K
        AATX = A * (F ** B)
        APTF = AATX * F
        PTX = APTF − APTF1
        PRINT 200, K, AATX, APTF, PTX
200     FORMAT (I10, 3F10.0)
        APTF1 = APTF
 77     CONTINUE
```

The values of AATX for one and two units are stored in the locations AV(1) and AV(2) by including the following statements after establishing AATX values in the above DO loop:

```
        IF (K.GT.2) GO TO 120
        AV(K) = AATX
120     CONTINUE
```

The learning rate in percent is calculated and printed out by the statements:

```
        P = AV(2)/AV(1)
        RATE = P * 100.0
        PRINT 140, RATE
140     FORMAT (10X, 5HRATE=, F6.2)
```

Table 8.1 is considered in our illustration as the input data. Therefore,

    APTF(1) = 240,000   QTY(1) = 2
    APTF(2) = 614,400   QTY(2) = 8

The program and the printout are presented in Figures 8.4 and 8.5, respectively. For the first 10 units, the AATX, APTF, and PTX values together with the learning rate for the production process are printed out.

```
                    DIMENSION APTX(2)
                    DIMENSION QTY(2)
                    DIMENSION AATN(2)
                    DIMENSION AV(2)
                    DO 10 J =1,2
                    READ 101, APTX(J),QTY(J)
               101 FORMAT (2F10.0)
                    PRINT 101, APTX(J), QTY(J)
                10 CONTINUE
                    AATN(1) = APTX(1)/QTY(1)
                    AATN(2) = APTX(2)/QTY(2)
                    Y=AATN(1)/AATN(2)
                    X=QTY(1)/QTY(2)
                    B=ALOG(Y)/ALOG(X)
                    A=AATN(1)/(QTY(1)**B)
                    APTF1=0.0
                    DO 77 K=1,10
                    F=K
                    AATX=A*(F**B)
                    IF(K.GT.2) GO TO 120
                    AV(K)=AATX
               120 CONTINUE
                    APTF = AATX*F
                    PTX = APTF-APTF1
                    PRINT 200,K,AATX,APTF,PTX
               200 FORMAT (I10,3F10.0)
                    APTF1=APTF
                77 CONTINUE
                    P=AV(2)/AV(1)
                    RATE=P*100.0
                    PRINT 140, RATE
               140 FORMAT (10X,5HRATE=,F6.2)
                    STOP
                    END
```

FIGURE 8.4   Computer program for learning curve problems (Case 2).

```
240000.         2.
614400.         8.
        1   150000.    150000.    150000.
        2   120000.    240000.     90000.
        3   105316.    315947.     75947.
        4    96000.    384000.     68053.
        5    89346.    446728.     62728.
        6    84252.    505515.     58787.
        7    80173.    561214.     55699.
        8    76800.    614400.     53186.
        9    73942.    665482.     51082.
       10    71476.    714765.     49283.
        RATE= 80.00
```

FIGURE 8.5   Computer printout for the cargo ship production problem (Case 2).

Note that the printout values are the same for case 1 and 2 since the learning rate (80%) and also the time to produce the first unit (150,000 man-hours) of case 1 are equivalent to those of case 2.

THE USE OF LINE OF BALANCE

One of the major problems facing the management is obtaining information on the status of operations soon enough for taking effective action. The effective action is pertinent in quantity production in order to reasonably ensure the delivery of the products according to the contract schedule. The Line of Balance (LOB) technique has been developed by the U.S. Navy to control the production processes. This method can be used to provide periodic, graphic displays of the operations performances at various production control points, such as receiving materials, completing assemblies, and testing quality. The status of the operations is compared with the planned delivery schedules and then the problem areas are identified and exposed. The management can utilize this information to properly monitor the production process to ensure that the end products will be delivered at the appropriate time.

SOLUTION PROCEDURE OF LINE OF BALANCE

The line of balance method consists of the four elements or phases:

1. The objective delivery schedule
2. The program or production plan
3. The program progress
4. The line of balance

The Objective Delivery Schedule

The first step in a line of balance study is to obtain the objective delivery schedule. For developing the objective schedule, two kinds of delivery information are required:

(a) The expected delivery of the units representing the cumulative quantity of units planned to be delivered over time.
(b) The actual delivery of the units indicating the cumulative number of completed units really delivered.

TABLE 8.2   Contract Delivery Schedule and Actual Delivery of Electrical Motors

Time, Week No.	Units to Be Delivered	Cumulative Units to Be Delivered	Units Actually Delivered	Cumulative Units Actually Delivered
15	10	10	5	5
16	10	20	9	14
17	10	30	12	26
18	10	40		
19	10	50		
20	10	60		
21	10	70		
22	10	80		
23	10	90		
24	10	100		
25	20	120		
26	20	140		
27	20	160		
28	20	180		
29	20	200		

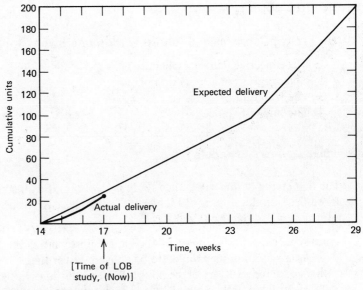

FIGURE 8.6   Objective chart.

Consider that the Warner Electrical Industries has received a contract to manufacture and deliver 200 units of electrical motors. Delivery of the units begins in the 15th week. The contract delivery schedule is shown in Table 8.2. The time of LOB study is the end of the 17th week. The actual number of units delivered until the 17th week is also given in Table 8.2. We note that until now (the end of the 17th week) 26 motors have been delivered, although the contract delivery is for 30 motors. The cumulative values for the expected delivery and the actual delivery are plotted on the objective chart, Figure 8.6.

The Program

The second step in the LOB study is to develop the program. The program denotes the manufacturer's planned production process of the product. In formulating the production plan, we consider three aspects:

(a) Establishing the activities to be performed and the key points (also called production control points).
(b) Establishing the sequence of activities.
(c) Establishing the activity times for each one of the operations.

For our example, the activities are presented in Figure 8.7.

Activity Identification	Description of Activity	Estimated Activity Time (Weeks)
1–4	Fabricate parts using the procured materials.	3
2–5	Assemble the components that are received from contractors and obtain subassembly A.	2
3–7	Assemble the purchased components to get subassembly B.	2
4–7	Assemble the fabricated parts and obtain subassembly B.	2
5–6	Inspect subassembly A.	1
6–8	Assemble the inspected works to obtain the final unit.	2
7–8	Assemble the final unit using subassembly B.	1
8–9	Inspect the final unit.	2
9–10	Deliver the tested unit.	1

FIGURE 8.7   Data for the production plan.

An activity is represented by a line connecting two control points. Each activity is identified by two control points, the starting and ending points. The program chart is drawn starting at the point of delivery and moving backwards to the beginning of the production process, from right to left.

The control points are numbered so that the starting point of any activity has a smaller index number than its ending index number. The starting point of the activity to be started earliest among all activities is given the index number 1. One way of numbering the events is to identify the control points in the program chart, as shown in Figure 8.8, starting from left to right and from top to bottom. Normally, symbols, colors, and letters are used to represent different types of operations such as purchasing materials, furnishing parts by contractors, assembly, and delivery. A horizontal scale drawn at the bottom of the program chart indicates the flow time factors from the delivery of the item to the beginning of the production process. The horizontal scale is numbered from right to left. Hence, the lead time for any control point denoting the number of time units from the final control point can be read on the time scale by observing where the vertical line through that control point intersects the scale. For example, in Figure 8.8, the lead time for the control point 7 is 4 weeks.

The Program Progress

The third step of the LOB study is to construct the program progress chart. It specifies the status of actual performance at the control points. The data for drawing the chart are obtained by taking a physical inventory of the quantities of materials, components, subassemblies, and finished units that have passed through the control points at the time of study. The heights of bars in the chart indicate the actual performances in cumulative number of completed units for the key points in the production plan.

The vertical axis of the chart is set adjacent to the objective chart, as shown in Figure 8.9. The vertical axis is drawn for the same scale as that of the objective chart. The horizontal axis denotes the control points.

For our illustration, the status of actual performance at the end of the 17th week is the following:

Control Point	Units Completed (Cumulative)
1	150
2	100
3	90
4	85
5	100
6	80
7	80
8	55
9	42
10	26

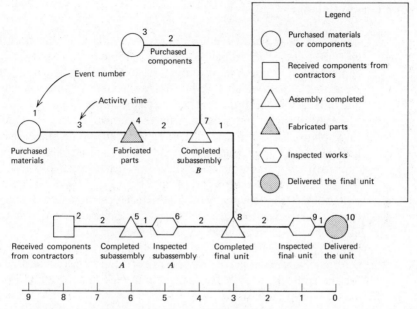

FIGURE 8.8   Program chart.

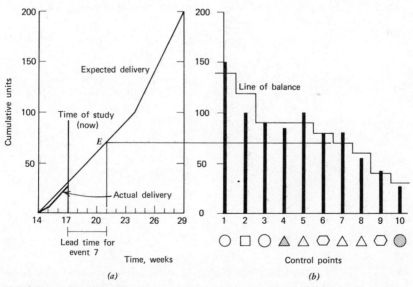

FIGURE 8.9   (a) **Objective chart and** (b) **program progress chart for the line of balance study.**

The Line of Balance

The last step is to draw the line of balance on the program chart. It represents the number of completed units that should have passed through each control point at the time of study in order to deliver the completed units according to the contract schedule. For each control point, the value for the line of balance that indicates the expected completed units is determined.

Let us illustrate the procedure for the control point 7 (completed subassembly B). We determine the lead time for key point 7 from the production plan, shown in Figure 8.8. We note that the lead time is 4 weeks. The cumulative units completed at the point 7 at the end of the 17th week would be delivered by the end of the 21st week (17 + 4). The number of units planned for delivery by the 21st week is determined from the objective chart. A vertical line is drawn for 21st week in the objective chart, and it intersects the expected delivery line at the point E in Figure 8.9. For the point E, the expected delivery is 70 units. Next, we draw a horizontal line through E and, thus, perceive its position on the progress chart, Figure 8.9, for the control point 7. The production performance for the point 7 is ahead of expected quantity by 10 units (= 80 − 70). This procedure is utilized for all other control points. The values are shown in Table 8.3 and in Figure 8.9.

TABLE 8.3   Comparison of Actual Performance with Contract Delivery Schedule

Control Point	Planned for Delivery, Week Number	Cumulative Units to Be Delivered (Planned)	Units Actually Completed	Deviation from Planned Units
1	26	140	150	+10
2	25	120	100	−20
3	23	90	90	0
4	23	90	85	−5
5	23	90	100	+10
6	22	80	80	0
7	21	70	80	+10
8	20	60	55	−5
9	18	40	42	+2
10	17	30	26	−4

We observe that the performance of control points 2, 4, 8, and 10 are behind schedule; hence, we need to look at them more closely and take corrective action.

A Computer Program

We have developed the solution procedure for the line of balance problem in the last section. We shall formulate in this section a computer model for the LOB problems. We might begin the program by giving the following information as input data:

LEVENT = total number of events or control points for the problem
LACTS = total number of activities or operations for the problem
TLOB = time of LOB study

Each one of LACTS activities is identified by its starting and ending key points; also, its estimated processing time is given by the following variables:

IEVENT(K) = starting key point of activity K
JEVENT(K) = ending key point of activity K
TIME(K) = estimated time for activity K

The variables IEVENT( ), JEVENT( ), and TIME( ) are one-dimensional arrays of size LACTS elements.

For each one of LEVENT key points, the actual performance accomplished in cumulative units at time TLOB is stored in the memory location:

ACTUAL(J) = actual performance for key point J

The variable ACTUAL( ) is a one-dimensional array with LEVENT elements.

With the following FORTRAN IV statements we might initialize the program

```
      READ 700, LEVENT, LACTS, TLOB
      DO 10 K = 1, LACTS
      READ 701, IEVENT(K), JEVENT(K), TIME(K)
  10  CONTINUE
      DO 20 J = 1, LEVENT
      READ 750, K, ACTUAL(J)
  20  CONTINUE
 700  FORMAT ( )
 701  FORMAT ( )
 750  FORMAT ( )
```

The estimated delivery for various time points are computed by SUBROUTINE SCHEME, and it is presented later in this section.

The lead-time values for the control points might be determined by the following instructions:

```
K = LEVENT - 1
NEVENT = LEVENT
FINISH(LEVENT) = 0.00
DO 30 J = 1, K
NEVENT = NEVENT - 1
DO 40 I = 1, LACTS
IF (IEVENT(L).NE.NEVENT) GO TO 40
Y = TIME(L)
NODEJ = JEVENT(L)
X = FINISH(NODEJ) + Y
GO TO 44
40  CONTINUE
44  FINISH(NEVENT) = X
30  CONTINUE
```

We note that the lead time for the last event, FINISH (LEVENT) is initialized with a zero. Then, the lead time of event (LEVENT-1) is calculated and subsequently that of events (LEVENT-2), (LEVENT-3), . . . , 2, 1 are determined. The variable FINISH( ) is a one-dimensional array with LEVENT elements.

The SUBROUTINE SCHEME is formulated for computing the expected delivery of end products planned for the time period, TPLAN. For our illustration of the Warren Electrical Industries, the following subroutine might be used to determine PLANED values for any given time period

```
SUBROUTINE SCHEME (TPLAN, PLANED)
X = TPLAN
IF (X.GT.14.0) GO TO 20
PLANED = 0
GO TO 100
20  IF (X.GT.24.0) GO TO 40
PLANED = (X - 14.0) * 10.0
GO TO 100
40  IF (X.GT.29) GO TO 60
PLANED = 100.0 + (X - 24.0) * 20.0
GO TO 100
60  PLANED = 200.0
100  CONTINUE
RETURN
END
```

For any key point M, the expected number of units to be completed, PLAN(M), at time TLOB as well as the deviation, DEVIN(M), of its actual performance from the expected value might be computed by the following statements:

```
TPLAN = TLOB + FINISH(M)
CALL SCHEME (TPLAN, PLANED)
PLAN(M) = PLANED
DEVIN(M) = ACTUAL(M) − PLAN(M)
```

The DEVIN(M) values for each one of LEVENT key points is determined by placing the above segment of statements within the DO loop

```
      DO 50 M = 1, LEVENT
50    CONTINUE
```

The variables PLAN( ) and DEVIN( ) are one-dimensional arrays with LEVENT elements.

The input data and the computed values can be printed out by the following PRINT instructions.

```
      PRINT 700, LEVENT, LACTS, TLOB
      DO 52 L = 1, LACTS
      PRINT 701, IEVENT(L), JEVENT(L), TIME(L)
52    CONTINUE
      DO 54 NEVENT = 1, LEVENT
      PRINT 702, NEVENT, FINISH (NEVENT)
54    CONTINUE
      PRINT 720
      PRINT 100
100   FORMAT (6X, 3HKEY, 18X, 5HUNITS)
      PRINT 110
110   FORMAT (5X, 5HPOINT, 5X, 7HPLANNED, 5X,
      16HACTUAL, 5X, 9HDEVIATION)
      DO 60 J = 1, LEVENT
      PRINT 731, J, PLAN(J), ACTUAL(J), DEVIN(J)
60    CONTINUE
702   FORMAT ( )
731   FORMAT ( )
```

Our complete program is shown in Figure 8.10.

```
         DIMENSION IEVENT(9) , JEVENT(9) , TIME(9)
         DIMENSION ACTUAL(10) , PLAN(10) , DEVIN(10) , FINISH(10)
C THE READ AND PRINT FORMAT STATEMENTS ARE CALLED OUT.
   699 FORMAT(1H1)
   700 FORMAT(2I10,F11.0)
   701 FORMAT(2I10,F11.0)
   750 FORMAT(I10,F12.0)
   702 FORMAT(5X,I3,F5.0)
   720 FORMAT (//)
   731 FORMAT(5X,I5,3F12.0)
C THE VALUES FOR LEVENT, LACTS,TLOB ARE READ IN.
C LEVENT = TOTAL NUMBER OF EVENTS OR CONTROL POINTS.
C LACTS = TOTAL NUMBER OF ACTIVITIES.
C TLOB = TIME OF LOB STUDY.
         READ 700, LEVENT,LACTS,TLOB
         DO 10 K=1,LACTS
C THE VALUES FOR IEVENT,JEVENT,TIME ARE READ IN.
         READ 701, IEVENT(K), JEVENT(K), TIME(K)
    10 CONTINUE
         DO 20 J=1,LEVENT
         READ 750,K,ACTUAL(J)
    20 CONTINUE
C THE LEAD TIME VALUES FOR THE CONTROL POINTS ARE DETERMINED BY THE
C        FOLLOWING POINTS.
         K=LEVENT-1
         NEVENT=LEVENT
C THE FOLLOWING STATEMENTS WILL INITIALIZE THE PROGRAM.
         FINISH (LEVENT)=0.00
C THE LEAD TIME AND ESTIMATED TIME ARE ADDED TOGEATHER.
         DO 30 J=1,K
         NEVENT=NEVENT-1
         DO 40 L=1,LACTS
         IF (IEVENT (L).NE.NEVENT) GO TO 40
         Y= TIME (L)
         NODEJ=JEVENT (L)
         X=FINISH (NODEJ)+Y
         GO TO 44
    40 CONTINUE
    44 FINISH (NEVENT)=X
    30 CONTINUE
         DO 50 M=1,LEVENT
         TPLAN=TLOB + FINISH (M)
C THE SUBROUTINE SCHEME IS CALLED FOR.
         CALL SCHEME (TPLAN,PLANED)
         PLAN(M)=PLANED
C DEVIN = THE DEVIATION.
         DEVIN (M)=ACTUAL (M) - PLAN(M)
    50 CONTINUE
C THE GIVEN AND COMPUTED VALUES ARE PRINTED OUT.
         PRINT 699
         PRINT 700,LEVENT,LACTS,TLOB
         DO 52 L=1,LACTS
         PRINT 701, IEVENT(L),JEVENT(L),TIME(L)
    52 CONTINUE
         DO 54 NEVENT=1,LEVENT
         PRINT 702, NEVENT,FINISH(NEVENT)
    54 CONTINUE
         PRINT 720
         PRINT 100
   100 FORMAT (6X,3HKEY,18X,5HUNITS)
         PRINT 110
   110 FORMAT (5X,5HPOINT,5X,7HPLANNED,5X,6HACTUAL,5X,9HDEVIATION)
         DO 60 J=1,LEVENT
         PRINT 731,J,PLAN(J),ACTUAL(J),DEVIN(J)
    60 CONTINUE
         STOP
         END
```

FIGURE 8.10   Computer model for the LOB problems (illustrated for Warner Electrical
Industries problem).

```
            SUBROUTINE SCHEME (TPLAN,PLANED)
            X=TPLAN
            IF (X.GT.14.0) GO TO 20
            PLANED=0
            GO TO 100
         20 IF (X.GT.24.0) GO TO 40
            PLANED= (X-14.0) * 10.0
            GO TO 100
         40 IF (X.GT.29.0) GO TO 60
            PLANED = 100.0 + (X-24.0) * 20.0
            GO TO 100
         60 PLANED = 200.0
        100 CONTINUE
            RETURN
            END
```

FIGURE 8.10 (*Continued*)

An Illustration of the Computer Program

Let us use the computer program shown in Figure 8.10 to solve our numerical example of Warner Electrical Industries. There are nine activities and 10 control points for the problem. Hence, we prepare the DIMENSION statements as follows:

DIMENSION IEVENT(9), JEVENT(9), TIME(9)
DIMENSION ACTUAL(10), PLAN(10), DEVIN(10), FINISH(10)

The first data card might be read according to the FORMAT statement:

700   FORMAT (2I10, F11.0)

The values of IEVENT(K), JEVENT(K), and TIME(K) for each activity may be read employing the following FORMAT statement:

701   FORMAT (2I10, F11.0)

There are nine cards in the data deck; each card gives information for one activity. The next set of data cards indicates the control indexes and their associated actual performances. These values may be read according to the FORMAT statement:

750   FORMAT (I10, F12.0)

In this set there are 10 data cards, each one giving the values for one event.
   The lead-time values of each control point might be printed out employing the FORMAT statement:

702   FORMAT (5X, I3, F5.0)

The values of expected completed units, actual performance, and the deviation from the target quantities might be printed out according to the FORMAT statement:

731   FORMAT (5X, I5, 3F12.0)

The resulting data deck for our problem has the following appearance:

10	9	17.
1	4	3.
2	5	2.
3	7	2.
4	7	2.
5	6	1.
6	8	2.
7	8	1.
8	9	2.
9	10	1.
1	150.	
2	100.	
3	90.	
4	85.	
5	100.	
6	80.	
7	80.	
8	55.	
9	42.	
10	26.	

10	9	17.
1	4	3.
2	5	2.
3	7	2.
4	7	2.
5	6	1.
6	8	2.
7	8	1.
8	9	2.
9	10	1.

1	9.
2	8.
3	6.
4	6.
5	6.
6	5.
7	4.
8	3.
9	1.
10	0

KEY POINT	PLANNED	UNITS ACTUAL	DEVIATION
1	140.	150.	10.
2	120.	100.	-20.
3	90.	90.	0
4	90.	85.	-5.
5	90.	100.	10.
6	80.	80.	0
7	70.	80.	10.
8	60.	55.	-5.
9	40.	42.	2.
10	30.	26.	-4.

FIGURE 8.11   Computer printout for the Warner Electrical Industries problem.

Figure 8.11 shows the resulting computer printout for the Warner Electrical Industries problem. Note that the computer printout values are the same as those obtained by our hand-computational method shown in Table 8.3.

PROBLEMS

8.1   An aircraft manufacturing company produces helicopters of a particular model. The amount of labor time required to produce the first unit was 2000 man-hours. The learning rate is 90 percent. Determine the learning curve formula. Calculate the estimated production time for the sixth unit.

8.2   The Reinhold Enterprises has been assembling color television sets since the third week of April. The data for the first two weeks of production are given below:

Time of Production (April)	Television Sets Assembled	Man-Hours of Labor
Third Week	150	710
Fourth Week	200	700

What is the learning curve equation for the assembly process? For the month of May, if the expected manhours is 3000, how many sets will be assembled?

8.3 A corporation has produced a total number of 500 refrigerators by the end of December, utilizing a total of 3750 man-hours. The production process has a learning rate of 83 percent. For each month in the following year (January to December), the available labor time is 7500 man-hours. Write a FORTRAN program to compute and print out the expected completed units for each month in the following year.

8.4 The Alpha Epsilon Industries has received a contract for the manufacture and delivery of castings to XYZ Company. The estimated direct labor time for the first unit is 480 man-hours with a learning rate of 78 percent. For storing each completed unit, 300 square feet of warehouse floor space are required. The shipping department wishes to determine the requirement of warehouse floor space for each week from the first through the twentieth week. The castings are shipped to XYZ company at the end of each week. The direct labor time available during each week is expected to be 1000 man-hours. Prepare a FORTRAN program for computing and printing out the number of units to be produced and the warehouse space required each week for the 20-week planning period.

8.5 An aircraft manufacturing corporation wishes to prepare its bid for the supply of 10 units of aircrafts to the ABC airlines. The estimated cost of the Xth unit is computed by applying the following expression:

ECXTH = DLC + MC + FBC

where

ECXTH = estimated cost of Xth unit
DLC = direct labor cost for Xth unit
MC = material cost
FBC = factory burden cost

The three cost components are obtained as follows:

DLC = (direct labor hours for the Xth unit). (Direct labor hourly rate) By applying expression 8.7, we can get the direct labor hours. The production time for the first unit is 95,000 hours and the learning rate is 90 percent. Direct labor hourly rate = $5.

MC = $100,000 per unit
FBC = 20 percent of DLC

The sales price of each unit is the average estimated cost plus 25 percent.

Write a FORTRAN program that will cause the computer to calculate and print out the ECXTH value for the 10 units, the average estimated cost per unit, and the unit sales price.

8.6 The Fournier Industries has received a contract for supplying 500 mobile homes. The delivery schedule is to supply 20 units each week beginning the 10th week and continuing through the 34th week. The production plan is shown in the figure.

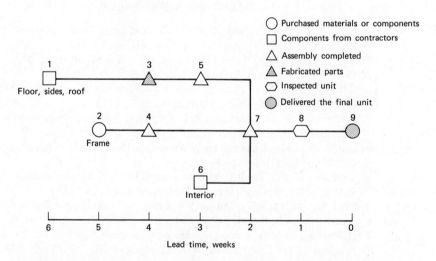

An LOB study is to be performed at the end of the 16th week. The number of completed units passed through each control point is as follows:

Control Point	Units
1	228
2	259
3	212

Control Point	Units
4	223
5	198
6	215
7	190
8	140
9	130

Draw the objective chart, the program progress chart, and the line of balance for the LOB chart. Determine the deviation of units for the key points.

8.7 Modify the computer model presented in Figure 8.10 so that the computer will determine the deviation values for the control points of Problem 8.6.

8.8 Write a program for the data of Problem 8.6 to determine the planned units and cumulative units for the key points for each week beginning with the third week and continuing through the 35th week.

―――――――――――――― **SELECTED REFERENCES** ――――――――――――――

Andress, F. J., "The Learning Curve As a Production Tool," *Harvard Business Review*, Vol. 32, No. 1, January–February 1954, pp. 87–97.

Blair, C., "The Learning Curve Gets an Assist from the Computer," *Management Review*, Vol. 57, No. 8, August 1968, pp. 31–37.

Department of the Navy, *Line of Balance Technology*, Office of Naval Material, Government Printing Office, Washington, D.C., 1962.

Fabrycky, W. J., P. M. Ghare, and P. E. Torgersen, *Industrial Operations Research*, Prentice-Hall, Englewood Cliffs, N.J., 1972.

Faught, A. N., "Managements' Early Warning System," *Business Automation*, Vol. 16, No. 3, March 1969, pp. 46–51.

Hirschmann, W. B., "Profit from the Learning Curve," *Harvard Business Review*, Vol. 42, No. 1, January–February 1964, pp. 125–139.

Holtz, J. N., *An Analysis of Major Scheduling Techniques in the Defense Systems Environment*, The RAND Corporation, RM-4697-PR, October 1966. Reprinted as Reading 32 of *Systems, Organizations, Analysis, Management—A Book of Readings*, D. I. Cleland and W. R. King, McGraw-Hill, New York, 1969.

Iannone, A. L., *Management Program Planning and Control with PERT, MOST and LOB*, Prentice-Hall, Englewood Cliffs, N.J., 1967.

Riggs, J. L., *Production Systems: Planning, Analysis, and Control*, Wiley, New York, 1970.

CHAPTER 9

inventory planning-deterministic models

To INSURE the smooth functioning of operating systems, the needed materials must be delivered in the right quantity at the right time. The availability of materials to be planned and controlled includes the following three kinds:

1. Purchased components and raw materials
2. In-process materials
3. Finished products

Most business enterprises invest a large amount of their capital in the inventories of these materials. For example, the Eastman Kodak Company had about $570,000,000 invested at the end of 1971 (this amount is approximately 33 percent of its assets), and the Republic Steel Corporation had over $280,000,000 of its capital tied up in inventories (this is about 50 percent of its assets) at the end of 1971. Because of the huge capital investment, management often faces the problem of planning and controlling inventories to minimize the total inventory cost.

THE NATURE OF INVENTORY PROBLEMS

The objectives of various divisions associated with the inventory system conflict with one another in planning the inventories. For the raw materials or components to be procured from outside sources, the purchasing division desires to procure these materials in large quantities so that the number of orders to be prepared and followed up will be minimal. For the components and finished products manufactured within the company, the production division wishes to produce large lot sizes in each production run so that the total time expended to set up the production line for the planning period is relatively small. On the other hand, the finance division would like to place orders or produce the items in small quantities so that the total capital outlay in inventories is small. The marketing division desires to have a large amount of units stocked in the warehouse so that the chance of running out of the materials is negligible. Therefore, in establishing inventory plans, the costs incurred, related to inventory, by these divisions are to be considered. It is important to develop inventory plans that will be desirable to the whole organization. We shall first describe the cost items and then present solution procedures to the inventory problems.

The cost items pertinent to inventory management can be classified into three groups:

1. Costs depending on number of lots.
2. Costs of carrying inventory.
3. Costs of not carrying inventory

The Costs Depending on the Number of Lots   These include costs associated with procuring the goods. The cost items taken into account in this group are those that are required in every order:

Clerical costs of filing and reviewing orders, as well as paying the bills.
Cost of related paper works, such as inspection and quality control documents.
Material handling costs.
Machine setup costs and the related expenses.

These costs depend on the number of orders placed over the given planning period, say one year, and do not normally depend on the size of orders. This cost group is also known as *procurement costs* or *ordering costs*.

The Costs of Carrying Inventory   These costs vary with the amount of inventory. The following cost items are included in this group:

Cost of capital invested in the inventory
Insurance
Taxes
Storage costs such as heating, maintenance, and supplies
Obsolescence and depreciation

Spoilage
Pilferage and other similar types of costs

This cost group is also called *the holding costs.*

The Costs of Not Carrying Inventory These occur whenever the supply of materials runs out before the demand for these products is satisfied. Two types of costs are incurred for not carrying inventory. If the products are not available when demanded by the customers or production division, the required products may be procured through rush orders. The additional costs of making them available on a short notice belongs to the group. On the other hand, if the required products cannot be obtained through emergency procedures, the customers are dissatisfied. These dissatisfied customers may go to a competitor and change their loyalty to the competitor's product. In such cases, not only are profits from the demanded products not realized, but future sales are also lost. If the requirements of the production division are not satisfied, the production is interrupted; idle labor and machine costs are incurred. These costs are to be included in the costs of not carrying inventory. This cost group is called *stockout costs.* Because of stockout, the opportunity to deliver the goods to the customers or production division is lost; hence, this group is also called the *opportunity costs.*

Many suppliers encourage their buyers to order in large lot sizes by offering *quantity discounts.* In inventory planning, the benefits of utilizing quantity discounts should be evaluated.

THE BASIC INVENTORY MODEL

We shall develop the basic inventory model in which the daily demand for the product is known and stockout of the goods is not allowed. It is also assumed that the unit cost of the items is constant for any lot size; this indicates that quantity discounts are not available. To illustrate the computational procedure, let us consider the following facts for freezers.

The Scheiper Enterprises procures and sells freezers, Model F2020.

1. Ordering cost per order is $180.00.
2. Unit cost of the freezer is $1000.00.
3. Holding cost per freezer is equal to 25 percent of the average yearly inventory.
4. Total demand per year is 2500 units. Number of working days per year is 250. Daily demand is constant.

5. Procurement lead time is two days. The procurement lead time is the period between the time when an order is placed and the time when the shipment of the ordered goods is received.
6. Stockout of freezers is not permitted.

The management of Scheiper Enterprises wishes to determine the inventory policy that minimizes the total yearly inventory cost for freezers.

The total yearly inventory cost can be computed in the following way:

$$\text{Total yearly inventory cost} = \text{Total ordering cost per year}$$
$$+ \text{Total holding cost per year} \quad (9.1)$$

We have mentioned the term "inventory policy" without really stating what it is. The inventory policy for any material should give information for the two questions in terms of time and quantity:

(a) *When* should the inventory be replenished?
(b) *How much* of the material should be ordered whenever an order is placed?

These two questions of "when?" and "how much?" are usually answered in the following ways:

(a) When? The material should be ordered when the amount in inventory reaches a level of JROPT units. JROPT is the reorder point at optimum.
(b) How much? The quantity to be ordered is KQOPT units of the product. KQOPT is the optimum lot size.

The optimum values correspond to the inventory decision alternative whose associated total inventory cost is lower than that of any other alternative. Let us illustrate the calculation steps for obtaining the total inventory cost for the alternative with a lot size of 50 units.

The inventory level of freezers for this alternative is shown in Figure 9.1. The reorder point is set at such an inventory level so that when the ordered units arrive, the inventory is zero. This will ensure that the average inventory is small, resulting in a lower total inventory holding cost.

$$\text{Procurement lead time} = 2 \text{ days}$$

$$\text{Daily demand} = \frac{\text{Total yearly demand}}{\text{Number of working days per year}}$$

$$= \frac{2500}{250} = 10 \text{ units}$$

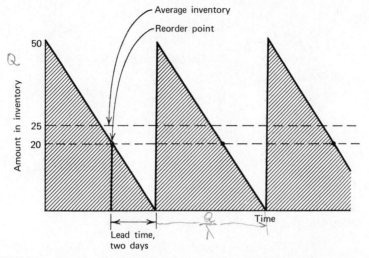

FIGURE 9.1 Graphical representation of inventory fluctuations for the basic model.

Therefore, demand during lead time or reorder point (JROPT)

$$= \text{Daily demand} \times \text{Procurement lead time} \quad (9.2)$$
$$= 10 \times 2 = 20 \text{ units}$$

$$\text{Number of orders placed in a year} = \frac{\text{Total demand per year}}{\text{Lot size}} \quad (9.3)$$

Total ordering cost per year can be written:

$$= (\text{Number of orders placed in a year}) (\text{Ordering cost per order}) \quad (9.4)$$

For the alternative of 50 units lot size,

$$\text{Number of orders placed in a year} = \frac{2500}{50} = 50 \text{ orders}$$

$$\text{Total ordering cost per year} = (50 \text{ orders})(\$180.00)$$
$$= \$9000.00$$

Total holding cost per year can be expressed as follows:

$$= \begin{pmatrix} \text{Average} \\ \text{inventory} \end{pmatrix} \cdot \begin{pmatrix} \text{Unit} \\ \text{cost} \end{pmatrix} \cdot \begin{pmatrix} \text{Inventory holding cost} \\ \text{as a decimal value} \end{pmatrix} \quad (9.5)$$

The average inventory, as shown in Figure 9.1, can be expressed:

$$\text{Average inventory} = \frac{\text{Maximum inventory} + \text{Minimum inventory}}{2} \quad (9.6)$$

Since

$$\text{Maximum inventory} = \text{Lot size,}$$

and

$$\text{Minimum inventory} = 0$$

we have

$$\text{Average inventory} = \frac{\text{Lot size}}{2} \qquad (9.7)$$

For our alternative of 50 units lot size,

$$\text{Total holding cost per year} = \left(\frac{50}{2}\right) \cdot (\$1000.00)(0.25)$$

$$= \$6250.00$$

Substituting in (9.1) we obtain

$$\text{Total yearly inventory cost} = \$9000.00 + \$6250.00$$

$$= \$15,250.00$$

Table 9.1 gives the total inventory costs for various alternatives of lot sizes, and Figure 9.2 shows the cost values as a function of lot sizes.

TABLE 9.1   Total Inventory Cost per Year as a Function of Various Lot Sizes

Lot Size, Units	Number of Orders Per Year	Total Ordering Cost per Year, $	Average Inventory, Units	Total Holding Cost per Year, $	Total Inventory Cost, $	
10	250	45,000	5	1,250	46,250	
20	125	22,500	10	2,500	25,000	
30	83.3	15,000	15	3,750	18,750	
40	62.5	11,250	20	5,000	16,250	
50	50	9,000	25	6,250	15,250	
60	41.7	7,500	30	7,500	15,000	←Minimum
70	35.7	6,428.57	35	8,750	15,178.57	
80	31.3	5,625	40	10,000	15,625	
90	27.8	5,000	45	11,250	16,250	
100	25	4,500	50	12,500	17,000	
110	22.7	4,090.91	55	13,750	17,840.91	
120	20.8	3,750	60	15,000	18,750	
130	19.2	3,461.54	65	16,250	19,711.54	

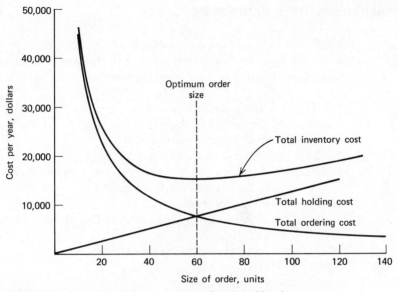

FIGURE 9.2   Cost patterns for freezers as a function of lot sizes.

We observe the following characteristics of the cost values:

When the lot size increases, the number of orders placed per year decreases. As a consequence, the total ordering cost decreases. However, when the lot size increases, the average inventory also increases. As a result, the total inventory holding cost is increased. The total inventory cost continues to decrease until some order quantity is reached; beyond this point the total cost increases.

The total inventory cost is minimum at the lot size of 60 units and the related inventory cost is \$15,000.00; hence the inventory policy will be:

The optimum reorder point is 20 units (JROPT = 20) and the optimum lot size is 60 units (KQOPT = 60).

We have established the optimum inventory policy by calculating the total inventory costs for numerous values of lot sizes and selecting the one with the smallest total cost. We shall develop the two methods to determine the optimum policy and formulate computer programs for each one of them:

Method 1: General solution procedure
Method 2: Solution through the economic lot size formula

GENERAL SOLUTION PROCEDURE

We have seen that the total inventory cost for any lot size can be computed utilizing the mathematical relationships (9.1), (9.4), and (9.5). The following notations will be used in developing the models:

NEEDY = yearly demand in units
KDAYS = number of working days per year
OCOST = ordering cost per order
UCOST = unit cost of the product
FACTOR = inventory holding cost per unit expressed as a decimal
of the average yearly inventory
Q   = lot size in units
TCOSTO = total inventory cost computed in previous iteration (for
$Q - 1$ units)
TCOST = total inventory cost for Q units
TOCOST = total ordering cost
THCOST = total holding cost
KQOPT = optimum lot size in units
JROPT = reorder point at optimum
TCOPT = total inventory cost at optimum
LTIME = procurement lead time in days

From (9.3) we get:

$$\text{Number of orders placed in a year} = \left(\frac{\text{NEEDY}}{Q}\right)$$

We obtain the total ordering cost per year from (9.4),

$$\text{TOCOST} = \left(\frac{\text{NEEDY}}{Q}\right) \cdot (\text{OCOST})$$

The total holding cost per year can be obtained utilizing (9.5) and (9.7),

$$\text{THCOST} = \left(\frac{Q}{2}\right) \cdot (\text{UCOST}) \cdot (\text{FACTOR})$$

Now, the total inventory cost by (9.1) is

$$\text{TCOST} = \text{TOCOST} + \text{THCOST}$$

We have developed the steps to determine the total inventory cost for a given lot size. The characteristic of the total inventory cost is that it continues to decrease until the optimum order size is reached; beyond this amount the total cost increases. Therefore, the computational procedure can be stated as follows:

STEP 1. Obtain the values of NEEDY, OCOST, UCOST, FACTOR, KDAYS, and LTIME.

STEP 2. Initialize the values of the two variables as follows:

$$TCOSTO = 999999.99 \text{ (a very large value)}$$

$$Q = 10 \text{ (a small order quantity)}$$

STEP 3. Compute TCOST for Q.

STEP 4. Compare TCOST with TCOSTO. If the value of TCOSTO is less than TCOST, go to Step 6. Otherwise go to Step 5.

STEP 5. Set TCOSTO = TCOST. Increase Q by 1. Go to Step 3.

STEP 6. Set KQOPT = Q − 1 and TCOPT = TCOSTO.

STEP 7. Compute JROPT from (9.2).

STEP 8. Print out the input and computed values.

A flowchart for the computational steps is shown in Figure 9.3. A FORTRAN program that would do the calculations and print out for us is shown in Figure 9.4. The input data card for our problem has the following appearance:

```
        2500  180.00    .25 250   21000.00
```

The computer printout is given below:

```
     2500 180.00    .25 250  21000.00


OPTIMUM INVENTORY POLICY
        REORDER POINT                  =    20 UNITS
        ORDER QUANTITY                 =    60 UNITS
        TOTAL ANNUAL INVENTORY COST=$      15000.00
```

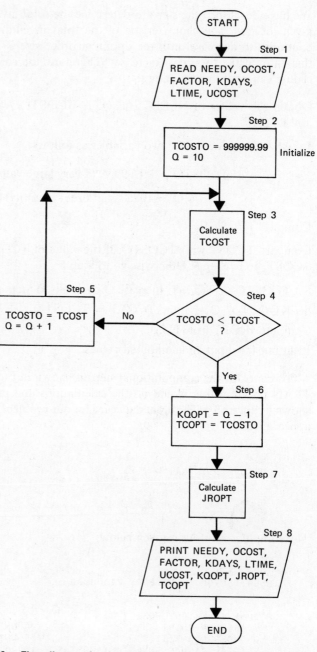

FIGURE 9.3   Flow diagram for the general solution procedure of the basic inventory model.

```
C   READ THE GIVEN VALUES
        READ   700, NEEDY,OCOST,FACTOR,KDAYS,LTIME,UCOST
        TCOSTO = 999999.99
        Q = 10
        YNEED = NEEDY
C
C   CALCULATE TCOST FOR ASSIGNED Q VALUE
    100 TCCOST =(YNEED/Q)* OCOST
        HCOST = UCOST * FACTOR
        THCOST =(Q/2.0)* HCOST
        TCOST =TOCOST + THCOST
        IF (TCOSTO .LT. TCOST ) GO TO 150
        TCOSTO = TCOST
        Q = Q+1.0
        GO TO 100
    150 KQOPT = Q - 1.0
        TCOPT= TCOSTO
C   VALUES OF KQOPT AND TCOPT HAVE BEEN DETERMINED
C
C   CALCULATE JROPT VALUE
        F=LTIME
        DAYS = KDAYS
        JROPT =(YNEED/DAYS)* F
C
C   PRINTOUT THE GIVEN AND COMPUTED VALUES
        PRINT 700, NEEDY,OCOST,FACTOR,KDAYS,LTIME,UCOST
        PRINT 830
        PRINT 860, JROPT
        PRINT 840, KQOPT
        PRINT 880, TCOPT
C
    700 FORMAT ( 5X,I5,F7.2,F6.2,I4,I3,F7.2 )
    830 FORMAT ( //, 5X,25HOPTIMUM INVENTORY POLICY )
    840 FORMAT ( 10X, 14HORDER QUANTITY, 13X, 1H=,I6, 6H UNITS )
    860 FORMAT ( 10X, 13HREORDER POINT, 14X, 1H=, I6, 6H UNITS )
    880 FORMAT ( 10X, 29HTOTAL ANNUAL INVENTORY COST=$, F11.2)
        STOP
        END
```

FIGURE 9.4 Computer program for the general solution procedure of the basic inventory model.

SOLUTION THROUGH THE ECONOMIC LOT SIZE FORMULA

Expression 11.1 for the total inventory cost is:

TCOST = TOCOST + THCOST

Substituting the expressions (9.4) and (9.5) in the above relationship, we get

$$TCOST = \left(\frac{NEEDY}{Q}\right) \cdot (OCOST) + \left(\frac{Q}{2}\right) \cdot (UCOST) \cdot (FACTOR) \quad (9.8)$$

In this function, all of these are parameters except TCOST and Q. TCOST and Q are the variables whose values are to be determined; Q is an independent variable and, TCOST is a dependent variable. The general shape of the function has the appearance shown in Figure 9.2. The point on the curve corresponding to the minimum total inventory cost indicates the optimum lot size. This lot size (KQOPT) can be determined by differentiating the function (9.8) with respect to Q, setting it equal to zero, and solving for Q.

$$\frac{d(\text{TCOST})}{dQ} = -\frac{(\text{NEEDY})(\text{OCOST})}{Q^2} + \frac{(\text{UCOST})(\text{FACTOR})}{2}$$

$$-\frac{(\text{NEEDY})(\text{OCOST})}{Q^2} + \frac{(\text{UCOST})(\text{FACTOR})}{2} = 0$$

$$Q = \sqrt{\frac{2 \cdot (\text{NEEDY}) \cdot (\text{OCOST})}{(\text{UCOST})(\text{FACTOR})}}$$

Hence

$$\text{KQOPT} = \sqrt{\frac{2(\text{NEEDY}) \cdot (\text{OCOST})}{(\text{UCOST}) \cdot (\text{FACTOR})}} \tag{9.9}$$

Expression 9.9 is called the *economic lot size formula*. By substituting the value of KQOPT for Q in expression 9.8, we can calculate the minimum total inventory cost. Figure 9.5 depicts a FORTRAN program that would compute and print out the optimum inventory policy utilizing the economic lot size formula (9.9).

The data card is the same as that of method 1. The computer printout is shown below:

```
2500   180.00   .25   250   21000.00
OPTIMUM INVENTORY POLICY
      REORDER POINT                 =   20 UNITS
      ORDER QUANTITY                =   60 UNITS
      TOTAL ANNUAL INVENTORY COST = $15000.00
```

In method 1, the total inventory cost is calculated for numerous order sizes. For each one of the incremented lot sizes, the total inventory cost is decreasing and this continues until we reach the optimum order size. For the next larger order size, the total cost increases, indicating that the previous order size is the optimum. However, in method 2, we directly

```
C   READ THE GIVEN VALUES
        READ  700, NEEDY,OCOST,FACTOR,KDAYS,LTIME,UCOST
C
C   CALCULATE KQOPT VALUE
        HCOST = UCOST * FACTOR
        YNEED = NEEDY
        V = (2.0*YNEED*OCOST) / HCOST
        KQOPT = SQRT (V)
C
C   CALCULATE TCOPT VALUE
        Q = KQOPT
        TOCOST =(YNEED/Q)* OCOST
        THCOST = (Q/2.0) * HCOST
        TCOST =TOCOST + THCOST
        TCOPT = TCOST
C
C   CALCULATE JROPT VALUE
        F=LTIME
        DAYS = KDAYS
        JROPT =(YNEED/DAYS)* F
C
C   PRINTOUT THE GIVEN AND COMPUTED VALUES
        PRINT 700, NEEDY,OCOST,FACTOR,KDAYS,LTIME,UCOST
        PRINT 830
        PRINT 860, JROPT
        PRINT 840, KQOPT
        PRINT 880, TCOPT
C
    700 FORMAT ( 5X,I5,F7.2,F6.2,I4,I3,F7.2 )
    830 FORMAT ( //, 5X,25HOPTIMUM INVENTORY POLICY )
    840 FORMAT ( 10X, 14HORDER QUANTITY, 13X, 1H=,I6, 6H UNITS )
    860 FORMAT ( 10X, 13HREORDER POINT, 14X, 1H=, I6, 6H UNITS )
    880 FORMAT ( 10X, 29HTOTAL ANNUAL INVENTORY COST=$, F11.2)
        STOP
        END
```

FIGURE 9.5   Computer program for the basic inventory model through the economic lot size formula approach.

obtain the optimum order size by utilizing the economic lot size formula. Therefore, method 2 requires less computational effort, and it should be useful especially when a company is establishing optimum inventory policies for numerous materials.

It is given in the basic inventory model that the unit cost is a constant value, and quantity discounts are not allowed. Consequently, the total annual cost of the items is a constant value. The lot sizes have no effect on it. However, if quantity discounts are offered, the total cost of the material will not be a constant anymore. It depends upon the lot sizes. By placing orders for large lot sizes, the quantity discounts can be obtained, and it will give rise to low total annual cost for the items. When quantity discounts are encountered, the cost of the materials should be included with other costs in the analysis. The inventory model with quantity discounts is the subject of discussion in the following section.

QUANTITY DISCOUNTS

It may be recalled that in calculating the total yearly inventory cost in our basic inventory model, according to the expression 9.1, we have included the total ordering cost and the total holding cost. In the basic model, the cost of materials per year has not been taken into account because a single unit price prevails for all order quantities and, therefore, the total material cost is a constant amount. However, for quantity discounts, low unit prices are offered for large order sizes. Consequently, the total annual material cost is not a constant and it depends upon the order size. When quantity discounts are offered, the total yearly inventory cost comprises, therefore, the ordering cost—the holding cost together with the material cost:

$$
\begin{aligned}
\text{Total yearly inventory cost} = \ &\text{Total ordering cost per year} \\
&+ \text{Total holding cost per year} \\
&+ \text{Total material cost per year} \quad (9.10)
\end{aligned}
$$

Analysis and Solution Procedure

Let us consider the single-price break schedule: If $0 < Q < b_1$, the unit price is C_1 and if $b_1 \leqq Q$, the unit price is C_2 and also that $C_1 > C_2$ where b_1 is the price break point. Expression 9.10 can be stated thus:

$$
\text{TCOST} = \left(\frac{\text{NEEDY}}{Q}\right) \cdot (\text{OCOST}) + \left(\frac{Q}{2}\right) \cdot (\text{UCOST}) \cdot (\text{FACTOR})
$$
$$
+ (\text{UCOST}) \cdot (\text{NEEDY}) \quad (9.11)
$$

The general shape of the TCOST curves for the unit prices of C_1 and C_2 is shown in Figure 9.6. For various order sizes, the TCOST can be calculated applying expression 9.11; the TCOST curves are then drawn. The feasible region for each of these prices refers to the range such that the order sizes fall within the limits specified in the price break schedule. In Figure 9.6, the feasible region is shown by a continuous line, and the infeasible portions are depicted by dotted lines. The minimum total inventory cost for a single-price break system may occur as any one of the four alternatives shown in Figure 9.6. It may be the lowest total inventory cost point at price C_1 (Figure 9.6a), the lowest total inventory cost point at price C_2 (Figure 9.6b), the price break point b, (Figure 9.6c), or the lowest total inventory cost point at price C_1 as well as the price break point b_1 (Figure 9.6d).

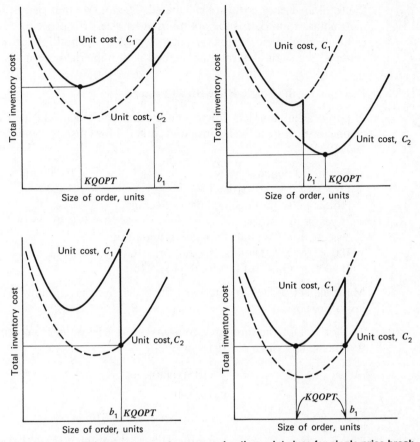

FIGURE 9.6   **Total inventory cost curves and optimum lot sizes for single-price break systems.**

The decision process for determining the optimum size in the presence of the price discounts is quite complex. The total inventory costs for numerous order quantities can be calculated, and the cost curves are drawn on a graph; then, utilizing the graph the optimum order quantity can be established. However, such an approach would be laborious, if not impracticable, especially in cases in which policies are to be determined for many products with quantity discounts.

An efficient method is to analyze the minimum total cost points for various unit prices as well as the price break points. The economic order quantities for each one of the unit prices are directly computed through expression 9.9. Among these quantities, the total inventory costs are determined only for the feasible lot sizes. Also, for the price break points, the total inventory costs are calculated. The lot size that corresponds to the smallest total cost is the optimum order size. This procedure is applicable to single-price break systems as well as to multiple-price break systems. We shall illustrate the method with numerical examples.

Example for the Single-Price Break System

A corporation markets home and garden equipment through several chain stores. Here is the information gathered for finishing sanders Model 1007:

Estimated demand per year	$= 10,500$ units
Ordering cost per order	$= \$40.00$
Holding cost as a decimal of the average yearly inventory	$= 0.30$
Assume that the daily demand is a constant	
If $0 < Q < 2000$ units, the unit price (C_1)	$= \$10.00.$
If $2000 \leq Q$, the unit price $(C_2) = \$9.50.$	

Management wishes to determine the order size that minimizes the trial inventory cost.

The lot size for the minimum total inventory cost point from expression 9.9 for the unit price of $10.00 is

$$KQOPT = \sqrt{\frac{2(10,500)(40)}{(10.00)(0.30)}}$$

$$= 530 \text{ units per order at } \$10.00 \text{ per unit}$$

This is a feasible order quantity. Hence, the total inventory cost from (9.11) is

$$TCOST = \left(\frac{10,500}{530}\right) 40 + \left(\frac{530}{2}\right)(10)(0.30) + (10)(10,500)$$

$$= \$106,590.00$$

For the unit price of $9.50, from (9.9) we have

$$KQOPT = \sqrt{\frac{2(10,500)(40)}{(9.50)(0.30)}}$$

$$= 543 \text{ units per order at } \$9.50 \text{ per unit}$$

The order size of 543 units is less than the price-break point (2000 units); hence, it is an infeasible plan. Therefore, we do not calculate the total cost for this lot size. The total cost for the price-break point from (9.11) is

$$TCOST = \left(\frac{10,500}{2,000}\right)40 + \left(\frac{2000}{2}\right)(9.50)(0.30) + (9.50)(10,500)$$

$$= \$102,660.00$$

Comparing the values of TCOST for Q = 530 and 2000 units, we conclude that the optimum order size is 2000 units. We observe that the nature of the cost curve would be similar to the curve shown in Figure 9.6c.

Example for the Multiple-Price Break System

The Sheppard Enterprises procures and sells hardware goods. Data for the dinnerware, Model 207 are given below:

Expected sales per year = 2500 units
Ordering cost per order = \$12.50
Holding cost as a decimal of
  the average yearly inventory = 0.25
Number of working days per year = 250
Lead time, days = 3

The product can be bought according to any of the three prices and the price schedule is:

Unit Price	Lot Size, Units
\$4.00 ($=C_1$)	1–259
\$3.60 ($=C_2$)	260–999
\$3.24 ($=C_3$)	1000 and above

Daily demand can be considered a constant value. Determine the inventory policy that will yield a minimum total inventory cost.

For the unit price of \$4.00, we obtain from (9.9)

$$KQOPT = \sqrt{\frac{2(2500)(12.5)}{(4.00)(0.25)}} = 250 \text{ units}$$

This is a feasible solution since the suppliers will be willing to sell the lot size of 250 units at \$4.00 unit price. The total inventory cost from expression 9.11 is

$$TCOST = \frac{2500}{250}(12.5) + \left(\frac{250}{2}\right)(4.00)(0.25) + (4.00)(2500)$$

$$= \$10,250.00$$

For the unit price of $3.60, we obtain from (9.9)

$$KQOPT = \sqrt{\frac{2(2500)(12.5)}{(3.60)(0.25)}} = 263 \text{ units}$$

This is also a feasible inventory plan and therefore, the associated total inventory cost from expression 9.11 is

$$TCOST = \frac{2500}{263}(12.5) + \left(\frac{263}{2}\right)(3.60)(0.25) + (3.60)(2500)$$

$$= \$9237.17$$

For the unit price of $3.24, we obtain from (9.9)

$$KQOPT = \sqrt{\frac{2(2500)(12.5)}{(3.24)(0.25)}} = 277 \text{ units}$$

The suppliers of the product will not accept the order with the lot size of 277 units for the unit price of $3.24. This is an infeasible inventory plan. Hence, we do not compute the total inventory cost for $Q = 277$ units. There are two price-break points for this product, namely, 260 units and 1000 units.

For the lot size of 260 units, we find the total inventory cost from (9.11).

$$TCOST = \left(\frac{2500}{260}\right)(12.5) + \left(\frac{260}{2}\right)(3.60)(0.25) + (3.60)(2500)$$

$$= \$9237.19$$

For the lot size of 1000 units, the total inventory cost from (9.11) is

$$TCOST = \left(\frac{2500}{1000}\right)(12.5) + \left(\frac{1000}{2}\right)(3.24)(0.25) + (3.24)(2500)$$

$$= \$8536.25$$

The computed values are depicted in Figure 9.7.

The total inventory cost is the smallest for the lot size of 1000 units. The order quantity at optimum, therefore, is 1000 units. The reorder point is obtained from expression 9.2.

$$\text{Reorder point} = \left(\frac{2500}{250}\right)(3) = 30 \text{ units}$$

The optimum inventory policy is: Whenever the inventory reaches 30 units, place an order for 1000 units of the dinnerware.

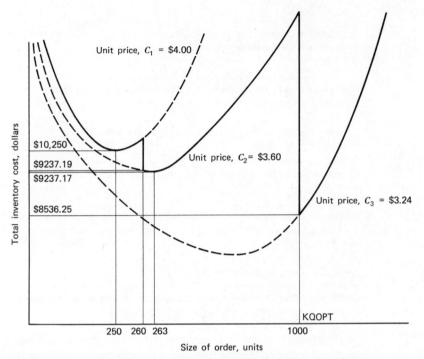

FIGURE 9.7 Total inventory cost curve for the multiple-price break problem.

A Computer Program

We shall formulate a computer program for the multiple-price break systems. This program is also applicable for the single-price break systems. The computational steps for establishing the optimum inventory policy are shown in Figure 9.8. Many variables used in the computer model are already introduced in the basic inventory model. Let us present the additional variables used in this computer model:

UCOST(K) = Kth highest unit price of the product.

IRANGE(K) = the smallest order quantity that can be procured at unit price UCOST(K). This is a price break point.

JRANGE(K) = the largest order quantity that can be purchased at unit price UCOST(K).

RANGEQ(K) = economic lot size for the unit price of UCOST(K).

RANGEC(K) = total inventory cost for the economic lot size of RANGEQ(K).

STEP 1

Calculate the economic lot size for each one of the unit prices. For the feasible lot sizes, determine the associated total inventory cost.

STEP 2

Choose the smallest total cost among those costs obtained in Step 1.

STEP 3

Compute the total inventory costs for the price break points.

STEP 4

Choose the smallest total cost among those costs obtained in Step 3.

STEP 5

Compare the two selected values, the one in Step 2 and the other in Step 4. Choose the smallest of them and its order quantity. This is the optimum order quantity.

STEP 6

Determine the optimum reorder point.

FIGURE 9.8 **Steps for determining the optimum inventory plan for quantity discount problems.**

$$CBREAK(K) = \text{total inventory cost for the price break point } IRANGE(K).$$

LEVELS = number of unit prices offered by the suppliers.

The variables UCOST( ), IRANGE( ), JRANGE( ), RANGEQ( ), RANGEC( ), and CBREAK( ) are one-dimensional arrays of size LEVELS elements.

TUCOST = total cost of the products per year.

We can read the input data by the FORTRAN IV statements.

```
        READ 700, NEEDY, OCOST, FACTOR, KDAYS, LTIME
        READ 720, LEVELS
        DO 10 K = 1, LEVELS
        READ 740, IRANGE(K), UCOST(K)
        IF (K.EQ.1) GO TO 10
        J = K - 1
        JRANGE(J) = IRANGE(K) - 1
   10   CONTINUE
        JRANGE(LEVELS) = 1000000
  700   FORMAT (5X, I5, F7.2, F6.2, I4, I3)
  720   FORMAT (10X, I5)
  740   FORMAT (I5, F5.2)
```

We observe that the JRANGE( ) values are computed from IRANGE( ) by the expression:

JRANGE(K − 1) = IRANGE(K) − 1

For the least unit price level, the variable JRANGE( ) is set a very large amount such as 1000000 so that the related economic lot size is less than this set amount.

We might initialize the value of RANGEC( ) with the following statements:

```
      DO 20 L = 1, LEVELS
      RANGEC(L) = 999999.99
  20  CONTINUE
```

The economic lot sizes for each one of the LEVELS prices and the total inventory costs, if feasible, are calculated by the following instructions:

```
      YNEED = NEEDY
      DO 30 J = 1, LEVELS
      HCOST = UCOST(J) * FACTOR
      V = (2.0 * OCOST * YNEED)/HCOST
      QC = SQRT(V)
      RANGEQ(J) = QC
      IF (QC.LT.IRANGE(J)) GO TO 30
      IF (QC.GT.JRANGE(J)) GO TO 30
      TOCOST = (YNEED/QC) * OCOST
      THCOST = (QC/2.0) * HCOST
      TUCOST = YNEED * UCOST(J)
      RANGEC(J) = TOCOST + THCOST + TUCOST
  30  CONTINUE
```

Note that if any economic lot size is not feasible, the total inventory cost is not calculated. Hence, for the infeasible economic lot sizes, the RANGE(J) values would still be the initialized values of 999999.99.

The statements below would cause the computer to select the price level having the economic lot size with the smallest total inventory cost.

```
      BIGVAL = 999999.99
      DO 40 N = 1, LEVELS
      IF (BIGVAL.LE.RANGEC(N)) GO TO 40
      BIGVAL = RANGEC(N)
      JRC = N
  40  CONTINUE
```

The economic lot size associated with the price level JRC has the smallest total inventory cost and the cost is BIGVAL.

The next step is to calculate the total inventory cost for the price break points and to choose the one that has the smallest total cost. The instructions for accomplishing this are given below:

```
        K = 2
 50     QNOW = IRANGE(K)
        HCOST = UCOST(K) * FACTOR
        TOCOST = (YNEED/QNOW) * OCOST
        THCOST = (QNOW/2.0) * HCOST
        TUCOST = YNEED * UCOST(K)
        CBREAK(K) = TOCOST + THCOST + TUCOST
        K = K + 1
        IF (K.LE.LEVELS) GO TO 50
        BIGNO = 999999.99
        DO 70 J = 2, LEVELS
        IF (BIGNO.LE.CBREAK(J)) GO TO 70
        BIGNO = CBREAK(J)
        JBC = J
 70     CONTINUE
```

The chosen price break point is IRANGE(JBC) and the smallest total cost is CBREAK(JBC).

The following statements would cause the computer to compare RANGEC(JRC) and CBREAK(JBC), select the smallest among them, and establish the values of KQOPT, TCOPT, and UCOPT:

```
        IF (RANGEC(JRC).GT.CBREAK(JBC)) GO TO 76
        KQOPT = RANGEQ(JRC)
        TCOPT = RANGEC(JRC)
        UCOPT = UCOST(JRC)
        GO TO 78
 76     KQOPT = IRANGE(JBC)
        TCOPT = CBREAK(JBC)
        UCOPT = UCOST(JBC)
 78     CONTINUE
```

We could calculate the reorder point thus:

```
        F = LTIME
        DAYS = KDAYS
        JROPT = (YNEED/DAYS) * F
```

The following instructions would output the input data and the desired results:

```
                PRINT 700, NEEDY, OCOST, FACTOR, KDAYS, LTIME
                PRINT 720, LEVELS
                DO 90 K = 1, LEVELS
                PRINT 760, IRANGE(K), JRANGE(K), UCOST(K)
        90  CONTINUE
                PRINT 802
                PRINT 800
                DO 100 L = 1, LEVELS
                LQ = RANGEQ(L)
                UCL = UCOST(L)
                TC = RANGEC(L)
                IF (RANGEC(L).EQ.999999.99) GO TO 95
                PRINT 780, LQ, UCL, TC
                GO TO 100
        95  PRINT 790, LQ, UCL
       100  CONTINUE
                PRINT 804
                PRINT 800
                DO 120 N = 2, LEVELS
                LQ = IRANGE(N)
                UCL = UCOST(N)
                TC = CBREAK(N)
                PRINT 780, LQ, UCL, TC
       120  CONTINUE
                PRINT 830,
                PRINT 860, JROPT
                PRINT 840, KQOPT
                PRINT 880, TCOPT
                PRINT 990, UCOPT

    C
       760  FORMAT (15X, 218, F7.2)
       780  FORMAT (5X, I8, 5X, F7.2, 5X, F9.2)
       790  FORMAT (5X, I8, 5X, F7.2, 5X, 12HNOT FEASIBLE)
       800  FORMAT (5X, 10HORDER SIZE, 3X, 9HUNIT COST,
                73X, 10HTOTAL COST)
       802  FORMAT (/, 4X, 25HCALCULATIONS FOR ECONOMIC,
                316H LOTSIZE POINTS)
```

```
804   FORMAT (/, 4X, 17HCALCULATIONS FOR,
      218HPRICE-BREAK POINTS)
830   FORMAT (//, 5X, 25HOPTIMUM INVENTORY POLICY)
840   FORMAT (10X, 14HORDER QUANTITY, 13X,
      51H =, I6, 6H UNITS)
860   FORMAT (10X, 13HREORDER POINT, 14X,
      41H =, I6, 6H UNITS)
880   FORMAT (10X, 22HTOTAL ANNUAL INVENTORY,
      67H COST = $, F11.2)
900   FORMAT (10X, 19HSELECTED UNIT PRICE, 8X, 2H = $,
      3F11.2)
      STOP
      END
```

A complete computer program composed of our previously designed program segments for solving our multiple-price break problem is shown in Figure 9.9.

For our multiple-price break system example problem, the data deck has the following appearance:

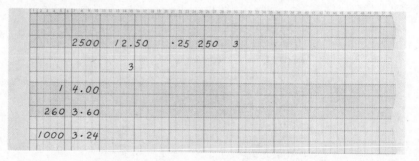

The resulting computer printout is shown in Figure 9.10. We have presented in this section inventory models in which the daily demand and the lead time are constant. This assumption is valid in many cases. However, in numerous realistic situations, the daily demand and the lead time are not constant and should be considered so as to take a range of daily demand and lead-time values. Such types of inventory models are discussed in Chapter 13.

```
      DIMENSION IRANGE(3),JRANGE(3),UCOST(3)
      DIMENSION RANGEQ(3),RANGEC(3),CBREAK(3
C
C  READ THE GIVEN VALUES
      READ  700, NEEDY,OCOST,FACTOR,KDAYS,LTIME
      READ 720, LEVELS
      DO 10 K = 1,LEVELS
      READ 740, IRANGE(K), UCOST(K)
C  CALCULATE JRANGE VALUES
      IF ( K .EQ. 1 ) GO TO 10
      J = K - 1
      JRANGE(J) =  IRANGE(K) - 1
   10 CONTINUE
      JRANGE(LEVELS) = 1000000
C
C  INITIALIZE RANGEC VALUES
      DO 20 L = 1,LEVELS
      RANGEC(L) = 999999.99
   20 CONTINUE
C
C  CALCULATE RANGEQ AND RANGEC VALUES
      YNEED = NEEDY
      DO 30 J = 1,LEVELS
      HCOST = UCOST(J) * FACTOR
      V = (2.0*OCOST*YNEED)/HCOST
      QC= SQRT(V)
      RANGEQ(J) = QC
      IF ( QC .LT. IRANGE(J) ) GO TO 30
      IF ( QC .GT. JRANGE(J) ) GO TO 30
      TOCOST = (YNEED/QC) * OCOST
      THCOST = (QC/2.0) * HCOST
      TUCOST = YNEED * UCOST(J)
      RANGEC(J) = TOCOST + THCOST + TUCOST
   30 CONTINUE
C
C  FIND OUT THE SMALLEST RANGEC VALUE
      BIGVAL = 999999.99
      DO 40 N= 1,LEVELS
      IF ( BIGVAL .LE. RANGEC(N)) GO TO 40
      BIGVAL  =  RANGEC(N)
      JRC = N
   40 CONTINUE
C
C  COMPUTE CBREAK VALUES
      K = 2
   50 QNOW = IRANGE(K)
      HCOST= UCOST(K) * FACTOR
      TOCOST= (YNEED/QNOW) * OCOST
      THCOST= (QNOW/2.0) * HCOST
      TUCOST= YNEED * UCOST(K)
      CBREAK(K)= TOCOST + THCOST + TUCOST
      K = K+1
      IF ( K.LE. LEVELS ) GO TO 50
```

FIGURE 9.9   Computer program for determining the optimum inventory plan for the Sheppard Enterprises problem.

```
C
C   FIND OUT THE SMALLEST CBREAK VALUE
        BIGNO = 999999.99
        DO 70 J = 2,LEVELS
        IF ( BIGNO.LE. CBREAK(J)) GO TO 70
        BIGNO = CBREAK(J)
        JBC = J
    70 CONTINUE
C
C   ESTABLISH KQOPT,TCOPT, AND UCOPT
        IF ( RANGEC(JRC).GT.CBREAK(JBC)) GO TO 76
        KQOPT = RANGEQ (JRC)
        TCOPT = RANGEC(JRC)
        UCOPT = UCOST(JRC)
        GO TO 78
    76 KQOPT = IRANGE(JBC)
        TCOPT = CBREAK(JBC)
        UCOPT = UCOST(JBC)
    78 CONTINUE
C
C   COMPUTE JROPT VALUE
        F = LTIME
        DAYS = KDAYS
        JROPT = (YNEED/DAYS)*F
C
C   PRINTOUT THE GIVEN VALUES
        PRINT 700, NEEDY,OCOST,FACTOR,KDAYS,LTIME
        PRINT720, LEVELS
        DO 90 K = 1,LEVELS
        PRINT 760, IRANGE(K), JRANGE(K), UCOST(K)
    90 CONTINUE
C
C   PRINTOUT THE CALCULATED VALUES
        PRINT 802
        PRINT 800
        DO 100 L = 1,LEVELS
        LQ = RANGEQ(L)
        UCL = UCOST(L)
        TC = RANGEC(L)
        IF ( RANGEC(L) .EQ. 999999.99 ) GO TO 95
        PRINT 780, LQ,UCL,TC
        GO TO 100
    95 PRINT 790, LQ,UCL
   100 CONTINUE
        PRINT 804
        PRINT 800
        DO 120 N = 2,LEVELS
        LQ = IRANGE(N)
        UCL = UCOST(N)
        TC = CBREAK(N)
        PRINT 780, LQ,UCL,TC
   120 CONTINUE
        PRINT 830
        PRINT 860, JROPT
        PRINT 840, KQOPT
        PRINT 880, TCOPT
        PRINT 900, UCOPT
```

FIGURE 9.9 (*Continued*)

```
C
   700 FCRMAT ( 5X,I5,F7.2,F6.2,I4,I3 )
   720 FCRMAT(10X,I5)
   740 FORMAT ( I5, F5.2 )
   760 FCRMAT(15X,2I8,F7.2)
   780 FCRMAT(5X,I8,5X,F7.2,5X,F9.2)
   790 FCRMAT ( 5X, I8, 5X, F7.2, 5X, 12HNOT FEASIBLE )
   800 FCRMAT(5X,10HORDER SIZE,3X,9HUNIT COST,3X,10HTOTAL COST)
   802 FCRMAT ( /, 4X, 41HCALCULATIONS FOR ECONOMIC LOTSIZE POINTS )
   804 FCRMAT ( /, 4X, 36HCALCULATIONS FOR PRICE-BREAK POINTS )
   830 FCRMAT ( //, 5X,25HOPTIMUM INVENTORY POLICY )
   840 FCRMAT ( 10X, 14HORDER QUANTITY, 13X, 1H=,I6, 6H UNITS )
   860 FCRMAT ( 10X, 13HREORDER POINT, 14X, 1H=, I6, 6H UNITS )
   880 FCRMAT ( 10X, 29HTOTAL ANNUAL INVENTORY COST=$, F11.2)
   900 FORMAT ( 10X, 19HSELECTED UNIT PRICE, 8X,2H=$,F11.2 )
       STOP
       END
```

FIGURE 9.9 (*Continued*)

```
   2500   12.50    .25 250   3
                3
                    1      259    4.00
                  260      999    3.60
                 1000  1000000    3.24

   CALCULATIONS FOR ECONOMIC LOTSIZE POINTS
     ORDER SIZE    UNIT COST    TOTAL COST
         250         4.00        10250.00
         263         3.60         9237.17
         277         3.24        NOT FEASIBLE

   CALCULATIONS FOR PRICE-BREAK POINTS
     ORDER SIZE    UNIT COST    TOTAL COST
         260         3.60         9237.19
        1000         3.24         8536.25


     OPTIMUM INVENTORY POLICY
         REORDER POINT              =      30 UNITS
         ORDER QUANTITY             =    1000 UNITS
         TOTAL ANNUAL INVENTORY COST=$     8536.25
         SELECTED UNIT PRICE        =$        3.24
```

FIGURE 9.10   Computer printout for the Sheppard Enterprises problem.

PROBLEMS

9.1   The data for product P106 is given below:

$\lambda =$ Annual usage $= 36,000$ units
Number of working days per year $= 250$
Lead time $= 3$ days
$a =$ Ordering cost $= \$25$ per order
$c =$ Unit cost of the product $= \$1$
$I =$ Holding cost $= 20$ percent of the average yearly inventory value

Use Golden Ratio   L ⟹ 0 - 56K , determine MIN Q.

Assume that the daily demand is a constant. Stockout of the product is not permitted. Calculate the total inventory cost per year for the 7 successive lot sizes, ranging from 1000 units per order through 7000 units per order by increments of 1000 units. Present your calculated values in the tabular form, as shown in Table 9.1. Determine the optimum inventory policy using the total inventory cost values.

9.2   Write a computer program for Problem 9.1.

9.3   Consider the data given in Problem 9.1. Establish the optimum inventory policy using the economic lot size formula.

9.4   Determine the optimum inventory policy for each of the two products:

	Product A	Product B
Ordering cost per order	$25	$20
Unit cost	$50	$10
Holding cost of the product per unit for one year as a percent of the unit cost	10%	20%
Lead time, days	4	3
Number of working days per year	300	300
Annual usage, units	27,300	500

9.5   Prepare a program that will cause the computer to calculate and print out the optimum inventory policy for each one of the five items:

IDNO	1007	8693	7200	2469	3218
NEEDY	4000	10,000	1000	6000	3000
KDAYS	250	250	250	250	250
OCOST	$120	$340	$600	$480	$430
UCOST	$50	$15	$120	$79	$60
FACTOR	0.10	0.25	0.20	0.15	0.20
LTIME	3 days	5 days	10 days	6 days	4 days

IDNO is the identification number of the item.

9.6 The Arichibald Enterprises procures product 623A and sells in Bowling Green. The annual demand is 18,000 units and the daily demand is a constant. Data for the product is the following:

Ordering cost = $10 per order
Holding cost per unit for one year as a percentage of the unit cost = 20%
Cost per unit = $5
Lead time = 2 days
Number of working days per year = 300

Shortage of the products is not allowed.

(a) The current inventory policy is:
    Lot size = 3000 units
    Reorder point = 120 units
    Determine the total inventory cost for the current policy.
(b) Establish the optimum inventory policy using the economic lot size formula.
(c) If the optimum inventory policy would be implemented, how much money would be saved per year?

9.7 For two products, the data are given below. Determine the optimum inventory policy for each of them.

Product K75:

Preparation cost per order = $50
Annual usage quantity = 8000 units
Holding cost = 70% of average inventory value
Cost per unit = $100 if lot size is less than 1000 units = $95 otherwise
Number of working days per year = 250
Lead time = 10 days

Product L809:

Preparation cost per order = $40
Annual usage quantity = 21,000 units
Holding cost = 30% of average inventory value
Cost per unit = $9.50 if lot size is 2000 units or above = 10.00 otherwise
Number of working days per year = 300
Lead time = 4 days

9.8 The Tooke Aluminum Corporation wants to use a systems approach to determine the inventory policy for the raw material ferrous sulphate. The raw material can be purchased according to the plan:

Cost per ton = \$100 if $Q < 1000$ tons
Cost per ton = \$95 if $Q \geq 1000$ tons

The usage amount is 10,000 tons per year at a uniform rate. If shortage of the raw material is not allowed, what is the optimum inventory policy.

Number of working days per year = 250
Lead time = 10 days
Ordering cost = \$200 per order
Holding cost = 100% of average inventroy value

Calculate the total inventory cost for the optimum plan.

9.9 The Markowitz Retail Stores procures EAZYKLEAN washing machines and sells in New England. Determine the inventory policy stating when and how much to replenish the product in the following situation:

Annual sales = 24,000 units
Ordering cost = \$250 per order
Inventory holding cost = 80% of average inventory value
Cost per unit = \$200 if $Q < 1000$
　　　　　　 = \$175 if $1000 \leq Q < 2000$
　　　　　　 = \$160 if $2000 \leq Q < 3500$
　　　　　　 = \$140 if $Q \geq 3500$
Lead time = 2 days
Number of working days per year = 300

9.10 For the drying machine model 6706M, the data are given below:

Annual sales = 22,500 units
Ordering cost = \$350 per order
Inventory holding cost = 30% of average inventory value
Cost per machine = \$150 if $Q < 2000$
　　　　　　　　 = \$140 if $2000 \leq Q < 4000$
　　　　　　　　 = \$125 if $4000 \leq Q < 7000$
　　　　　　　　 = \$120 if $Q \geq 7000$
Lead time = 3 days
Number of working days per year = 300

Determine the optimum inventory plan.

─────────────── **SELECTED REFERENCES** ───────────────

Ackoff, R. L., and M. W. Sasieni, *Fundamentals of Operations Research*, Wiley, New York, 1968.

Biegel, J. E., *Production Control—A Quantitative Approach*, Prentice-Hall, Englewood Cliffs, N.J., 1963.

Bowman, E. W., and R. B. Felter, *Analysis for Production and Operations Management*, Richard D. Irwin, Homewood, Ill., 1967.

Buffa, E. S., and W. H. Taubert, *Production-Inventory Systems: Planning and Control*, Richard D. Irwin, Homewood, Ill., 1972.

Felter, R. B., and W. C. Dalleck, *Decision Models for Inventory Management*, Richard D. Irwin, Homewood, Ill., 1961.

Groff, G. K., and J. F. Muth, *Operations Management: Analysis for Decisions*, Richard D. Irwin, Homewood, Ill., 1972.

Hadley, G., and T. M. Whitin, *Analysis of Inventory Systems*, Prentice-Hall, Englewood Cliffs, N.J., 1963.

McMillan C., and R. F. Gonzalez, *Systems Analysis: A Computer Approach to Decision Models*, Richard D. Irwin, Homewood, Ill., 1973.

Naddor, E., *Inventory Systems*, Wiley, New York, 1966.

Prince, T. R., *Information Systems for Management Planning and Control*, Richard D. Irwin, Homewood, Ill., 1966.

Wagner, H. N., *Statistical Management of Inventory Systems*, Wiley, New York, 1962.

Whitin, T. M., *The Theory of Inventory Management*, Princeton University Press, Princeton, N.J., 1953.

concepts of risk

CHAPTER 10

WHEN MANAGEMENT can make a perfect forecast of the future, decisions could then be made with certainty. If an accurate knowledge of forecast does not exist, which is true in many situations, the possible outcomes may be considered with a probability of occurrence for each outcome. For example, a businessman says that there is a 90 percent chance of selling 1000 Christmas trees during this Christmas season. When decisions are made concerning a situation for which the consequences of alternatives can be stated in a probabilistic frame of reference, the probability concepts can be of great help to the decision maker. The purpose of this chapter is to present the basic probability concepts and probability distributions that are widely used in the decision models.

PROBABILITY CONCEPTS

We can determine the numerical value of probabilities by the three methods:

1. A prior estimation
2. Experimental estimation
3. Subjective estimation

A Prior Estimation   When a fair coin is tossed, we know that the chance that a head will appear is 1/2. The answer is based on the concept that there is no reason to favor either a head or tail, as the possible outcome of a head or tail should be equally likely to occur. This concept of abstract reasoning is used in the prior estimation of probabilities. This approach is limited to situations where the symmetry of probabilities can be assumed. Other examples of prior estimation are tossing a fair die and drawing from a fair deck of cards.

Experimental Estimation   Another method of determining probabilities is the experimental approach. A large number of trials of the experiment for which probabilities are needed is performed. Then we can calculate the probability applying expression 10.1.

$$\left.\begin{array}{l}\text{Probability of the}\\\text{outcome}\end{array}\right\} = \frac{\text{Number of occurrences of the outcome}}{\text{Number of trials}} \quad (10.1)$$

It indicates that if a fair coin is tossed one million times, heads would appear in about 500,000 trials.

Subjective Estimation   The objective interpretation of probability has been employed in our previously discussed methods by using intuitive judgment and definitive historical information. However, historical information is not available for many management situations. For these situations, probabilities are assigned based on personal judgment. For example, consider that a company is planning to manufacture and market a new product. A machine must be purchased to produce the item. Two kinds of machines, machine A and machine B, fulfill the requirements. One of them, machine A, requires a large capital investment and a small labor expenditure. On the other hand, machine B requires a relatively small capital investment and a much higher labor cost. Among them, A is suitable when sales are high, and B is appropriate when sales are not so high. The company cannot estimate probabilities by the objective approach. In this case, the management may assign probabilities for the different possible sales in the future based on personal judgment or belief about the likelihood of the states of nature and other related actions.

Probability Terms

We shall illustrate the probability terms by numerical examples. Suppose that there are 2 green balls, 3 black balls, and 5 red balls in a box (Figure 10.1). In each trial, the 10 balls are mixed thoroughly, and one of them is picked at random. Because a ball is selected at random, each ball in the box has an equal chance of being picked. After noting which ball it is, it is returned to the box.

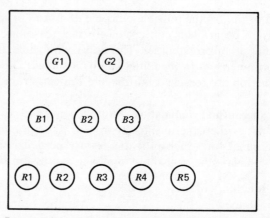

FIGURE 10.1   Box containing 2 green balls, 3 black balls, and 5 red balls.

The possible outcomes of a trial are the following 10 states:

G1	R1
G2	R2
B1	R3
B2	R4
B3	R5

In each trial, one and only one outcome can occur. At any time, two or more events cannot happen. If this condition is satisfied, we say that the events are *mutually exclusive*. We have seen that there are 10 states in the possible outcome. One of these states must occur in any trial and no other outcome is possible. When all the possible outcomes for a given action are listed, the list is said to be *collectively exhaustive*. The set of all possible outcomes of an experiment is often referred to as sample space. The elements of a sample space are the outcomes of the experiment. Each outcome is represented by a sample point within the sample space. Figure 10.2 shows the sample space for our example.

• G1	• R1
• G2	• R2
• B1	• R3
• B2	• R4
• B3	• R5

FIGURE 10.2   A sample space containing the possible outcomes of selecting a ball.

Basic Rules for Assigning Probabilities The three important rules of probability theory are:

RULE 1. The probability of any event is a real number on the interval from 0 to 1.

$$0 \leq P(A) \leq 1 \qquad \text{for each event } A \qquad (10.2)$$

The probability of an event that is certain to occur is 1. The probability of an impossible event is 0.

RULE 2. The sum of probabilities assigned to a set of mutually exclusive and collectively exhaustive events must total 1.

We have for our example

$$\left[\begin{array}{l} P(G1) + P(G2) + P(B1) + P(B2) + P(B3) \\ + P(R1) + P(R2) + P(R3) + P(R4) + P(R5) \end{array} \right] = 1 \quad (10.3)$$

Since each ball has equal likelihood of being selected, we obtain

$$P(G1) = 0.10 \qquad P(R1) = 0.10$$
$$P(G2) = 0.10 \qquad P(R2) = 0.10$$
$$P(B1) = 0.10 \qquad P(R3) = 0.10$$
$$P(B2) = 0.10 \qquad P(R4) = 0.10$$
$$P(B3) = 0.10 \qquad P(R5) = 0.10$$

Note that the probabilities satisfy Rule 1.

RULE 3. The probability of an event composed of two or more mutually exclusive outcomes is equal to the sum of the probabilities of the individual outcomes.

The probability of drawing a green ball shall be

$$P(G) = P(G1) + P(G2)$$
$$= 0.10 + 0.10 = 0.20$$

Likewise, the probability of drawing a black ball or red ball is

$$P(B) = P(B1) + P(B2) + P(B3)$$
$$= 0.30$$

$$P(R) = P(R1) + P(R2) + P(R3) + P(R4) + P(R5)$$
$$= 0.50$$

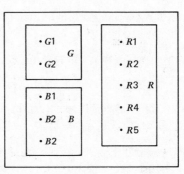

FIGURE 10.3   **A sample space containing the three events of drawing a green ball, a black ball, or a red ball.**

The three events, drawing a green ball, drawing a black ball, or drawing a red ball, are shown in Figure 10.3.

Statistical Independence

If the probability of an event is in no way affected by the occurrence of any other event, the events are said to be independent. For example, in the ball drawing experiment, the probability of drawing a red ball in the second trial is 0.5 regardless of the outcome of the first trial. Therefore, the events that occur in the consecutive trials are independent.

If two or more events are independent, the probability that all these events will occur together or in succession is equal to the product of the individual probabilities. If events A and B are independent of one another, then

$$P(AB) = P(A) \times P(B) \tag{10.4}$$

where

$$P(AB) = \text{probability of events } A \text{ and } B \text{ occurring}$$
$$\text{together or in succession}$$

Expression 10.4 is called *the equation for joint probability*. Consider that we throw a coin and a die simultaneously. The 12 possible outcomes are shown in Figure 10.4. The outcome $H4$ denotes the occurrence of head on the coin and 4 on the die. Each outcome is equally likely. Therefore, we can assign the probability 1/12 for each of them. Assume

$$A = \text{event of occurrence head on the coin}$$

$$B = \text{event of occurrence 5 on the die}$$

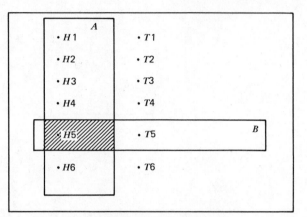

FIGURE 10.4 A sample space containing the possible outcomes of throwing a coin and a die simultaneously.

Figure 10.4 shows event AB.

$$P(AB) = P(H5) = \frac{1}{12}$$

From (10.4)

$$P(AB) = P(H) \times P(5)$$

$$= \left(\frac{1}{2}\right) \times \left(\frac{1}{6}\right) = \frac{1}{12}$$

For our ball drawing example, the probability of drawing a green ball in the first draw, a red ball in the second draw, and a black ball in the third draw $[P(GRB)]$ is depicted in the probability tree of Figure 10.5. From (10.4)

$$P(GRB) = P(G) \times P(R) \times P(B)$$

$$= 0.2 \times 0.5 \times 0.3$$

$$= 0.03$$

In like manner,

$$P(BB) = P(B) \times P(B)$$

$$= 0.3 \times 0.3 = 0.09$$

and

$$P(RG) = P(R) \times P(G)$$

$$= 0.5 \times 0.2 = 0.10$$

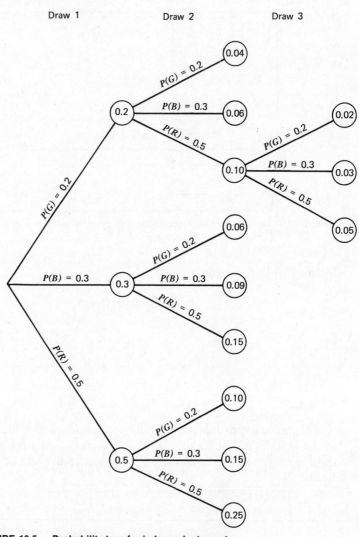

FIGURE 10.5   Probability tree for independent events.

Statistical Dependence

If the probability of an event is affected by the occurrence of any other event, the events are said to be dependent. For example, consider that two machines, $M1$ and $M2$, in a factory produce exactly the same parts and are packed in the same type of boxes. It is known that 20 percent and 3 percent of the produced parts are defective for machines $M1$ and

*M*2, respectively. The experiment consists of selecting a box at random first and then drawing a part from the selected box. The part is tested to see if it is defective. The probability of drawing a defective part is influenced by the previous selection. The events of selecting a box and drawing a defective part are dependent.

Let us illustrate the probability theory for the statistically dependent events by a numerical example. Suppose there are three boxes that contain color balls as follows:

Box 1 ... 2 red and 8 white
Box 2 ... 3 green and 7 yellow
Box 3 ... 6 green and 4 yellow

Assume we draw a ball from box 1 at random. If the ball drawn from box 1 is red, we draw a ball from box 2. If the ball drawn from box 1 is white, we draw a ball from box 3. Figure 10.6 shows the experiment. The probability tree for the game is shown in Figure 10.7.

One of the important relationships is the conditional probability of a dependent event. The relationship, expressed as an equation, is the probability of event *B* given that the event *A* has happened; it is equal to the joint probability of *A* and *B*, divided by the probability of event *A*.

$$P(B/A) = \frac{P(AB)}{P(A)} \tag{10.5}$$

The joint probability of two dependent events is obtained by multiplying both sides of equation 10.5 by $P(A)$ and rewriting it as follows:

$$P(AB) = P(B/A) \cdot P(A) \tag{10.6}$$

Suppose a ball is drawn from box 1 and it is red. Then a ball is drawn from box 2. What is the probability that it is green? Because there are

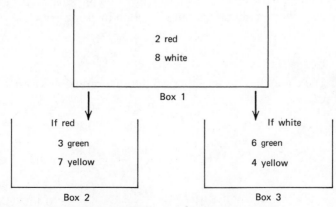

FIGURE 10.6   Three boxes indicating the rules of the experiment.

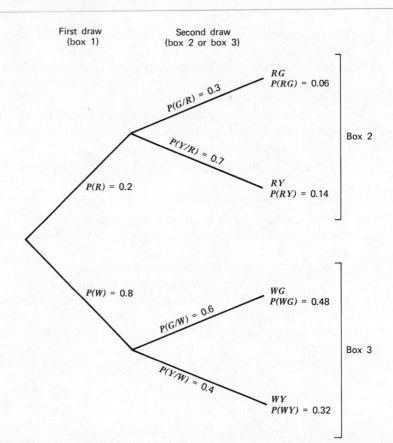

FIGURE 10.7 **Probability tree for the first draw from box 1 and the second draw from boxes 2 and 3.**

3 green balls in a total of 10 balls, we conclude that the probability is 0.30. We can obtain the same solution by using expression 10.5.

$$P(G/R) = \frac{P(RG)}{P(R)} = \frac{0.06}{0.20} = 0.30$$

The computations for other conditional probabilities are as follows:

$$P(Y/R) = \frac{P(RY)}{P(R)} = \frac{0.14}{0.20} = 0.70$$

$$P(G/W) = \frac{P(WG)}{P(W)} = \frac{0.48}{0.80} = 0.60$$

$$P(Y/W) = \frac{P(WY)}{P(W)} = \frac{0.32}{0.80} = 0.40$$

The joint probability values that are given in Table 10.1 are computed by employing equation 10.6.

$$P(RG) = P(G/R) \cdot P(R) = 0.3 \times 0.2 = 0.06$$

$$P(RY) = P(Y/R) \cdot P(R) = 0.7 \times 0.2 = 0.14$$

$$P(WG) = P(G/W) \cdot P(W) = 0.6 \times 0.8 = 0.48$$

$$P(WY) = P(Y/W) \cdot P(W) = 0.4 \times 0.8 = 0.32$$

The marginal probabilities are the unconditional probabilities, and they are given in the margins of the joint probability table. In calculating the marginal probability values, it is assumed that each event stands alone and that the occurrence of any event is not conditional on the occurrence of any other event. $P(R)$ and $P(W)$ are examples of marginal probabilities.

TABLE 10.1   Joint Probability Table

	Second Draw		Marginal Probability of Outcome on
First Draw	G	Y	First Draw
R	$P(RG) = 0.06$	$P(RY) = 0.14$	0.20
W	$P(WG) = 0.48$	$P(WY) = 0.32$	0.80
Marginal probability of outcome on second draw	0.54	0.46	1.00

Example 10.1

An automobile dealer sells three models of automobiles. The number of cars sold are shown in the following table:

Kind of transmission	Model C	Model T	Model W	Total
Standard (S)	100	150	150	400
Automatic (A)	400	120	80	600
Total	500	270	230	1000

If it is known that a car was sold, the probability that it was model C is

$$P(C) = \frac{500}{1000} = 0.50$$

The probability that it had standard transmission is

$$P(S) = \frac{400}{1000} = 0.40$$

The probability that it was model C, given that it had standard transmission is

$$P(C/S) = \frac{100}{400} = 0.25$$

The probability that it had standard transmission, given that it was model C is

$$P(S/C) = \frac{100}{500} = 0.20$$

The probability that it was model C and it had standard transmission is

$$P(CS) = P(S/C) \cdot P(C) = 0.20 \times 0.50 = 0.10$$

or

$$P(CS) = P(C/S) \cdot P(S) = 0.25 \times 0.40 = 0.10$$

PERMUTATIONS AND COMBINATIONS

When a set of items are given, the process of determining the possible number of ways of arranging the items becomes complex. The techniques of counting the arrangements of a set of objects are discussed in this section.

Permutations

To illustrate the arrangement of items, let us consider the example: Three locations $L1$, $L2$, and $L3$ are available in a store to display the three products $P1$, $P2$, and $P3$. The six possible ways of arranging the products are shown in the tree diagram of Figure 10.8. Any one of the three products can be selected for location $L1$. Then two products are available for $L2$. Any one of them may be chosen for $L2$. After the selection for $L2$, the remaining one product is arranged in $L3$. Thus, there are $3 \cdot 2 \cdot 1 = 6$ arrangements or permutations. Generally speaking, if the first operation can be performed in n_1 ways, the second in n_2 ways, the third in n_3 ways, and so on for K operations, then the total of K operations can be performed in $n_1 \cdot n_2 \cdot n_3 \cdots n_K$ ways.

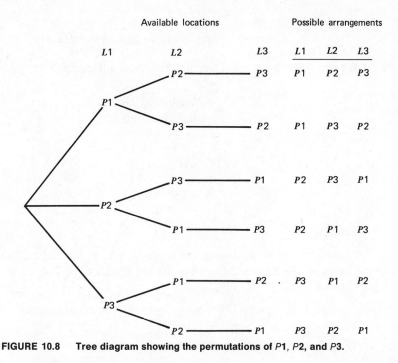

FIGURE 10.8   Tree diagram showing the permutations of P1, P2, and P3.

For example, if there are four routes between cities A and B, and two routes between cities B and C, the number of routes between A and C is $4 \times 2 = 8$, as shown in Figure 10.9.

The number of possible permutations of n different things taken r at a time is

$$_nP_r = \frac{n!}{(n-r)!} \tag{10.7}$$

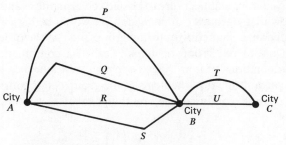

FIGURE 10.9   Highway routes between City A and City C.

where

$$n! = n(n - 1)(n - 2) \cdots (3)(2)(1)$$

We define 0! to be 1. The symbol $n!$ is read n factorial.

Possible Routes
between A and C

PT	PU
QT	QU
RT	RU
ST	SU

Example 10.2

The number of permutations of the four letters A, B, C, and D taken three at a time can be determined employing equation 10.7.

$$_4P_3 = \frac{4!}{(4 - 3)!} = \frac{4 \cdot 3 \cdot 2 \cdot 1}{1} = 24 \text{ arrangements}$$

The arrangements are

ABC	ACB	BCA	BAC	CAB	CBA
ABD	ADB	BDA	BAD	DAB	DBA
ACD	ADC	CAD	CDA	DAC	DCA
BCD	BDC	CBD	CDA	DBC	DCB

Combinations

In some counting situations, order of the items in the groups is not important. Consider that a company has a complex design problem to be solved. Five engineers are to be selected from the available seven engineers. How many groups can be made? We would be basically interested in the individuals that constitute any group and not the order within the group. An unordered group of items is called a combination. The number of combinations of four different items taken four at a time is one because the arrangement of the items in the group is not considered. The number of possible combinations of n things taken r at a time is

$$_nC_r = \binom{n}{r} = \frac{n!}{r!(n - r)!} \tag{10.8}$$

Example 10.3

A trucking company wishes to ship five boxes A, B, C, D, and E from New York to Chicago. Assume all boxes have the same weight and the same sizes. There is place for only three boxes in the truck that leaves today. How many combinations of three boxes are possible?

Using the equation 10.8 we have

$$_5C_3 = \binom{5}{3} = \frac{5!}{3!(5-3)!} = \frac{5 \cdot 4 \cdot 3 \cdot 2 \cdot 1}{(3 \cdot 2 \cdot 1) \cdot (2 \cdot 1)} = 10$$

The combinations are

$$\begin{array}{ccccc} ABC & ABD & ACD & BCD & ABE \\ ACE & BCE & ADE & BDE & CDE \end{array}$$

DECISION TREES

We have used the probability trees to calculate the probabilities of different events. The decision trees are employed to determine the preferred course of actions. In analyzing the various alternatives, we estimate the economic outcomes and the probability factors. We shall represent the decision points by circles and the outcomes by squares in the decision tree. The dotted lines denote the alternatives available and the solid lines signify the effects of selecting the alternatives. The initial decision point is represented on the left side of the decision tree and the subsequent alternatives branch out beginning at the first event. There would be two or more alternatives available at each decision point. The expected value of different actions at each decision point are computed using expression 10.9.

$$\text{Expected value } (EV) = P_1E_1 + P_2E_2 + \cdots + P_jE_j + \cdots + P_nE_n \quad (10.9)$$

where

P_j = probability of occurring the event j

E_j = estimated economic outcome for the event j

and

$$P_1 + P_2 + \cdots + P_n = 1$$

Example 10.4

The Giant Supply Corporation has the option of distributing either product A or product B in Canada. The possible outcomes and probability factors are as follows:

Product	Outcome	Sales Value (in 10⁶ dollars)	Probability
A	Large national demand	10	0.3
	Average national demand	5	0.5
	Limited national demand	1	0.2
			1.0
B	Large national demand	10	0.1
	Average national demand	5	0.7
	Limited national demand	1	0.2
			1.0

Which one of the two products would yield the maximum expected sales?

The calculations depicted in Figure 10.10 indicate that the expected sales are \$5.7 and \$4.7 millions for A and B, respectively. Hence, we conclude that it would be better to distribute product A.

We have considered in Example 10.4 one stage only. When a decision tree has two or more stages, the expected values at each outcome point are calculated. Subsequently, the course of action for each decision point that yields the maximum expected value is determined. This process is called the averaging out and folding back procedure.

Example 10.5

The Barnhart Corporation operates many stores from coast to coast. It wishes to buy and market a particular kind of toy. The decision tree shown in Figure 10.11 shows the data gathered for the two alternatives, buying 25,000 boxes of toys and buying 40,000 boxes of toys. Which of the two alternatives must be recommended? The objective is to implement the alternative that yields the maximum profit.

$$\text{Expected profit of } O_3 = (0.8 \times 800,000) + [0.2 \times (-400,000)]$$
$$= \$560,000$$

$$\text{Expected profit of } O_4 = (0.6 \times 700,000) + (0.4 \times 500,000)$$
$$= 620,000$$

At the decision point D_2, one of the two alternatives must be selected:
Advertise in TV (O_3). This yields an expected profit of \$560,000.
or
Advertise in newspapers (O_4). This yields an expected profit of \$620,000.

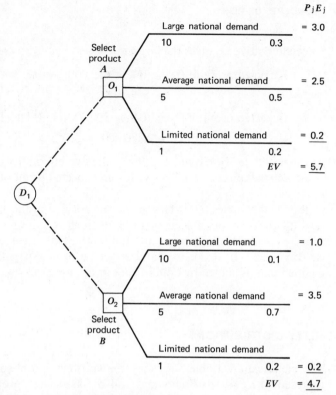

		$P_j E_j$
Large national demand		= 3.0
10	0.3	
Average national demand		= 2.5
5	0.5	
Limited national demand		= 0.2
1	0.2	
	EV	= 5.7

Large national demand		= 1.0
10	0.1	
Average national demand		= 3.5
5	0.7	
Limited national demand		
1	0.2	= 0.2
	EV	= 4.7

Select product A — O_1

Select product B — O_2

D_1

FIGURE 10.10   Decision tree for Example 10.4.

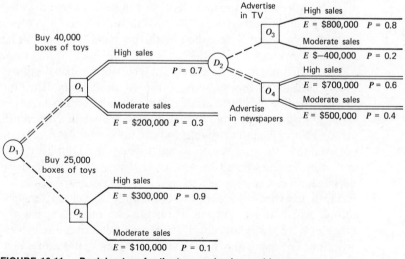

FIGURE 10.11   Decision tree for the toy purchasing problem.

Since alternative O_4 gives the largest expected value, it is selected.

$$\text{Expected profit of } O_1 = (0.7 \times 620{,}000) + (0.3 \times 200{,}000)$$
$$= \$494{,}000$$

$$\text{Expected profit of } O_2 = (0.9 \times 300{,}000) + (0.1 \times 100{,}000)$$
$$= \$280{,}000$$

At the initial decision point D_1, one of the two alternatives is selected:
Buy 40,000 boxes (O_1). This yields an expected profit of \$494,000.
or
Buy 25,000 boxes (O_2). This yields an expected profit of \$280,000.
Among the two possible courses of actions, O_1 gives the maximum expected profit. Hence, choose alternative O_1. Thus, the recommendation is: Buy 40,000 boxes of toys and advertise in newspapers. The recommended plan is shown by double lines in Figure 10.11.

PROBABILITY DISTRIBUTIONS

The significant variables in many management problems can be best represented by a set of outcomes and the associated probability of occurrence. Some examples are demand for inventory items, number of customers serviced in an hour, and number of failed machines in a shift. A probability distribution represents the pattern of the distribution of probabilities over all possible events. It can be employed to determine the likelihood of the occurrence of all possible outcomes. The variable, whose pattern of variation is described by the probability distribution, is called a random variable.

We shall present some important probability distributions that are widely used in systems analysis. A discrete probability distribution is associated with a random variable that can take only a finite number of values. On the other hand, a continuous probability distribution is associated with a random variable that can take on values over a continuum. For example, the number of defective components in a lot size of 1000 units can be represented by a discrete distribution, while the life of an electric bulb can be represented by a continuous distribution.

Mean and variance are the two commonly used characteristics of probability distributions. The mean (arithmetic mean, expected value, or average) is a measure of central tendency, it describes the values of a distribution. The mean may be defined as the sum of the values of a variable divided by the number of values.

$$m = \frac{\sum\limits_{i=1}^{i=n} x_i}{n} \qquad (10.10)$$

where

m = mean of the set of n values

$\sum\limits_{i=1}^{i=n} x_i$ = sum of the values of the variable

n = number of values being averaged

x_i = a given set of n values, such as $x_1, x_2, x_3, \ldots, x_n$

The variance is a measure of dispersion, and it describes the variability of the values in a distribution. The variance is defined as the arithmetic mean of the squared deviations of the individual values from their arithmetic mean.

$$v = \frac{\sum\limits_{i=1}^{i=n} (x_i - m)^2}{n} \qquad (10.11)$$

where

v = variance of the set of n values

m = arithmetic mean of the set of n values

$\sum\limits_{i=1}^{i=n} (x_i - m)^2$ = sum of the squared deviations of the n values from their mean (m)

x_i = a given set of n values, such as $x_1, x_2, x_3, \ldots, x_n$

n = number of values whose variance is determined

Discrete Probability Distributions

Binomial Distribution   For many decision problems, there are only two possible outcomes, such as success or failure, yes or no, and accept or reject. When the probability of either outcome is constant from trial to trial, and the occurrence of outcomes does not adhere to any fixed pattern, we can use the binomial probability distribution. Let

p = probability of success

q = probability of failure = $(1 - p)$

$$n = \text{number of trials considered}$$

$$x = \text{number of successes}$$

$$x = 0, 1, 2, \ldots, n$$

The probability of obtaining exactly x successes in n trials is given as

$$P(x) = {}_nC_x \cdot p^x q^{n-x} \tag{10.12}$$

where

$${}_nC_x = \frac{n!}{x!(n-x)!}$$

$$\text{Mean} = np$$

$$\text{Variance} = npq$$

Example 10.6

A milling machine produces gears that are consistently 20% defective. A sample of six gears is selected at random from the output of the machine. The probability of defective gears occurring in the sample can be computed as follows:

n and p are known as the parameters of the distribution.

Number of trials, n $= 6$

Probability of a success (defective gear), $p = 0.20$

Probability of a failure (acceptable gear), $q = 0.80$

The probability of obtaining exactly two defective items is

$$P(2) = \frac{6!}{2!(6-2)!}(0.2)^2(0.8)^{6-2}$$

$$= 0.2458$$

The likelihood of other outcomes are

$$P(0) = \frac{6!}{0!(6-0)!}(0.2)^0(0.8)^{6-0} = 0.2621$$

$$P(1) = \frac{6!}{1!(6-1)!}(0.2)^1(0.8)^{6-1} = 0.3932$$

Likewise, we can calculate the probability of the events $x = 3, 4, 5,$ and 6 and obtain

$$P(3) = 0.0819 \qquad P(5) = 0.0015$$

$$P(4) = 0.0154 \qquad P(6) = 0.0001$$

Note that the sum of the probabilities equals

$$P(0) + P(1) + P(2) + \cdots + P(6) = 1.000$$

The binomial probability distribution is shown as a histogram in Figure 10.12. The probability of at least three successes is

$$
\begin{aligned}
P(x \geq 3) &= P(3) + P(4) + P(5) + P(6) \\
&= 0.0819 + 0.0154 + 0.0015 + 0.0001 \\
&= 0.0989
\end{aligned}
$$

The probability of more than three successes is

$$
\begin{aligned}
P(x > 3) &= P(4) + P(5) + P(6) \\
&= 0.0154 + 0.0015 + 0.0001 \\
&= 0.0170
\end{aligned}
$$

The probability of less than three successes is

$$
\begin{aligned}
P(x < 3) &= P(0) + P(1) + P(2) \\
&= 0.2621 + 0.3932 + 0.2458 \\
&= 0.9011
\end{aligned}
$$

$$
\begin{aligned}
or\ P(x < 3) &= 1 - P(x \geq 3) \\
&= 1 - 0.0989 \\
&= 0.9011
\end{aligned}
$$

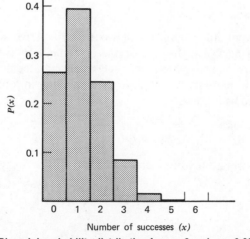

FIGURE 10.12　Binomial probability distribution for $n = 6$ and $p = 0.20$

Poisson Distribution When the probability of an event occurring is extremely small in each trial of an experiment, the Poisson distribution can be found useful. It is assumed that the probability of the event occurring is constant from trial to trial and is independent of the outcomes of any past or future trials. The Poisson distribution can be employed in various areas that include insurance involving the number of accidents in a time period, counting the number of defects in a given length of cable, and the analysis of waiting lines at service facilities.

Let

p = probability of occurring the event

λ = expected or average number of occurrences of the event in n trials ($= np$)

x = number of occurrences

$$x = 0, 1, 2, \ldots, \infty$$

The probability of exactly x occurrences of the event is given as

$$P(x) = \frac{(\lambda)^x e^{-\lambda}}{x!} \tag{10.13}$$

where

$$e = 2.718282 = \text{base of natural logarithms}$$

$$\text{Mean} = \lambda$$

$$\text{Variance} = \lambda$$

Example 10.7

A recent study indicates that customers arrive at a checkout stand of a supermarket at the rate of one customer per five minutes. Assume that the arrival rate pattern is closely approximated by a Poisson distribution. The probability of two customers arriving in four minutes can be determined as follows:

$$\lambda = np = 4(1/5)$$
$$= 0.8 \text{ customers}$$

The probability of exactly two customers arriving at the stand is

$$P(2) = \frac{(0.8)^2 (2.718282)^{-0.8}}{2!}$$

$$= 0.1438$$

The calculation of probability according to the binomial distribution is laborious when n is large and p is small. In those situations, the Poisson distribution may be applied as an approximation to the binomial distribution, especially for $pn < 5$.

Continuous Probability Distributions

Exponential Distribution   We have shown in the previous section that a Poisson distribution can be employed to find the likelihood of the occurrence of a specified number of events (x) in a finite interval (n) of time or space. The exponential distribution is applicable where we study the time or space between two successive occurrences of the event. The random variable (t) in the exponential distribution represents the time or space interval that separates two successive events.

The exponential probability distribution is given by

$$f(t) = \frac{1}{a} e^{-t/a} \tag{10.14}$$

where

$$a = \frac{1}{\text{Average number of successes per interval}}$$

$e = 2.718282 = $ base of natural logarithms

$t = $ number of intervals $0 \leqq t \leqq \infty$

Mean $= a$

Variance $= a^2$

The probability that an event will not occur in t intervals can be obtained by using the relationship (10.15)

$$P(t) = e^{-t/a} \tag{10.15}$$

Example 10.8

The Lipson Company manufactures minicomputers. The product is very reliable, with an average of only two breakdowns every 10,000 hours of use. The breakdowns are known to be Poisson distributed. A high school planning to purchase a minicomputer is interested in the probability distribution of the length of time the product is expected to operate before a breakdown occurs.

The probability distribution of the time interval between breakdowns can be represented by an exponential distribution. Consider one time interval equals 10,000 hours.

$$\text{Mean} = a = \tfrac{1}{2}$$

Therefore, the probability distribution function is

$$f(t) = 2e^{-t/0.5}$$

The ordinates of the curve displayed in Figure 10.13 denote the associated $f(t)$ values. The $f(t)$ values are computed as follows:

Time Interval t	Elapsed Time (Hours) $t \times 10{,}000$	Ordinate (height) $f(t) = 2(2.7183)^{-t/0.5}$	
0.0	0	$2(2.7183)^{-0.0/0.5}$	$= 2.00000$
0.5	5,000	$2(2.7183)^{-0.5/0.5}$	$= 0.73576$
1.0	10,000	$2(2.7183)^{-1.0/0.5}$	$= 0.27068$
1.5	15,000	$2(2.7183)^{-1.5/0.5}$	$= 0.09958$
2.0	20,000	$2(2.7183)^{-2.0/0.5}$	$= 0.03664$

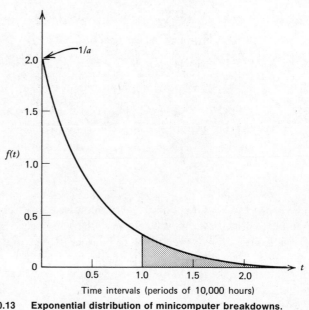

FIGURE 10.13   **Exponential distribution of minicomputer breakdowns.**

The probability that the minicomputer will not breakdown in the first 10,000 hours ($t = 1$) is shown in Figure 10.13. The associated probability is obtained by using expression 10.15.

$$P(t) = e^{-1/0.5}$$
$$= 0.13534$$

Normal Distribution   The normal distribution is one of the most important of all distributions. The probability density function of the normal distribution is given by

$$f(x) = \frac{1}{\sqrt{2\pi\sigma^2}} e^{-(x-\mu)^2/2\sigma^2} \tag{10.16}$$

$$-\infty \leq x \leq +\infty$$

where

$$\pi = 3.1416$$

$$e = 2.718282 = \text{base of natural logarithms}$$

$$\sigma = \text{standard deviation}$$

$$\text{Mean} = \mu$$

$$\text{Variance} = \sigma^2$$

The general form of the normal distribution is symmetrical about the mean. A unit or standard distribution is just a normal distribution with zero mean and unit variance. The probability density is

$$p(t) = \frac{1}{\sqrt{2\pi}} e^{-t^2/2} \tag{10.17}$$

$$-\infty \leq t \leq +\infty$$

The parameters for the unit normal distribution are

$$\text{Mean} = 0$$

$$\text{Variance} = 1$$

Areas under the unit normal distribution can be read from tables. Most statistics texts present these tables. The random variable is usually represented as a deviation from the mean expressed in terms of standard deviations. The deviation of the random variable x from μ is

$$Z = \frac{x - \mu}{\sigma} \tag{10.18}$$

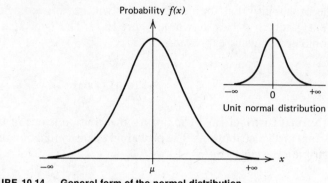

FIGURE 10.14   General form of the normal distribution.

where Z is the standard normal variate. The general form of the normal distribution is shown in Figure 10.14.

A summary of the probability distributions is shown in Table 10.2.

TABLE 10.2   A Summary of Probability Distributions

Distribution	Density Function	Mean	Variance
Binomial	$P(x) = \dfrac{n!}{x!(n-x)!} p^x q^{n-x}$	np	npq
Poisson	$P(x) = \dfrac{\lambda^x e^{-\lambda}}{x!}$	λ	λ
Exponential	$f(t) = \dfrac{1}{a} e^{-t/a}$	a	a^2
Normal	$f(x) = \dfrac{1}{\sqrt{2\pi\sigma^2}} e^{-(x-\mu)^2/2\sigma^2}$	μ	σ^2

PROBLEMS

10.1   A box contains six yellow balls, 12 red balls, and two green balls. In each trial, a ball is selected at random, its color is noted, and then it is returned to the box.

(a) What is the probability of drawing a red ball in the first draw?

(b) What is the probability of drawing a red ball in the first draw and a yellow ball in the second draw?

(c) What is the probability of drawing a red ball in the first draw, a yellow ball in the second draw, and a green ball in the third draw?

(d) What is the probability of drawing a red ball in the second trial given that the outcome of the first trial is a green ball?

10.2 The balls contained in three boxes are as follows:

Box	Number of Green balls	Number of Yellow balls	Number of Red balls	Total
A	8	12	0	20
B	4	4	2	10
C	3	2	20	25

The first draw is made from box *A*. If it is a green ball, the next draw is made from box *C*. If the outcome of the first trial is a yellow ball, the next draw is made from box *B*.

(a) What is the probability of getting a red ball in the second trial, given that the outcome of the first trial is a green ball?

(b) What is the probability of getting a yellow ball in the first draw and a red ball in the second draw?

10.3 A warehouse receives a particular product from three plants. The number of acceptable and defective items received in a shipment from each plant is depicted in the following table:

Kind of Item	Plant P1	Plant P2	Plant P3	Total
Acceptable (A)	900	1420	480	2800
Defective (D)	100	80	20	200
	1000	1500	500	3000

(a) If a unit is chosen at random, what is the probability that it is defective?

(b) What is the probability that the item was manufactured in plant *P3*?

(c) What is the probability that it was manufactured in plant *P1* given that it was defective?

(d) What is the probability that it was defective and it was manufactured in plant *P2*?

10.4 Two automatic machines are used to produce hexagonal bolts. The units produced during a particular hour are tested and the resulting values are given in the following table:

Kind of Bolt	Machine M1	Machine M2	Total
Acceptable bolts (A)	460	665	1125
Defective bolts (D)	40	35	75
	500	700	1200

(a) A bolt is selected at random. What is the probability that it was defective, given that it was produced on machine M2?

(b) What is the probability that it was defective and that it was produced on machine M1?

10.5 If automobile license plates are to have four digits preceded by a letter, how many distinct plates can be developed? Assume that letter O cannot be used?

10.6 A car dealer wishes to exhibit five new cars in the show room. However, there are only three locations available. Determine in how many ways the cars can be arranged.

10.7 A corporation has eight executives available to be assigned to five divisions. How many distinct assignments can be made?

10.8 A construction company wishes to form a group of six workers from a pool of 10 painters. How many groups are possible?

10.9 A business has chosen seven perspective locations in a city to build warehouses. It desires to build three warehouses. In any location, at most one warehouse can be constructed. In how many ways can the selection be made?

10.10 The Rosenthal Enterprises is examining the manufacturing method to be implemented to produce a new product. Any one of the three methods can be selected: M1, M2, or M3.

Manufacturing Method	Resulting Sales Price/Unit
M1	$4.50
M2	$4.70
M3	$5.00

The planning department has assigned the following probabilities and estimated values:

Sales Price per Unit	Possible Outcomes	Probability	Estimated Value*
$4.50	High sales	0.70	$850,000
	Moderate sales	0.30	−200,000
$4.70	High sales	0.60	700.000
	Moderate sales	0.40	300,000
$5.00	High sales	0.55	400,000
	Moderate sales	0.45	150,000

* A + value denotes a profit and a − value a loss.

(a) Draw a decision tree similar to Figure 10.11.
(b) The planning department desires to recommend the method that yields the largest expected profit. Determine which method gives the maximum expected profit.

10.11 It is known that 10% of the families of a given suburb of a city own two cars per family. In the suburb, seven families are randomly selected, and each of them is asked how many cars they own. What is the probability that the number of families owning two cars will be
(a) Exactly two?
(b) Less than two?
(c) More than three?

10.12 In the head office of a corporation, 30% of the personnel are engineers. If a sample of 10 personnel is taken, what is the probability that the sample will include
(a) No engineers?
(b) Exactly one engineer?
(c) Two or more engineers?

10.13 Automobiles arrive at a car wash in a Poisson manner at the rate of 10 per hour. In a nine-minute interval, what is the probability that the number of arrivals will be
(a) Exactly three?
(b) Less than three?
(c) Less than one?
(d) Either two or three?

10.14 The number of defects per 100 feet of wire is considered a Poisson process. The mean number of defects is 2 per 100 feet. In a 175 feet of the wire,

what is the probability that the number of defects will be
(a) Exactly one?
(b) At least two?
(c) Less than three?

10.15 The time required to load trucks is exponentially distributed, with a mean of 90 minutes.
(a) Calculate mean and variance.
(b) What is the probability that a truck will be loaded in more than 90 minutes?
(c) What is the probability that a truck will be loaded in more than three hours?
(d) Draw the exponential distribution of truck service times.

─────────────── **SELECTED REFERENCES** ───────────────

Berman, S. M., *The Elements of Probability*, Addison-Wesley, Reading, Mass., 1969.

Bierman, Jr. H., C. P. Bonini, L. E. Fouraker, and R. K. Jaedicke, *Quantitative Analysis for Business Decisions*, Richard D. Irwin, Homewood, Ill., 1965.

Clark, C. T., and L. L. Schkade, *Statistical Methods for Business Decisions*, South-Western Publishing, Cincinnatti, 1969.

Fabrycky, W. J., P. M. Ghare, and P. E. Torgersen, *Industrial Operations Research*, Prentice-Hall, Englewood Cliffs, N.J., 1972.

Freund, J. E., and F. J. Williams, *Elementary Business Statistics: The Modern Approach*, Prentice-Hall, Englewood Cliffs, N.J., 1964.

Guenther, W. C., *Concepts of Statistical Inference*, McGraw-Hill, New York, 1965.

Hodges, Jr., J. L., and E. L. Lehmann, *Basic Concepts of Probability and Statistics*, Holden-Day, San Francisco, Second Edition, 1970.

Magee, J. F., "Decision Trees for Decision Making," *Harvard Business Review*, July–August, 1964

Magee, J. F., "How to Use Decision Trees in Capital Investment," *Harvard Business Review*, September–October, 1964.

Parzen, E., *Modern Probability and Its Applications*, Wiley, New York, 1960.

Riggs, J. L., *Economic Decision Models for Engineers and Managers*, McGraw-Hill, New York, 1968.

Schlaifer, R., *Probability and Statistics for Business Decisions*, McGraw-Hill, New York, 1959.

Thierauf, R. J., and R. A. Grosse, *Decision Making Through Operations Research*, Wiley, New York, 1970.

Thompson, Jr. W. A., *Applied Probability*, Holt, Rinehart and Winston, New York, 1969.

Wine, R. L., *Statistics for Scientists and Engineers*, Prentice-Hall, Englewood Cliffs, N.J., 1964.

CHAPTER 11

project planning - cpm and pert

NUMEROUS PROJECTS have been performed by organizations and individuals since the beginning of civilization. Some examples of projects are the construction of a building, planning and launching a new product, design and production of a new product, manufacturing and assembly of a large generator, and modifications of an existing plant or equipment. In undertaking a project, one is concerned with the successful completion of the whole project by a certain time and staying within the budget. The basic managerial functions of planning, scheduling, and control of projects impose severe burdens upon the managers because of the special problems such as nonrepetitive nature of work, complexity of the project, problem of scheduling, and effective use of the available resources, as well as the problem of controlling the activities and rescheduling the project where necessary. The Critical Path Method (CPM) and the Program Evaluation and Review Technique (PERT) are two powerful tools that can be used in diverse situations of project planning, scheduling, and control.

The CPM is used in projects such as construction and manufacturing where the company has some experience in executing similar projects and the uncertainty surrounding the time estimates for each one of the tasks forming the project is considered insignificant. On the other hand, PERT is applicable for projects such as research and development as well as planning and launching a new product, in which the performance times of their activities are subject to a considerable degree of uncertainty. The CPM uses a deterministic

377

time for each task of the project, while the PERT is based on three-time estimates of the performance time for each task; namely, the optimistic time, the most likely time, and the pessimistic time.

A project is to have the following characteristics for applying the CPM or PERT:

1. The project consists of a collection of activities. Commonly used terms synonymous with "activity" are "job," "task," "operation," or "work."
2. The activities are performed in a given sequence and can be started and stopped independently.
3. The activities are executed according to the technological requirement.

NETWORK CONSTRUCTION

A project is represented by a *network*, also called *arrow diagram*, to enable one to analyze the activities and their interrelationships. A network consists of two basic elements: Activity and Event.

Activity

An activity is a basic task of a project that must be performed in the course of executing the project. Consider the following set of activities pertaining to the construction of a wall.

	Activity	*Estimated Time*
A.	Build wall	10 days
B.	Plaster wall	3 days
C.	Paint wall	4 days

The project has a collection of the three activities: *A*, *B*, and *C*. The technological requirement is that activity *A* (build wall) should be done first and activity *B* (plaster wall) can be started upon completion of activity *A*. Activity *B* (plaster wall) must be done before one can start task *C* (paint wall). The wall should be built before plastering; hence, we say that "building the wall" is the *immediate predecessor* of plastering wall. In turn, plastering wall is the immediate predecessor of painting wall. Task *B* is also known as the *postrequisite* task of task *A* and, similarly, task *C* is the postrequisite task of task *B*. Provided that these technological conditions are satisfied, the tasks can be started and stopped independently. This construction project is represented by an arrow diagram in Figure 11.1.

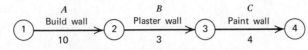

FIGURE 11.1   Arrow diagram for a wall construction project.

The activities are shown by lines in the arrow diagram. The arrow placed on an activity indicates the direction of time flow. It is conventional to draw the network so that time flows from left to right. Descriptions and time estimates of the tasks are usually written along the lines. The lengths of the lines do not indicate activity times.

Event

The starting point and the ending point of any activity are known as *events*. The events, as shown in Figure 11.1, are represented by circles. Commonly used terms synonymous with "event" are "node," "milestone," "stage," or "special point." The major difference between activities and events is that the activities involve the passage of time while events are milestones in time. Some of the events for the wall construction project are: begin building wall, complete building wall, end painting wall, and finishing the project.

Numbering the Events

Each event is indexed with an integer number such that the starting event (i) of any activity K has a smaller number than the ending event (j) of the same activity ($i < j$).

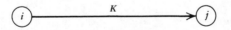

The event denoting the beginning of the project is numbered 1, and the other events are numbered with consecutively increasing values, without omitting any integer number in between. An activity can also be referred to by its starting node and ending node. For instance, the task K is also called "$i - j$." A number can be used in indexing the nodes at most once. The abovementioned conditions on numbering the events are usually required in utilizing the computer routines. An illustration of numbering the events for a milling machine rebuilding project is shown in Figure 11.2. An event is called a *merge event* if it represents the completion of more than one activity. A merge event may denote the beginning of any number of activities (Figure 11.3). An event is called a *burst event* if it represents

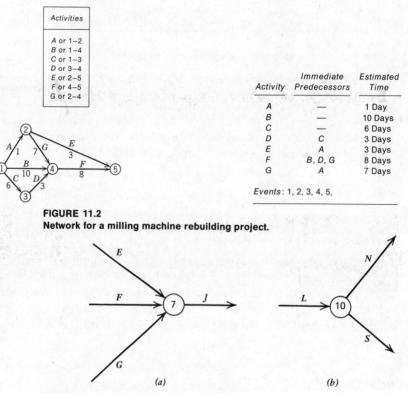

Activities
A or 1–2
B or 1–4
C or 1–3
D or 3–4
E or 2–5
F or 4–5
G or 2–4

Activity	Immediate Predecessors	Estimated Time
A	—	1 Day
B	—	10 Days
C	—	6 Days
D	C	3 Days
E	A	3 Days
F	B, D, G	8 Days
G	A	7 Days

Events: 1, 2, 3, 4, 5,

FIGURE 11.2
Network for a milling machine rebuilding project.

FIGURE 11.3 *(a)* **Merge event and** *(b)* **burst event.**

the beginning of more than one task. A burst event may be the completion of any number of tasks (Figure 11.3). In Figure 11.2, events 4 and 5 are the merge events, and nodes 1 and 2 are the burst events.

Dummy Activity

We have drawn two networks, one for the wall construction project and the other for the milling machine rebuilding project. It is not always easy to draw a network. The difficulty of plotting the network arises under two situations:

1. Numbering the nodes
2. Fulfilling the technological requirements

Dummy activities are employed for these two cases in drawing the networks. Dummy activities are considered activities, and they do not require any time or resource. The time needed for a dummy task is zero. Dummy activities are represented by broken lines in the arrow diagram.

Numbering the Nodes

When a node i is the starting node for two or more tasks, and each of these activities have the same single node j as their ending node, confusion arises in identifying these tasks. Consider the wall painting project shown in Figure 11.4. We note in Figure 11.4a that event 2 is the starting point for activities B and C, and these have event 3 as their ending point. Therefore, when we say activity $2-3$, it is not certain if it means activity B or activity C. Hence, it is invalid. A valid network is shown in Figure 11.4b. It is observed that the introduction of the dummy task $3-4$ has enabled us to identify the tasks by the event numbers. Now, task B is $2-4$, task C is $2-3$, and task $3-4$ is a dummy activity.

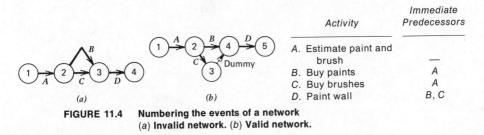

Activity	Immediate Predecessors
A. Estimate paint and brush	—
B. Buy paints	A
C. Buy brushes	A
D. Paint wall	B, C

FIGURE 11.4 **Numbering the events of a network**
(a) **Invalid network.** (b) **Valid network.**

Fulfilling the Technological Requirements

When drawing the networks, care should be exerted to insure that the network really represents the predecessor requirements. Let us illustrate the difficulty that may arise through the example shown in Figure 11.5. In the network shown in Figure 11.5a, we note that task M has the predecessors K and L. This is not true according to the predecessor requirements. A dummy activity is introduced in Figure 11.5b, and now this network satisfies all the requirements.

Any number of dummy activities can be employed in a network, and these do not alter the final solution.

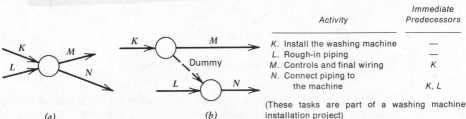

Activity	Immediate Predecessors
K. Install the washing machine	—
L. Rough-in piping	—
M. Controls and final wiring	K
N. Connect piping to the machine	K, L

(These tasks are part of a washing machine installation project)

FIGURE 11.5 **Dummy activity for fulfilling the technological requirements.**
(a) **Invalid network.** (b) **Valid network.**

CRITICAL PATH METHOD

The critical path of a project is the longest path through its network. Suppose that the longest path for a project is 30 months. This indicates that the project can be completed in 30 months, and that the project cannot be finished in less than 30 months without decreasing some activity times. In this section, we shall present the computational steps for determining the critical path and its implications. Figure 11.6 summarizes the steps of the CPM.

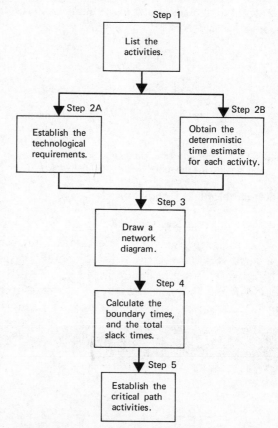

FIGURE 11.6   Summary of steps in applying the CPM.

We shall illustrate the computational procedure by a house-building project. The first step is to obtain the list of activities. The following list is furnished:

List of Activities

A. Excavate

B. Order and deliver materials

C. Foundation

D. Install outside water, gas, sewer, and electric power lines

E. Wooden frame

F. Brickwork

G. Roofing

H. Install plumbing, ducts, and wiring

I. Paint interior

J. Grade and pave outside area

After Step 1, we can do Steps 2*A* and 2*B* simultaneously. In this step, the technological requirements as well as the time estimates for the activities are compiled. We can determine the technological requirements by inquiring of each task the following set of questions.

What tasks must be completed immediately before starting this task?
What tasks can be started on completion of this task?
What tasks can be performed during the same time?

Here are the data for our example.

Activity	Immediate Predecessors	Estimated Time
A. Excavate	—	3 weeks
B. Order and deliver materials	A	2 weeks
C. Foundation	A	5 weeks
D. Install outside water, gas, sewer and electric power lines	B, C	3 weeks
E. Wooden frame	B, C	2 weeks
F. Brickwork	E	4 weeks
G. Roofing	F	3 weeks
H. Install plumbing, ducts, and wiring	D, G	2 weeks
I. Paint interior	H	1 week
J. Grade and pave outside area	D, G	2 weeks

In Step 3, we draw the network diagram. The network diagram for our example is shown in Figure 11.7. The boundary values for each activity are calculated in Step 4. The four components of the boundary times for any activity (i, j) are:

1. The Earliest Start Time, $ES(i, j)$
2. The Earliest Finish Time, $EF(i, j)$

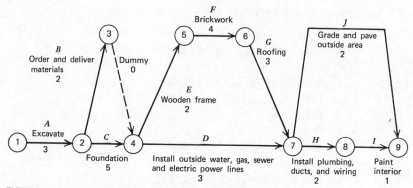

FIGURE 11.7   Network for the house building project.

3. The Latest Finish Time, $LF(i, j)$
4. The Latest Start Time, $LS(i, j)$

The *earliest start time* is the earliest time at which an activity can be started. It is the time at which all related immediately preceding activities can be completed assuming that all of these preceding activities are also started at their earliest start times. The mathematical expression for determining the ES of any activity is:

$$ES(i, j) = \text{Maximum}\left[ES(c, i) + T(c, i)\right]$$

$$\text{for all immediately preceding tasks of } (i, j) \qquad (11.1)$$

where

$$T(c, i) = \text{estimated time of task } (c, i). \qquad \text{Also, } ES \text{ of event } i = ES(i, j)$$

$$(11.2)$$

It is assumed that the ES of event 1 is zero. After the CPM calculations, if the computed values are to be modified to represent the actual dates, the change is done by simple additions. If the beginning of a project is the 12th week, namely the ES of node 1 = 12, a calculated value 4 would denote the 16th week ($= 12 + 4$).

The ES values for our example are shown in Table 11.1. For instance, the immediately preceding tasks of 4–5 are 2–4 and 3–4. Therefore,

$$ES(4, 5) = \text{Max}\begin{bmatrix} ES(2, 4) + T(2, 4) \\ ES(3, 4) + T(3, 4) \end{bmatrix}$$

$$= \text{Max}(3 + 5 = 8, 5 + 0 = 5)$$

$$= 8$$

The ES value or the earliest arriving time for the last event, d, or for that matter for any event is:

ES of event d = maximum $[ES(i, d) + T(i, d)]$

$$\text{for all entering activities} \qquad (11.3)$$

For event 9, the entering tasks are (7, 9) and (8, 9).

$$ES \text{ of event } 9 = \text{maximum} \begin{bmatrix} ES(7, 9) + T(7, 9) \\ ES(8, 9) + T(8, 9) \end{bmatrix}$$

$$= \text{maximum} (17 + 2 = 19, 19 + 1 = 20)$$

$$= 20$$

The ES of the last event specifies the total project time. The total project time is thus 20 weeks.

The *earliest finish time* is the earliest time at which an activity can be completed. It is obtained by adding the estimated time to the earliest start time.

$$EF(i, j) = ES(i, j) + T(i, j) \qquad (11.4)$$

The EF values for our example are given in Table 11.1.

TABLE 11.1   Calculation of ES and EF Values

Activity	Computations for ES Value	ES	EF
1–2	0 (assumed)	0	0 + 3 = 3
2–3	0 + 3 = 3	3	3 + 2 = 5
2–4	0 + 3 = 3	3	3 + 5 = 8
3–4	3 + 2 = 5	5	5 + 0 = 5
4–5	Max $\begin{cases} \text{for 2–4, } 3 + 5 = 8 \\ \text{for 3–4, } 5 + 0 = 5 \end{cases}$	8	8 + 2 = 10
5–6	8 + 2 = 10	10	10 + 4 = 14
4–7	Max $\begin{cases} \text{for 2–4, } 3 + 5 = 8 \\ \text{for 3–4, } 5 + 0 = 5 \end{cases}$	8	8 + 3 = 11
6–7	10 + 4 = 14	14	14 + 3 = 17
7–8	Max $\begin{cases} \text{for 4–7, } 8 + 3 = 11 \\ \text{for 6–7, } 14 + 3 = 17 \end{cases}$	17	17 + 2 = 19
7–9	Max $\begin{cases} \text{for 4–7, } 8 + 3 = 11 \\ \text{for 6–7, } 14 + 3 = 17 \end{cases}$	17	17 + 2 = 19
8–9	17 + 2 = 19	19	19 + 1 = 20

The *latest finish time* is the latest time at which an activity can be completed, given that the project must be completed at the *ES* of its terminal event. The *LF* value can be calculated by applying the expression:

$$LF(i, j) = \text{minimum } [LF \text{ for event } e - T(j, e)]$$

$$\text{for all the post activities of } (i, j) \quad (11.5)$$

Also,

$$LF \text{ for event } e = LF(f, e), \quad (11.6)$$

The calculations are done backwards starting with the last activity and moving toward the activity denoting the beginning of the project. An activity (i, j) is selected each time for establishing its *LF* value such that the *LF* values of all of its postrequisite activities are known. This procedure is continued until the *LF* values are computed for all the tasks. The *LF* values for our example are given in Table 11.2. For instance, the postrequisite activities for task 6–7 are 7–8 and 7–9. Therefore,

$$LF(6, 7) = \text{Min} \begin{bmatrix} LF \text{ for event } 8 - T(7, 8) \\ LF \text{ for event } 9 - T(7, 9) \end{bmatrix}$$

$$= \text{Min} \begin{bmatrix} LF(7, 8) - T(7, 8) = 19 - 2 = 17 \\ LF(8, 9) - T(7, 9) = 20 - 2 = 18 \end{bmatrix}$$

$$= 17$$

The *LF* value for event 1 that indicates the beginning of the project is calculated applying the expression:

$$LF \text{ of event } 1 = \text{Minimum } [LF(1, j) - T(1, j)]$$

$$\text{for all the tasks bursting out from event 1} \quad (11.7)$$

Now, the only task that bursts out of event 1 is 1–2. Thus,

$$LF \text{ of event } 1 = \text{Minimum } [LF(1, 2) - T(1, 2)]$$

$$= 3 - 3 = 0$$

We observe that both the *ES* and *LF* of event 1 equal zero. This information can be utilized to check the validity of our *ES* and *LF* calculations.

The *latest start time* is the latest time at which an activity can be started, given that the project is completed at the *ES* of its terminal event. The *LS* of an activity is obtained by subtracting the estimated task time from its latest finish time.

$$LS(i, j) = LF(i, j) - T(i, j) \quad (11.8)$$

See Table 11.2 for the *LS* values of our example.

TABLE 11.2   Calculation of *LF* and *LS* Values

Activity	Computations for *LF* Value	*LF*	*LS*
8–9	20 (= *ES* of event 9)	20	20 − 1 = 19
7–9	20 (= *ES* of event 9)	20	20 − 2 = 18
7–8	20 − 1 = 19	19	19 − 2 = 17
6–7	Min $\begin{cases} \text{for } 7\text{–}8,\ 19 - 2 = 17 \\ \text{for } 7\text{–}9,\ 20 - 2 = 18 \end{cases}$	17	17 − 3 = 14
4–7	Min $\begin{cases} \text{for } 7\text{–}8,\ 19 - 2 = 17 \\ \text{for } 7\text{–}9,\ 20 - 2 = 18 \end{cases}$	17	17 − 3 = 14
5–6	17 − 3 = 14	14	14 − 4 = 10
4–5	14 − 4 = 10	10	10 − 2 = 8
3–4	Min $\begin{cases} \text{for } 4\text{–}7,\ 17 - 3 = 14 \\ \text{for } 4\text{–}5,\ 10 - 2 = 8 \end{cases}$	8	8 − 0 = 8
2–4	Min $\begin{cases} \text{for } 4\text{–}7,\ 17 - 3 = 14 \\ \text{for } 4\text{–}5,\ 10 - 2 = 8 \end{cases}$	8	8 − 5 = 3
2–3	8 − 0 = 8	8	8 − 2 = 6
1–2	Min $\begin{cases} \text{for } 2\text{–}3,\ 8 - 2 = 6 \\ \text{for } 2\text{–}4,\ 8 - 5 = 3 \end{cases}$	3	3 − 3 = 0

Next, we compute the total slack time for each activity. The maximum time that an activity can be delayed beyond its *ES* time for starting it, still enabling the project to be completed at the *ES* of the terminal event, is called the *total slack of the activity*. The total slack of an activity is often called "total float of the activity."

The total float for the task (i, j) is:

$$TF(i, j) = LS(i, j) - ES(i, j)$$

or

$$TF(i, j) = LF(i, j) - EF(i, j) \tag{11.9}$$

The calculations are shown in Table 11.3.

The last step is to establish the critical path. The critical path activities are identified by zero total float. The critical path tasks for our house building project as can be seen in Table 11.3 are 1–2, 2–4, 4–5, 5–6, 6–7, 7–8, and 8–9. This is the longest path in the network and the project duration is 20 weeks.

TABLE 11.3 **Boundary Times and *TF* Values for the House Building Project**

Activity	T	ES	EF	LS	LF	TF
1–2	3	0	3	0	3	0*
2–3	2	3	5	6	8	3
2–4	5	3	8	3	8	0*
3–4	0	5	5	8	8	3
4–5	2	8	10	8	10	0*
5–6	4	10	14	10	14	0*
4–7	3	8	11	14	17	6
6–7	3	14	17	14	17	0*
7–8	2	17	19	17	19	0*
7–9	2	17	19	18	20	1
8–9	1	19	20	19	20	0*

* Critical path activity.

We can determine the boundary times as well as the total float times (Figure 11.6, Step 4) by the matrix method presented below.

Matrix Method

We shall illustrate the matrix method with the aid of the numerical example shown in Figure 11.2. Here are the computational steps and the calculations:

STEP 4a. Construct a square matrix of size *M* rows by *M* columns, where *M* equals (2 + total number of events in the network). The matrix is labeled as shown in Table 11.4.

STEP 4b. Enter the *T* value of each activity in the respective box of the matrix. (Table 11.5). There are many empty boxes in the matrix. An empty box situated, for instance, in row *p* and column *q* indicates that task (*p*, *q*) does not exist.

STEP 4c. The *ES* values are determined in this step. Set the *ES* of event 1 equals zero. Then consecutively *ES* of event 2, event 3, . . . , etc. are calculated as follows. To find the *ES* of event *k*, draw a horizontal line through starting event *k* until it intersects the diagonal line, and then erect a vertical line. For the completed boxes through which the vertical line

TABLE 11.4   Construction and Labeling the Matrix for the Milling Machine Rebuilding Project

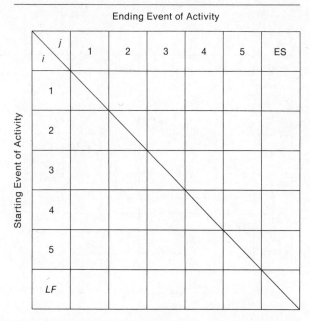

The box that lies at the intersection of row i and column j is used to record the values pertaining to activity (i, j). In the boxes, the values of T, EF, LS, and TF of the related tasks will be entered in the four corners.

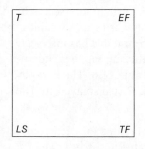

has passed compute their $(T + ES)$ value. Select the largest computed value. This denotes the ES of event k. The method of calculating the ES value of starting event 4 is shown in Table 11.5, which also gives the ES values for other events. Note that $ES(i, j) = ES$ of event i from the expression 11.2.

TABLE 11.5   Calculation of _ES_ Values

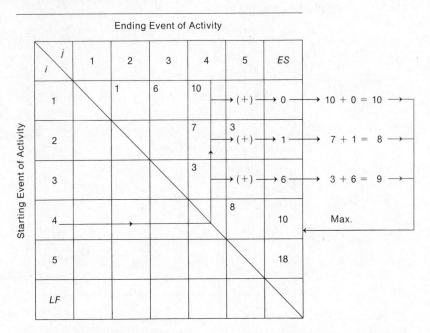

Ending Event of Activity

STEP 4d. Enter the _LF_ of the terminal event as the _ES_ of the same event. The _LF_ values are calculated backwards starting with the terminal event moving towards event 1. To find the _LF_ of event _n_, draw a vertical line through ending event _n_ until it intersects the diagonal line, and then draw a horizontal line toward the _ES_ values. For the completed boxes through which the horizontal line has intersected, calculate their $(LF - T)$ values. The smallest value among these denotes the _LF_ of event _n_. The _LF_ values are shown in Table 11.6. The computational procedure for determining the _LF_ of event 2 is displayed in Table 11.6. Recall that $LF(i, j) = LF$ of event _j_.

STEP 4e. Compute the _EF_ value for each completed box applying expression 11.4, and enter the value in the box. The _EF_ values for our example are shown in Table 11.7.

STEP 4f. Calculate the _LS_ value for each task applying expression 11.8 and write the value in the related box. Table 11.7 gives the _LS_ values for our example.

STEP 4g. Applying expression 11.9, determine the _TF_ values and put them in the associated boxes. Activities having a _TF_ of zero specify critical path activities.

TABLE 11.6   **Calculation of** *LF* **Values**

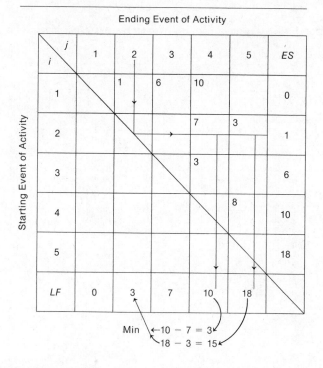

TABLE 11.7   **Calculation of** *EF* **and** *LS* **Values**

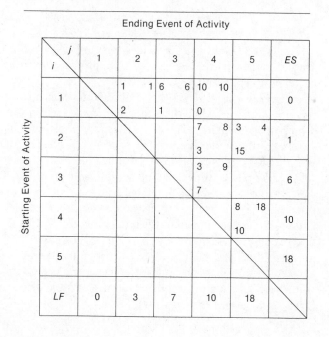

TABLE 11.8 Boundary Times and *TF* Values for the Milling Machine Rebuilding Project

Ending Event of Activity

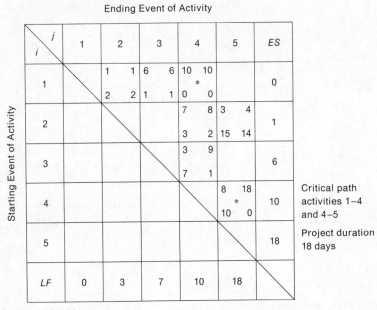

j \ i	1	2	3	4	5	ES
1		1 1 / 2 2	6 6 / 1 1	10 10 * / 0 0		0
2				7 8 / 3 2	3 4 / 15 14	1
3				3 9 / 7 1		6
4					8 18 * / 10 0	10
5						18
LF	0	3	7	10	18	

Critical path activities 1–4 and 4–5

Project duration 18 days

* Critical path activity.

Table 11.8 shows the TF values for our problem. Tasks 1–4 and 4–5 are the critical path activities, and the project duration is 18 days. It is pertinent to state that for any activity (i, j), its ES is given in row i, its LF in column j, and its T, EF, LS, and TF values in box (i, j) of Table 11.8. The boundary times and the total float calculations for our house building project are presented in Table 11.9. It can be observed that these computed values are the same as those displayed in Table 11.3. The critical path activities for our example are shown in Figure 11.8.

The computed values may be considered as the planned scheme if the calculated project time is approved as the target date. In case the project time must be decreased, more resources such as men, money, materials, and machines are applied to the tasks in an efficient manner. This topic is discussed by Wiest and Levy (1969) and Davis (1966). After obtaining an acceptable project plan, the next phase is to perform the activities according to the plan. The earliest start schedule graph can be used as a valuable aid during the implementation stage and also for controlling the tasks whenever necessary.

TABLE 11.9 **Calculations for the House Building Project by Matrix Method**

Ending Event of Activity

i \ j	1	2	3	4	5	6	7	8	9	ES
1		3 3 * 0 0								0
2			2 5 6 3	5 8 * 3 0						3
3				0 5 8 3						5
4					2 10 * 8 0		3 11 14 6			8
5						4 14 * 10 0				10
6							3 17 * 14 0			14
7								2 19 * 17 0	2 19 18 1	17
8									1 20 * 19 0	19
9										20
LF	0	3	8	8	10	14	17	19	20	

Starting Event of Activity

* Critical path activity.

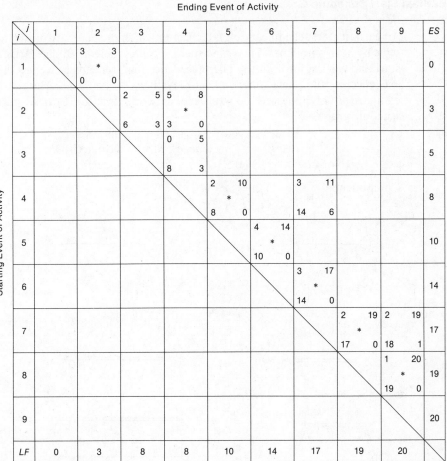

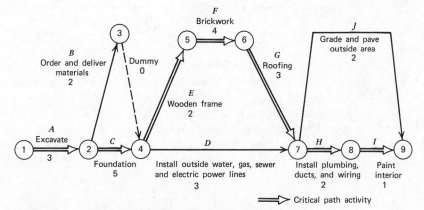

FIGURE 11.8 **Critical path activities for the house building project.**

Earliest Start Schedule Graph

An earliest start schedule graph is a chart in which the tasks are drawn to scale by horizontal lines. Their placement specifies their scheduled earliest start and earliest finish times. The precedence requirements as well as the existence of total slack times are displayed by broken lines. Here is a comparison of notations between the network and the earliest start schedule graph.

Network   Earliest Start Schedule Graph
(Not drawn to scale)   (Drawn to scale)

Components:

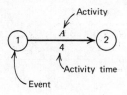
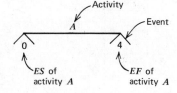

Critical path activity:

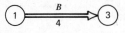
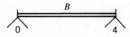

Example of a network: Equivalent graph:

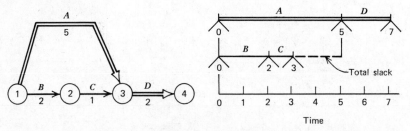

For our house construction project, the schedule graph is depicted in Figure 11.9. The schedule graph can be utilized to determine the planned progress of the project and the beginning, execution, and ending of tasks. We observe in Figure 11.9 that at the end of week 3, task A is expected to be completed, and consecutively at the beginning of week 4 tasks B and C can be started. Among these two tasks, C is a critical task while B is not a critical task. Thus C must be started at the beginning of week 4; on the

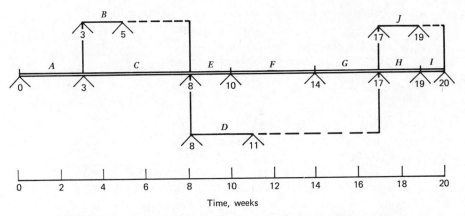

FIGURE 11.9 Earliest start schedule graph for the house building project.

other hand, if it is so desired, the starting of B can be delayed until the end of week 6 and still it can be completed at the end of week 8.

The flexibility available in starting and ending the noncritical jobs is an invaluable aid in scheduling the project. Let us consider the illustration: At the end of week 9 (now) actually tasks B and C were completed. According to the schedule graph, their latest completion times are the end of week 8. Hence, the project is behind the planned schedule by one week. Now, tasks E and D may be started. The project manager wishes to take effective action so that the project can still be completed in 20 weeks. What corrective action is available to the manager? Activity E is a critical job and hence it must be started now. Job D is not a critical task, and it has a total float of six weeks. Therefore, D need not be started now. It may be possible to transfer the resources from D to E. The consequent effect is that task E can be speeded up.

If a project has a few activities, the CPM calculations can be performed manually. However, in large projects with complex sequence relationships, the hand calculations become more difficult and are subject to errors. The computer program developed in the following section would be of great value, especially in applying the CPM for large projects.

COMPUTER MODEL FOR CPM

We shall formulate a computer program for the CPM in this section. Let us establish the following list of symbols.

NODES   = total number of nodes in the network
KACTS   = total number of activities in the network
INODE(K) = starting node of activity K
JNODE(K) = ending node of activity K
TIME(K)   = estimated time for activity K
ES(N)   = earlist start time of event N
YLF(N)   = latest finish time of event N
YLS(K)   = latest start time for activity K
EF(K)   = earliest finish time for activity K
FLOAT(K) = total float time for activity K
E   = earliest start time for the particular activity
Y   = latest finish time for the particular activity

The variables INODE( ), JNODE( ), TIME( ), YLS( ), EF( ), and FLOAT( ) are one-dimensional arrays of size KACTS elements, and the variables ES( ) and YLF( ) are one-dimensional arrays of size NODES elements.

The values for NODES, KACTS, and for each task INODE(K), JNODE(K), TIME(K) are the input data for the program. We might initialize the program with the following statements:

```
     READ 200, NODES, KACTS
     DO 10 K = 1, KACTS
     READ 210, INODE(K), JNODE(K), TIME(K)
 10  CONTINUE
200  FORMAT ( )
210  FORMAT ( )
```

The earliest start time for each event has been computed in the manual method by expressions 11.1 and 11.2. The flow chart is shown in Figure 11.10. The ES values might be determined by the following instructions:

```
     ES(1) = 0
     DO 20 J = 2, NODES
     SMALV = -999999.99
     DO 30 K = 1, KACTS
     IF (JNODE(K).NE.J) GO TO 30
     N = INODE(K)
     P = ES(N) + TIME(K)
     IF (SMALV.GE.P) GO TO 30
     SMALV = P
 30  CONTINUE
     ES(J) = SMALV
 20  CONTINUE
```

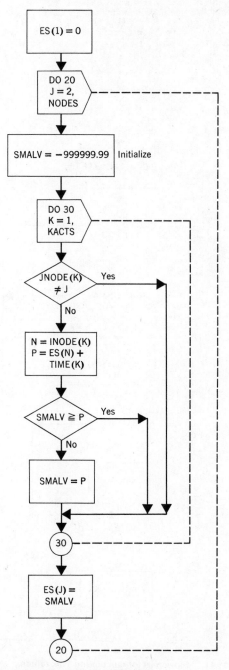

FIGURE 11.10 Flowchart for calculating *ES* values.

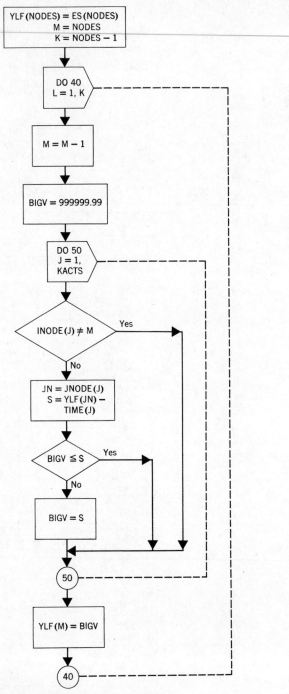

FIGURE 11.11   Flowchart for calculating *YLF* values.

The latest finish time for each event can be established by applying expressions 11.5 and 11.6. Figure 11.11 displays the flowchart for calculating YLF values. The set of statements would cause the computer to compute YLF values:

```
      YLF(NODES) = ES(NODES)
      M = NODES
      K = NODES − 1
      DO 40 L = 1, K
      M = M − 1
      BIGV = 999999.99
      DO 50 J = 1, KACTS
      IF (INODE(J).NE.M) GO TO 50
      JN = JNODE(J)
      S = YLF(JN) − TIME(J)
      IF (BIGV.LE.S) GO TO 50
      BIGV = S
   50 CONTINUE
      YLF(M) = BIGV
   40 CONTINUE
```

For determining the earliest finish time, the latest start time, and the total float for each task we can employ the instructions:

```
      DO 60 K = 1, KACTS
      I = INODE(K)
      EF(K) = ES(I) + TIME(K)
      J = JNODE(K)
      YLS(K) = YLF(J) − TIME(K)
      FLOAT(K) = YLS(K) − ES(I)
   60 CONTINUE
```

We might print out the input data, the boundary times, and the total float values with the statements:

```
      PRINT 202, NODES
      PRINT 204, KACTS
      DO 70 N = 1, KACTS
      E = EF(N) − TIME(N)
      Y = YLS(N) + TIME(N)
      PRINT 300, INODE(N), JNODE(N), TIME(N), E, EF(N),
    3 YLS(N), Y, FLOAT(N)
```

```
 70  CONTINUE
202  FORMAT ( )
204  FORMAT ( )
300  FORMAT ( )
```

The following instructions would cause the computer to print out the critical path activities and the project time:

```
     DO 820 K = 1, KACTS
     IF (FLOAT(K).NE.0.0) GO TO 820
     PRINT 825, INODE(K), JNODE(K)
820  CONTINUE
     PRINT 830, YLF(NODES)
825  FORMAT ( )
830  FORMAT ( )
```

An Illustration of the Computer Model We shall demonstrate the application of the computer model by the house-building project example. There are nine events and 11 activities in the network. Thus we have

```
DIMENSION INODE(11), JNODE(11), TIME(11)
DIMENSION YLS(11), EF(11), FLOAT(11)
DIMENSION ES(9), YLF(9)
```

The FORMAT statements are designed as follows:

```
200  FORMAT (2I2)
202  FORMAT (5X, 16HNUMBER OF EVENTS, 5X, 1H = , I3)
204  FORMAT (5X, 20HNUMBER OF ACTIVITIES, 1X, 1H = , I3)
210  FORMAT (2I1, F3.1)
300  FORMAT (4X, 2I4, 4F6.1, 1X, 2F6.1)
825  FORMAT (20X, I2, 1H-, I2)
830  FORMAT (10X, 27HEXPECTED PROJECT DURATION = , F6.2)
```

Figure 11.12 shows the complete computer program for our example to which many comment statements and printout instructions are included to enhance readability of the model and the printout results.

The first card in the data deck is prepared according to FORMAT 200 in inputting NODES and KACTS values. The KACTS data cards that are arranged following the first card are prepared according to FORMAT 210 to transmit the data for each activity. The data deck has this appearance:

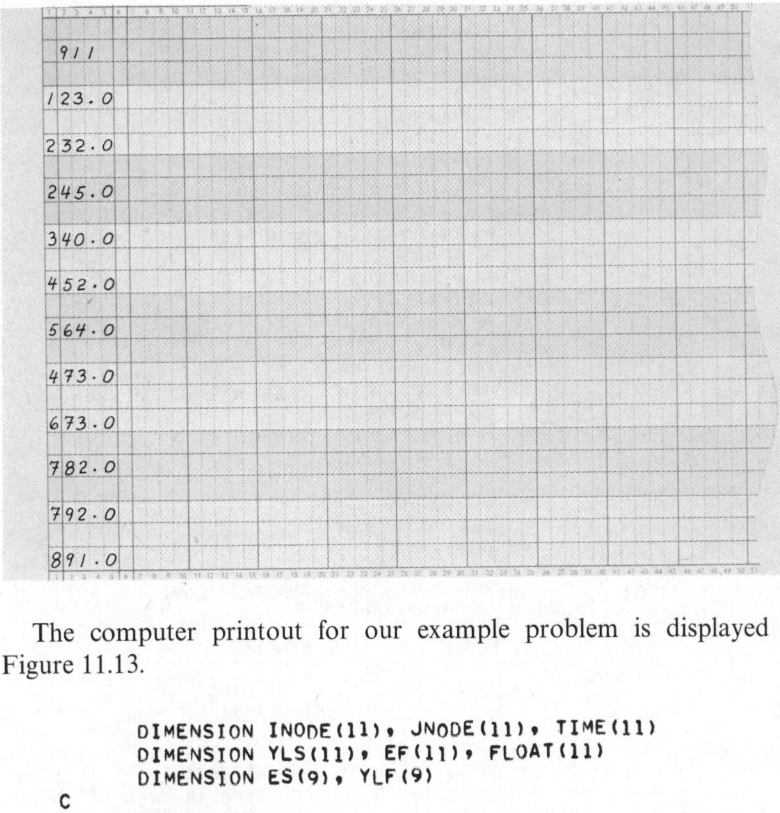

The computer printout for our example problem is displayed in Figure 11.13.

```
      DIMENSION INODE(11), JNODE(11), TIME(11)
      DIMENSION YLS(11), EF(11), FLOAT(11)
      DIMENSION ES(9), YLF(9)
C
C  PRINT THE TITLE
      PRINT 201
      PRINT 520
      PRINT 525
      PRINT 545
C
C  READ AND PRINTOUT SIZE OF NODES AND KACTS
      READ 200, NODES, KACTS
      PRINT 202, NODES
      PRINT 204, KACTS
C
C  READ INODE, JNODE AND TIME VALUE FOR EACH ACTIVITY
      DO 10 K=1, KACTS
      READ 210, INODE(K), JNODE(K), TIME(K)
   10 CONTINUE
```

FIGURE 11.12   Computer model for the CPM (illustrated for the house building project).

```
C
C    CALCULATE ES VALUE FOR EACH EVENT
     ES(1)=0
     DO 20 J=2,NODES
     SMALV= -999999.99
     DO 30 K=1,KACTS
     IF ( JNODE(K) .NE. J  )  GO TO 30
     N = INODE(K)
     P= ES(N) + TIME(K)
     IF ( SMALV .GE. P ) GO TO 30
     SMALV = P
  30 CONTINUE
     ES(J) = SMALV
  20 CONTINUE
C
C    CALCULATE LF VALUE FOR EACH EVENT
     YLF(NODES) = ES(NODES)
     M = NODES
     K = NODES-1
     DO 40 L=1,K
     M = M-1
     BIGV = 999999.99
     DO 50 J=1,KACTS
     IF ( INODE(J) .NE. M ) GO TO 50
     JN = JNODE(J)
     S = YLF(JN) - TIME(J)
     IF ( BIGV .LE. S ) GO TO 50
     BIGV = S
  50 CONTINUE
     YLF(M) = BIGV
  40 CONTINUE
C
C    CALCULATE EF, YLS, AND FLOAT VALUES
     DO 60 K=1,KACTS
     I = INODE(K)
     EF(K) = ES(I) + TIME(K)
     J = JNODE(K)
     YLS(K) = YLF(J) - TIME(K)
     FLOAT(K) = YLS(K)- ES(I)
  60 CONTINUE
C
C    CALCULATE E AND Y FOR THE ACTIVITIES
C    PRINTOUT THE GIVEN AND COMPUTED VALUES
     PRINT 525
     PRINT 530
     PRINT 535
     PRINT 525
     DO 70 N=1,KACTS
     E= EF(N)-TIME(N)
     Y=YLS(N)+TIME(N)
     PRINT 300, INODE(N),JNODE(N),TIME(N),E,EF(N),
    3YLS(N),Y,FLOAT(N)
  70 CONTINUE
     PRINT 525
     PRINT 545
     PRINT 800
     DO 820 K=1,KACTS
     IF (FLOAT(K) .NE. 0.0) GO TO 820
     PRINT 825, INODE(K), JNODE(K)
```

FIGURE 11.12 (*Continued*)

```
820 CONTINUE
    PRINT 830, YLF(NODES)
C
  200 FORMAT (2I2)
  201 FORMAT (1H1)
  202 FORMAT (5X,16HNUMBER OF EVENTS, 5X,1H=,I3)
  204 FORMAT (5X,20HNUMBER OF ACTIVITIES, 1X,1H=,I3)
  210 FORMAT (2I1,F3.1)
  300 FORMAT ( 4X,2I4,4F6.1, 1X, 2F6.1 )
  520 FORMAT ( 10X,35HCPM TIME ESTIMATES AND CALCULATIONS)
  525 FORMAT ( 5X, 45H------------------------------------------------)
  530 FORMAT ( 5X,8HACTIVITY,1X,4HTIME,3X,
     78HEARLIEST,6X,6HLATEST,4X,5HSLACK)
  535 FORMAT ( 7X,1HI,3X,1HJ,7X,5HSTART,1X,
     36HFINISH,1X,5HSTART,1X,6HFINISH,2X, 4HTIME )
  545 FORMAT (/)
  800 FORMAT (10X, 24HCRITICAL PATH ACTIVITIES )
  825 FORMAT (20X,I2,1H-,I2)
  830 FORMAT (10X, 27HEXPECTED PROJECT DURATION =, F6.2 )
      STOP
      END
```

FIGURE 11.12 (*Continued*)

```
          CPM TIME ESTIMATES AND CALCULATIONS
    ------------------------------------------------


    NUMBER OF EVENTS      =  9
    NUMBER OF ACTIVITIES  = 11
    ------------------------------------------------
    ACTIVITY TIME   EARLIEST      LATEST      SLACK
     I    J        START FINISH START FINISH  TIME
    ------------------------------------------------
     1    2   3.0      0    3.0     0    3.0      0
     2    3   2.0    3.0    5.0   6.0    8.0    3.0
     2    4   5.0    3.0    8.0   3.0    8.0      0
     3    4     0    5.0    5.0   8.0    8.0    3.0
     4    5   2.0    8.0   10.0   8.0   10.0      0
     5    6   4.0   10.0   14.0  10.0   14.0      0
     4    7   3.0    8.0   11.0  14.0   17.0    6.0
     6    7   3.0   14.0   17.0  14.0   17.0      0
     7    8   2.0   17.0   19.0  17.0   19.0      0
     7    9   2.0   17.0   19.0  18.0   20.0    1.0
     8    9   1.0   19.0   20.0  19.0   20.0      0
    ------------------------------------------------


          CRITICAL PATH ACTIVITIES
                 1- 2
                 2- 4
                 4- 5
                 5- 6
                 6- 7
                 7- 8
                 8- 9
          EXPECTED PROJECT DURATION = 20.00
```

FIGURE 11.13 Computer printout for the house building project.

PROGRAM EVALUATION AND REVIEW TECHNIQUE

The PERT approach was developed for projects in which uncertainties in activity times exist. The CPM and the PERT are similar in their logical framework. The CPM employs deterministic activity time estimates, while the PERT utilizes three time estimates for each activity. The three-time estimates are an optimistic time, a most likely time, and a pessimistic time. The *optimistic time* of an activity is the shortest conceivable time for completing the activity when everything goes right. The *pessimistic time* of an activity is the longest possible time required to complete the activity in case bad luck is encountered. The most probable time required to complete a task is known as the *most likely time.* The PERT reduces the three-time values to a single expected activity time. The expected activity times of the tasks are employed to determine the critical path of the given project.

A summary of computational steps for applying PERT is displayed in Figure 11.14. By comparing the steps of CPM (Figure 11.6) and PERT (Figure 11.14), we notice that the solution procedures are basically the same, except Steps 2*B* and 4 of PERT. We shall illustrate the PERT algorithm by a numerical example.

An Illustrative Example

Suppose that a company has received an order to design, manufacture, and supply a special piece of equipment in 22 months. Information pertaining to Step 1 of PERT is given below:

List of Activities

A. Design analysis
B. Design tooling
C. Design parts
D. Design castings
E. Manufacture castings
F. Buy tooling
G. Manufacture parts
H. Procure parts
I. Assemble
J. Test

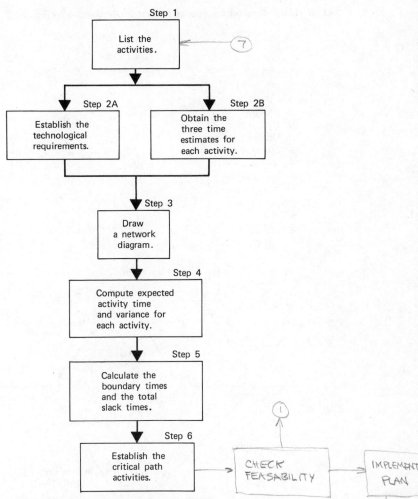

FIGURE 11.14   Summary of steps for employing the PERT.

Table 11.10 shows the technological requirements as well as the three-time estimates for each activity (Steps 2A and 2B).

In Step 3, we draw the network that is shown in Figure 11.15. In Step 4, the expected activity time and the variance applying the three-time estimates for each activity are calculated. The expected time and the variance for each task are obtained by assuming that the three-time estimates form part of a population obeying a beta distribution. The expected activity

TABLE 11.10   Data for the Special Equipment Production Project

Activity	Immediate Predecessors	Time Estimates (Months)		
		Optimistic	Most Likely	Pessimistic
A. Design analysis	–	3	4	5
B. Design tooling	A	1	2	3
C. Design parts	A	3	4	8
D. Design castings	A	1	2	9
E. Manufacture castings	D	4	5	12
F. Buy tooling	B	5	6	7
G. Manufacture parts	E, F	4	4	4
H. Procure parts	C	5	6	13
I. Assemble	G, H	4	4	10
J. Test	I	1	1	1

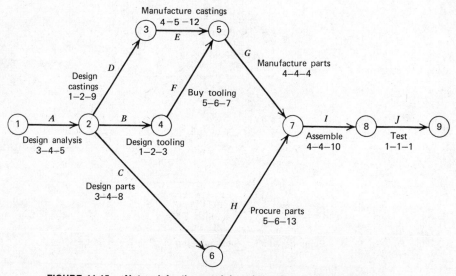

FIGURE 11.15   Network for the special equipment production project.

time and the variance for each task are calculated by applying the expressions:

$$T = \frac{A + 4(M) + B}{6} \tag{11.10}$$

$$V = \left[\frac{B - A}{6} \right]^2 \tag{11.11}$$

where

$$T = \text{expected activity time}$$

$$V = \text{variance of the activity}$$

$$A = \text{optimistic time}$$

$$M = \text{most likely time}$$

$$B = \text{pessimistic time}$$

TABLE 11.11   Calculation of T and V Values

| Task | Time Estimates | | | $T = [A + 4(M) + B]/6$ | $V = [(B - A)/6]^2$ |
	A	M	B		
1–2	3	4	5	$[3 + 4(4) + 5]/6 = 4$	$[(5 - 3)/6]^2 = 0.11$
2–3	1	2	9	$[1 + 4(2) + 9]/6 = 3$	$[(9 - 1)/6]^2 = 1.78$
2–4	1	2	3	$[1 + 4(2) + 3]/6 = 2$	$[(3 - 1)/6]^2 = 0.11$
2–6	3	4	8	$[3 + 4(4) + 8]/6 = 4.5$	$[(8 - 3)/6]^2 = 0.69$
3–5	4	5	12	$[4 + 4(5) + 12]/6 = 6$	$[(12 - 4)/6]^2 = 1.78$
4–5	5	6	7	$[5 + 4(6) + 7]/6 = 6$	$[(7 - 5)/6]^2 = 0.11$
5–7	4	4	4	$[4 + 4(4) + 4]/6 = 4$	$[(4 - 4)/6]^2 = 0$
6–7	5	6	13	$[5 + 4(6) + 13]/6 = 7$	$[(13 - 5)/6]^2 = 1.78$
7–8	4	4	10	$[4 + 4(4) + 10]/6 = 5$	$[(10 - 4)/6]^2 = 1.00$
8–9	1	1	1	$[1 + 4(1) + 1]/6 = 1$	$[(1 - 1)/6]^2 = 0$

Table 11.11 gives the T and V values for our example. The boundary times as well as the total slack times are determined either by applying expressions 11.1 to 11.9 or by the matrix method. The calculations for the matrix method are displayed in Table 11.12. The critical path activities as shown in Table 11.12 and Figure 11.16 are 1–2, 2–3, 3–5, 5–7, 7–8, and 8–9. The expected critical path time is 23 months. Because the activity time estimates are probability distributions in the PERT structure, the critical path may be considered a probabilistic critical path. Recall that the due date for the project is 22 months. The project manager may often wish to have an answer for the question, "what is the probability that the project will be completed by the desired due date?" The variance along the critical path is the sum of variances for the activities forming the path. Thus, the variance is $0.11 + 1.78 + 1.78 + 0 + 1 + 0 = 4.67$. Assuming that T is normally distributed, the number of standard deviations by which the due date deviates from the expected critical path time (Z) is computed

TABLE 11.12 Calculations for the Special Equipment Production Project

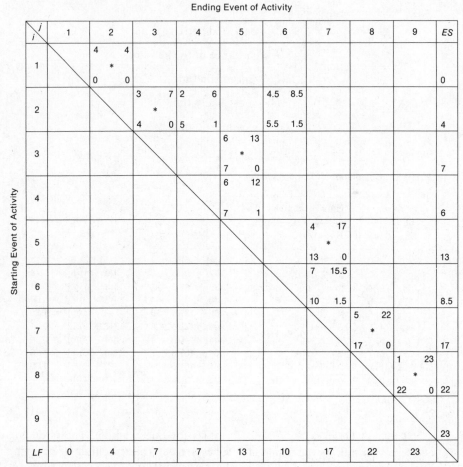

Ending Event of Activity

i \ j	1	2	3	4	5	6	7	8	9	ES
1		4 4 * 0 0								0
2			3 7 * 4 0	2 6 5 1		4.5 8.5 5.5 1.5				4
3					6 13 * 7 0					7
4					6 12 7 1					6
5							4 17 * 13 0			13
6							7 15.5 10 1.5			8.5
7								5 22 * 17 0		17
8									1 23 * 22 0	22
9										23
LF	0	4	7	7	13	10	17	22	23	

Starting Event of Activity

* Critical path activity.

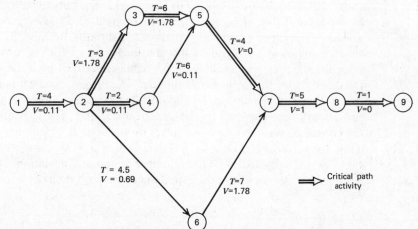

FIGURE 11.16 PERT network with expected activity times and variances.

by the expression

$$Z = \frac{\text{Due date—expected critical path time}}{\text{Standard deviation along the critical path}} \qquad (11.12)$$

where standard deviation along the critical path

$$= \sqrt{\text{Sum of variances for activities along the critical path}}$$

Thus

$$Z = \frac{22 - 23}{\sqrt{4.67}}$$

$$= -0.463$$

The probability of completing the project by the due date is read directly from Table 11.13.

TABLE 11.13   Probability of Completing a Project by the Due date

Z	Probability	Z	Probability	Z	Probability
3.0	0.9987	0.8	0.7881	−1.2	0.1151
2.8	0.9974	0.6	0.7257	−1.4	0.0808
2.6	0.9953	0.4	0.6554	−1.6	0.0548
2.4	0.9918	0.2	0.5793	−1.8	0.0359
2.2	0.9861	0.0	0.5000	−2.0	0.0228
2.0	0.9772	−0.2	0.4207	−2.2	0.0139
1.8	0.9641	−0.4	0.3446	−2.4	0.0082
1.6	0.9452	−0.6	0.2741	−2.6	0.0047
1.4	0.9192	−0.8	0.2119	−2.8	0.0026
1.2	0.8849	−1.0	0.1587	−3.0	0.0013
1.0	0.8413				

We can meet the due date of 22 months with the probability of 0.3446 (approximate) as shown in Table 11.13. For the due date of 23 months, $Z = (23 - 23)/\sqrt{4.67} = 0$, and it denotes that the probability of meeting the 23 month due date is 0.50. The probability values also indicate the following:

Probability (P)

$0 \leq P < 0.25$ The project cannot be completed reasonably by the due date with the given resources.

$0.25 \leq P \leq 0.60$ This is an acceptable level and, hence, the project can be started.

$0.60 < P \leqq 1.00$ This refers to more than acceptable level in many situations. It signifies that excess resources are applied and, hence, some of them may be transferred to other projects.

COMPUTER MODEL FOR PERT

We develop a computer model for the PERT in this section. The computer program is similar to the CPM computer model with a few modifications. In both CPM and PERT computer models the same variables and logic are used. Over and above the symbols established for CPM, we also utilize these variables:

A(K) = Optimistic time for activity K
TMOST(K) = Most likely time for activity K
B(K) = Pessimistic time for activity K
VARI(K) = Variance for activity K
SVARI = Sum of variances for activities along the critical path

The variables A( ), TMOST( ), B( ), and VARI( ) are one-dimensional arrays of size KACTS. The values of a network might be read in by these statements:

```
      READ 305, NODES, KACTS
      DO 10 K = 1, KACTS
      READ 310, INODE(K), JNODE(K), A(K), TMOST(K), B(K)
   10 CONTINUE
  305 FORMAT ( )
  310 FORMAT ( )
```

The following set of instructions would compute the expected activity time, TIME( ), and variance for each task and print out the computed values:

```
      DO 12 M = 1, KACTS
      C = A(M) + (4.0 * TMOST(M)) + B(M)
      TIME(M) = C/6.0
      D = (B(M) − A(M))/6.0
      VARI(M) = D ** 2.0
      PRINT 100, INODE(M), JNODE(M), A(M), TMOST(M),
     1B(M), TIME(M), VARI(M)
   12 CONTINUE
  100 FORMAT ( )
```

Next, the values of ES, YLF, EF, YLS, and FLOAT are computed by applying the related segments of the CPM computer model. Then, these values are printed out.

We shall illustrate the computer model by the equipment production project. There are nine events and 10 activities in our example. Thus, we have

```
DIMENSION INODE(10), JNODE(10), A(10), TMOST(10), B(10)
DIMENSION TIME(10), YLS(10), EF(10), FLOAT(10), VARI(10)
DIMENSION ES(9), YLF(9)
```

The format instructions utilized to input and output the data are

```
100   FORMAT (5X, 2I4, 1X, 4F6.1, 4X, F5.2)
202   FORMAT (5X, 16HNUMBER OF EVENTS, 5X, 1H =, I3)
204   FORMAT (5X, 20HNUMBER OF ACTIVITIES, 1X, 1H =, I3)
300   FORMAT (4X, 2I4, 4F6.1, 1X, 2F6.1)
305   FORMAT (2I2)
310   FORMAT (2I1, 3F3.0)
```

Our complete program is depicted in Figure 11.17. The data deck has the following appearance:

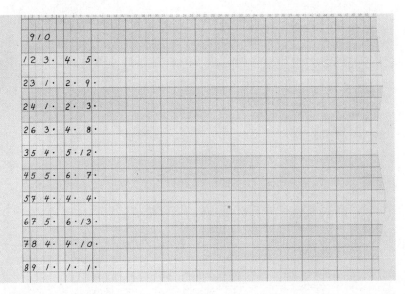

Among the 11 cards, the first gives the values of NODES and KACTS. The other 10 cards supply INODE, JNODE, A, TMOST, and B values for the 10 activities. The resulting computer printout is shown in Figure 11.18.

```
      DIMENSION INODE(10),JNODE(10),A(10),TMOST(10),B(10)
      DIMENSION TIME(10),YLS(10),EF(10),FLOAT(10),VARI(10)
      DIMENSION ES(9),YLF(9)
C
C     PRINT THE TITLE
      PRINT 201
      PRINT 570
      PRINT 525
      PRINT 545
C
C     READ AND PRINTOUT SIZE OF NODES AND KACTS
      READ 305,NODES,KACTS
      PRINT 202, NODES
      PRINT 204, KACTS
C
C     READ INODE, JNODE, A, TMOST, AND B VALUES FOR THE ACTIVITIES
      PRINT 525
      PRINT 575
      PRINT 580
      PRINT 525
      DO 10 K=1,KACTS
      READ 310,INODE(K),JNODE(K),A(K),TMOST(K),B(K)
   10 CONTINUE
C
C     CALCULATE TIME AND VARI VALUES
      DO 12 M=1,KACTS
      C = A(M)+(4.0*TMOST(M))+ B(M)
      TIME(M)= C/6.0
      D =(B(M)-A(M))/6.0
      VARI(M) = D**2.0
      PRINT 100, INODE(M),JNODE(M),A(M),TMOST(M),
     1B(M),TIME(M),VARI(M)
   12 CONTINUE
      PRINT 525
C
C     CALCULATE ES VALUE FOR EACH EVENT
      ES(1)=0
      DO 20 J=2,NODES
      SMALV= -999999.99
      DO 30 K=1,KACTS
      IF ( JNODE(K) .NE.  J  ) GO TO 30
      N = INODE(K)
      P= ES(N) + TIME(K)
      IF ( SMALV .GE. P ) GO TO 30
      SMALV = P
   30 CONTINUE
      ES(J) = SMALV
   20 CONTINUE
C
C     CALCULATE LF VALUE FOR EACH EVENT
      YLF(NODES) = ES(NODES)
      M = NODES
      K = NODES-1
      DO 40 L=1,K
      M = M-1
      BIGV = 999999.99
      DO 50 J=1,KACTS
      IF ( INODE(J) .NE. M ) GO TO 50
      JN = JNODE(J)
      S = YLF(JN) - TIME(J)
      IF ( BIGV .LE. S ) GO TO 50
      BIGV  =  S
```

FIGURE 11.17   Computer program for the PERT (illustrated for the special equipment production project).

```
   50 CONTINUE
      YLF(M) = BIGV
   40 CONTINUE
C
C  CALCULATE EF, YLS, AND FLOAT VALUES
      DO 60 K=1,KACTS
      I= INODE(K)
      EF(K) = ES(I) + TIME(K)
      J = JNODE(K)
      YLS(K) = YLF(J) - TIME(K)
      FLOAT(K) = YLS(K)- ES(I)
   60 CONTINUE
C
C  CALCULATE E AND Y FOR THE ACTIVITIES
C  PRINTOUT THE GIVEN AND COMPUTED VALUES
      PRINT 545
      PRINT 525
      PRINT 530
      PRINT 535
      PRINT 525
      DO 70 N=1,KACTS
      E= EF(N)-TIME(N)
      Y=YLS(N)+TIME(N)
      PRINT 300, INODE(N),JNODE(N),TIME(N),E,EF(N),
     3YLS(N),Y,FLOAT(N)
   70 CONTINUE
      PRINT 525
      SVARI = 0.0
      PRINT 545
      PRINT 800
      DO 820 K=1,KACTS
      IF (FLOAT(K) .GT. 0.0) GO TO 820
      PRINT 825, INODE(K), JNODE(K)
      SVARI = SVARI + VARI(K)
  820 CONTINUE
      PRINT 830, YLF(NODES)
      PRINT 835, SVARI
C
  100 FORMAT ( 5X,2I4,1X,4F6.1,4X,F5.2)
  201 FORMAT (1H1)
  202 FORMAT (5X,16HNUMBER OF EVENTS, 5X,1H=,I3)
  204 FORMAT (5X,20HNUMBER OF ACTIVITIES, 1X,1H=,I3)
  300 FORMAT ( 4X,2I4,4F6.1, 1X, 2F6.1 )
  305 FORMAT (2I2)
  310 FORMAT (2I1, 3F3.0)
  525 FORMAT ( 5X, 45H---------------------------------------------)
  530 FORMAT ( 5X,8HACTIVITY,1X,4HTIME,3X,
     78HEARLIEST,6X,6HLATEST,4X,5HSLACK)
  535 FORMAT ( 7X,1HI,3X,1HJ,7X,5HSTART,1X,
     36HFINISH,1X,5HSTART,1X,6HFINISH,2X, 4HTIME )
  545 FORMAT (/)
  570 FORMAT ( 8X,36HPERT TIME ESTIMATES AND CALCULATIONS)
  575 FORMAT ( 6X,8HACTIVITY,4X,4HTIME,1X,
     7 9HESTIMATES,2X,4HTIME,2X,8HVARIANCE)
  580 FORMAT ( 8X,1HI,3X,1HJ,5X,1HA,5X,1HM,5X,1HB)
  800 FORMAT (10X, 24HCRITICAL PATH ACTIVITIES )
  825 FORMAT (20X,I2,1H-,I2)
  830 FORMAT (10X, 27HEXPECTED PROJECT DURATION =, F6.2 )
  835 FORMAT (1X,48HSUM OF VARIANCES OF ACTIVITIES ON CRITICAL PATH=,
     8 F6.2 )
      STOP
      END
```

FIGURE 11.17 (Continued)

```
        PERT TIME ESTIMATES AND CALCULATIONS
--------------------------------------------------------


NUMBER OF EVENTS      =  9
NUMBER OF ACTIVITIES  = 10
--------------------------------------------------------
 ACTIVITY    TIME ESTIMATES   TIME   VARIANCE
   I   J     A    M    B
--------------------------------------------------------
   1   2    3.0  4.0  5.0   4.0      .11
   2   3    1.0  2.0  9.0   3.0     1.78
   2   4    1.0  2.0  3.0   2.0      .11
   2   6    3.0  4.0  8.0   4.5      .69
   3   5    4.0  5.0 12.0   6.0     1.78
   4   5    5.0  6.0  7.0   6.0      .11
   5   7    4.0  4.0  4.0   4.0       0
   6   7    5.0  6.0 13.0   7.0     1.78
   7   8    4.0  4.0 10.0   5.0     1.00
   8   9    1.0  1.0  1.0   1.0       0
--------------------------------------------------------



--------------------------------------------------------
ACTIVITY TIME    EARLIEST        LATEST      SLACK
  I   J       START FINISH   START FINISH    TIME
--------------------------------------------------------
  1   2   4.0     0    4.0     0    4.0      0
  2   3   3.0    4.0   7.0    4.0   7.0      0
  2   4   2.0    4.0   6.0    5.0   7.0     1.0
  2   6   4.5    4.0   8.5    5.5  10.0     1.5
  3   5   6.0    7.0  13.0    7.0  13.0      0
  4   5   6.0    6.0  12.0    7.0  13.0     1.0
  5   7   4.0   13.0  17.0   13.0  17.0      0
  6   7   7.0    8.5  15.5   10.0  17.0     1.5
  7   8   5.0   17.0  22.0   17.0  22.0      0
  8   9   1.0   22.0  23.0   22.0  23.0      0
--------------------------------------------------------



          CRITICAL PATH ACTIVITIES
                  1- 2
                  2- 3
                  3- 5
                  5- 7
                  7- 8
                  8- 9
          EXPECTED PROJECT DURATION = 23.00
   SUM OF VARIANCES OF ACTIVITIES ON CRITICAL PATH=  4.67
```

FIGURE 11.18   Computer printout for the special equipment production project.

PROBLEMS

11.1 Construct an arrow diagram for each set of technological requirements.

(a)

Activity	Postrequisite Activity
A	C,E
B	D
C	—
D	—
F	D
F	G
G	—

(b)

Activity	Postrequisite Activity
A	D
B	D,E,F
C	F
D	—
E	—
F	G
G	—

11.2 Draw a network diagram for the set of technological requirements.

Task	Immediate Predecessor	Estimated Time (Weeks)
A	—	6
B	A,D	7
C	A,D	10
D	—	8
E	D	2
F	A,D	5
G	B	1
H	B	6
I	B	3
J	C,G	4
K	E,F	6
L	E,F	7
M	E,F	10
N	I,J,K,L	8

11.3 For Problem 11.2, find the critical path of the network diagram.

11.4 Determine the critical path of the network shown below. How many time units will it take to complete this project.

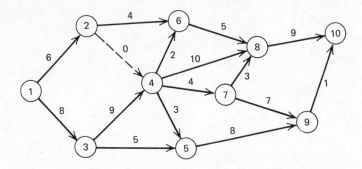

11.5 Prepare a computer program for the network shown in Problem 11.4 that will enable the computer to compute and print out the critical path activities and the expected project duration.

11.6 The network diagram shown below has been prepared for the installation of a milling machine in a factory. Determine the critical path activities and the estimated project duration.

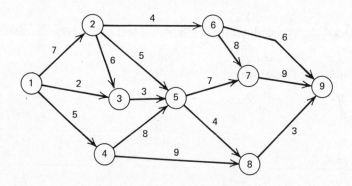

11.7 The following PERT diagram has been prepared for the salesmen training program in a company.

(a) Calculate the expected activity time and variance for each task.
(b) Determine the critical path activities, expected critical path time, and the variance along the critical path.
(c) If the due date is 47, find the probability of completing the project by the due date.

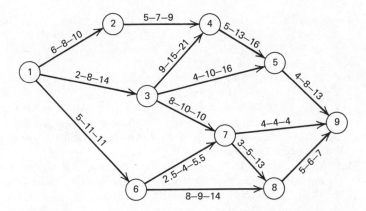

11.8 For a computer installation project, the interrelationships and the three-time estimates for each task are shown in the PERT network:

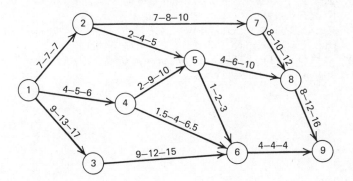

(a) Calculate the expected activity times and the variances of the activities.
(b) Determine the critical path, expected critical path time, and the variance along the critical path.
(c) The due date for the project is 38. Find the probability of completing the project by the due date.

———————————— SELECTED REFERENCES ————————————

Archibald, R. D., and R. L. Villonia, *Network—Based Management Systems*, Wiley, New York, 1967.

Biegel, J. H., *Production Control: A Quantitative Approach*, Prentice-Hall, Englewood Cliffs, N.J., Second Edition, 1971.

Buffa, E. S., and W. H. Taubert, *Production—Inventory Systems: Planning and Control*, Richard D. Irwin, Homewood, Ill., Revised Edition, 1972.

Davis, E. W., "Resource Allocation in Project Network Models—A Survey," *Journal of Industrial Engineering*, Vol. XVII, No. 4, April 1966, pp. 177–188.

Elmaghraby, S. E., "The Theory of Networks and Management Science, II," *Management Science*, Vol. 17, No. 2, October 1970, pp. 54–71.

Fabrycky, W. J., P. M. Ghare, and P. E. Torgersen, *Industrial Operations Research*, Prentice-Hall, Englewood Cliffs, N.J., 1972.

Levin, R. I., and C. A. Kirkpatrick, *Planning and Control with PERT/CPM*, McGraw-Hill, New York, 1966.

Levy, F. K., G. L. Thompson, and J. D. Wiest, "The ABC's of the Critical Path Method," *Harvard Business Review*, September–October 1963, pp. 98–108.

Lockyer, K. G., *An Introduction to Critical Path Analysis*, Sir Isaac Pitman & Sons, Ltd., London, Third Edition, 1969.

McMillan, C., and R. F. Gonzalez, *Systems Analysis: A Computer Approach to Decision Models*, Richard D. Irwin, Homewood, Ill., Third Edition, 1973.

Miller, R. W., "How to Plan and Control with PERT," *Harvard Business Review*, March–April 1962, p. 93.

Moder, J. J., and C. R. Phillips, *Project Management with CPM and PERT*, Reinhold, New York, 1964.

Riggs, J. L., *Economic Decision Models for Engineers and Managers*, McGraw-Hill, New York, 1968.

Shaffer, L. R., J. B. Ritter, and W. L. Meyer, *The Critical Path Method*, McGraw-Hill, New York, 1965.

Wiest, J. D., and F. K. Levy, *A Management Guide to PERT/CPM*, Prentice-Hall, Englewood Cliffs, N.J., 1969.

bayesian analysis

WHEN THE DECISION MAKER encounters a problem, two or more actions are available in most situations. Each possible action would lead to a particular event. If the resulting event can be any one of a set of events, and the probability of each event can be estimated, it is considered a risk. The possible events are usually called states of nature. Consider that an electronic manufacturer wishes to produce a newly developed calculator in the proposed plant at San Diego. He is about to determine the production capacity of the plant. The market demand must be considered in deciding the plant capacity. Some possible actions are to build the plant that will be capable to produce 50,000 units per year, 80,000 units per year, and so on. When the market demand can be predicted with certainty, the capacity can be easily decided. The manufacturer does not know what the demand for the calculator will be, but he must decide on some definite size for the plant capacity. Some states of nature are a demand for 45,000 units per year, 90,000 units per year, and so forth. The problems that are characterized as risk situations can be solved employing Bayesian analysis. We shall illustrate the Bayesian analysis by a numerical example.

ILLUSTRATION: AN OIL DRILLING PROBLEM

An oil wildcatter must decide whether or not he wants to drill in a particular location. He is not certain on the extent of oil

419

deposits at the location. Even though the quantity of oil deposits can assume continuous values, the oil drillers tend to think in discrete terms such as 100,000 and 500,000 barrels. Past records, preliminary analysis, and managerial judgment can be used to assign the probability of each state of nature. The states of nature and the probabilities for our example are:

State of Nature	Probability $P(N_i)$
N_1 = Dry hole	0.6
N_2 = 100,000 barrels	0.1
N_3 = 500,000 barrels	0.2
N_4 = 1,000,000 barrels	0.1
	1.0

$P(N_i)$ values are called *prior probabilities*. The actions available to the oil driller are:

$$A_1 = \text{drill}$$

$$A_2 = \text{don't drill}$$

Table 12.1 presents the monetary consequence for each action-state combination. If an oil well is drilled, any one of the four states of nature, N_1 to N_4, would be encountered. For events N_1 and N_2 the cost of drilling is more than the available oil deposits. The payoffs for N_1 and N_2 are a loss of $50,000 and $30,000, respectively. However, if the outcome is N_3 or N_4, the payoff is a profit of $100,000 or $350,000, respectively. When the decision is "Don't drill," the payoff is $0.

TABLE 12.1   Payoff Table for the Oil Drilling Problem

		Actions	
State of Nature		A_1 Drill	A_2 Don't Drill
Dry hole	N_1	$-\ 50,000	$0
100,000 barrels	N_2	$-\ 30,000	0
500,000 barrels	N_3	100,000	0
1,000,000 barrels	N_4	350,000	0

Information on the underlying geological structure would reveal whether the terrain below has closed structure (T_1), open structure (T_2), or no structure. The geological structure can be determined by conducting seismic soundings. The relationship between the observations of the structure and the estimated states of nature based upon the past experience is presented in Table 12.2. The relationships show that they are not the perfect indicators of the states of nature. Among the observations, T_1 corresponds to a promising site and T_3 is for the least preferred site. Assume that the oil driller wishes to maximize the expected gain. Determine:

(a) The optimal action without seismic information.
(b) The optimal action for each observation of experimentation.
(c) The expected gain for the optimal strategy.

When the experimental data is not available, the expected gain for each action is calculated using expression 11.9.

$$\text{Expected gain for } A_1 = 0.6(-50,000) + 0.1(-30,000)$$
$$+ 0.2(100,000) + 0.1(350,000)$$
$$= \$22,000$$

$$\text{Expected gain for } A_2 = -0.6(0) + 0.1(0) + 0.2(0) + 0.1(0)$$
$$= \$0$$

The expected gain of A_1 is the largest; hence, the optimum action is A_1 (drill). We can determine the solutions for (b) and (c) by employing either of the two approaches:

1. Enumeration of all strategies
2. Bayes' theorem

TABLE 12.2 **Conditional Probability $[P(T_j/N_i)]$ Table**

	Observations on the Geological Structure		
State of Nature	T_1 Closed	T_2 Open	T_3 No Structure
Dry hole N_1	0.1	0.1	0.8
100,000 barrels N_2	0.1	0.3	0.6
500,000 barrels N_3	0.2	0.5	0.3
1,000,000 barrels N_3	0.7	0.2	0.1

ENUMERATION OF ALL STRATEGIES

A strategy in the context of our oil drilling problem is a decision rule that tells the oil driller what action to take for each kind of experimental observation. For instance, one possible strategy would be to drill if the observation is T_1, not to drill if the observation is T_2, and to drill if the observation is T_3. If m actions are available for each observation, and k types of observations are the possible outcomes of experimentation, the total number of strategies (TS) is given by expression 12.1.

$$TS = m^k \qquad\qquad (12.1)$$

For our example, $m = 2$ and $k = 3$. Therefore, number of strategies = $2^3 = 8$. The eight possible strategies are listed in Table 12.3. Next, the action probabilities for each state of nature for a given strategy is calculated. We shall illustrate the computational method of determining the action probabilities for strategy $S2$. The decision rule for $S2$ is

If observation is T_1, selected action is A_1,
If observation is T_2, selected action is A_1,
If observation is T_3, selected action is A_2.

TABLE 12.3   Strategies and Action Probabilities

Strategy	Action Assigned to Each Observation				Action Probability for Each State of Nature			
	T_1	T_2	T_3		N_1	N_2	N_3	N_4
S1	A_1	A_1	A_1	A_1	1.0	1.0	1.0	1.0
				A_2	0.0	0.0	0.0	0.0
S2	A_1	A_1	A_2	A_1	0.2	0.4	0.7	0.9
				A_2	0.8	0.6	0.3	0.1
S3	A_1	A_2	A_1	A_1	0.9	0.7	0.5	0.8
				A_2	0.1	0.3	0.5	0.2
S4	A_1	A_2	A_2	A_1	0.1	0.1	0.2	0.7
				A_2	0.9	0.9	0.8	0.3
S5	A_2	A_1	A_1	A_1	0.9	0.9	0.8	0.3
				A_2	0.1	0.1	0.2	0.7
S6	A_2	A_1	A_2	A_1	0.1	0.3	0.5	0.2
				A_2	0.9	0.7	0.5	0.8
S7	A_2	A_2	A_1	A_1	0.8	0.6	0.3	0.1
				A_2	0.2	0.4	0.7	0.9
S8	A_2	A_2	A_2	A_1	0.0	0.0	0.0	0.0
				A_2	1.0	1.0	1.0	1.0

TABLE 12.4   Strategies and Expected Gains

Strategy	Action Assigned to Each Observation			Average Gain for Each State of Nature (in $1000 Units)				Expected Gain (in $1000 Units)
	T_1	T_2	T_3	N_1	N_2	N_3	N_4	
S1	A_1	A_1	A_1	−50	−30	100	350	22.00
S2	A_1	A_1	A_2	−10	−12	70	315	38.30
S3	A_1	A_2	A_1	−45	−21	50	280	8.90
S4	A_1	A_2	A_2	− 5	− 3	20	245	25.20
S5	A_2	A_1	A_1	−45	−27	80	105	− 3.20
S6	A_2	A_1	A_2	− 5	− 9	50	70	13.10
S7	A_2	A_2	A_1	−40	−18	30	35	−16.30
S8	A_2	A_2	A_2	0	0	0	0	0.00

If N_1 is the state of nature, we observe T_1 for 0.10 of the time, T_2 for 0.10 of the time, and T_3 for 0.80 of the time (see Table 12.2). Because A_1 is the selected action for observation T_1 and T_2, action A_1 would be implemented for $0.10 + 0.10 = 0.20$ of the time. A_2 is the selected action for observation T_3; therefore the action probability of A_2 for N_1 is 0.80. Likewise,

Action probability of A_1 for $N_2 = 0.1 + 0.3 = 0.4$
Action probability of A_2 for $N_2 = 0.6$
Action probability of A_1 for $N_3 = 0.2 + 0.5 = 0.7$
Action probability of A_2 for $N_3 = 0.3$
Action probability of A_1 for $N_4 = 0.7 + 0.2 = 0.9$
Action probability of A_2 for $N_4 = 0.1$

The action probabilities for other strategies are computed similarly. The action probabilities for all strategies are depicted in Table 12.3.

For strategy S2, if N_1 is the state of nature, action A_1 is taken for 0.20 of the time and action A_2 is taken for 0.80 of the time. Table 12.1 indicates that the outcome for actions A_1 and A_2 are $-50,000$ and $0, respectively. Therefore, the average gain for N_1 is

$$-50,000(0.2) + 0(0.8) = \$-10,000$$

Similarly,

Average gain for $N_2 = -30,000(0.4) + 0(0.6) = \$-12,000$
Average gain for $N_3 = 100,000(0.7) + 0(0.3) = \$\ 70,000$
Average gain for $N_4 = 350,000(0.9) + 0(0.1) = \$\ 315,000$

The average gains for other strategies are calculated similarly. The average gains for all strategies are shown in Table 12.4. How can we select a strategy among the eight possible strategies? In other words, which strategy yields the largest expected gain? We have determined the average gain of different states of nature for each strategy. We can calculate the expected gain for any strategy by adding the products of average gain and the corresponding state of nature probability for all states of nature.

$$EG(Sj) = P(N_1) \cdot AG(N_1) + P(N_2) \cdot AG(N_2) + P(N_3) \cdot AG(N_3)$$
$$+ P(N_4) \cdot AG(N_4) \tag{12.2}$$

where

$$EG(Sj) = \text{expected gain for strategy } Sj$$

$$P(N_i) = \text{prior probability of state of nature } N_i$$

$$AG(N_i) = \text{average gain for state of nature } N_i$$

For instance, the expected gain for strategy S2 is

$$EG(S2) = 0.6(-10,000) + 0.1(12,000) + 0.2(70,000) + 0.1(315,000)$$
$$\$38,300$$

Table 12.4 presents the expected gains for all strategies. Note that strategy S2 yields the maximum expected gain. Therefore, we conclude that S2 is the optimal strategy and the corresponding expected gain is \$38,300. The optimal decision rule is

If observation is T_1, select action A_1
If observation is T_2, select action A_1
If observation is T_3, select action A_2

The steps of the enumeration procedure are given by:

STEP 1. List the states of nature for the decision-making problem, and estimate prior probability for each state of nature.

STEP 2. List the strategies and the decision rule for each strategy.

STEP 3. Calculate the action probabilities of each state of nature for each strategy.

STEP 4. For each strategy, compute the average gain for each state of nature.

STEP 5. Determine the expected gain of each strategy. Select the strategy with the largest expected gain. The selected strategy is the optimal strategy.

In determining optimal strategy we find that the application of the enumeration procedure to a decision problem with a large number of possible strategies would require impressive, if not impossible, computational effort. The Bayes' theorem is used as an efficient decision aid, especially for large-size problems, in establishing the optimal decision rule.

BAYES' THEOREM

Let the events $N_1, N_2, \ldots, N_s$ be a set of mutually exclusive events. Suppose T is a dependent event; when T occurs, it must be accompanied or preceded by some N_i. That is,

$$P(T) = P(TN_1) + P(TN_2) + \cdots + P(TN_s) \tag{12.3}$$

Suppose T has occurred, and we wish to compute the probability that a given N_i also occurred, that is, $P(N_i/T)$. From (10.5) we obtain

$$P(N_i/T) = \frac{P(TN_i)}{P(T)} \tag{12.4}$$

Likewise,

$$P(T/N_i) = \frac{P(TN_i)}{P(N_i)} \tag{12.5}$$

Solving (12.4) and (12.5) for $P(TN_i)$, we have

$$P(N_i/T)P(T) = P(T/N_i)P(N_i)$$

$$P(N_i/T) = \frac{P(T/N_i)P(N_i)}{P(T)} \tag{12.6}$$

From (12.5) we get

$$P(TN_i) = P(T/N_i)P(N_i) \tag{12.7}$$

Substituting $P(TN_i)$ in (12.3) by its equivalent, given in (12.7), leads to

$$P(T) = P(T/N_1)P(N_1) + P(T/N_2)P(N_2) + \cdots + P(T/N_s)P(N_s) \tag{12.8}$$

Expression 12.6 can be rewritten to obtain the final statement of Bayes' theorem:

$$P(N_i/T) = \frac{P(T/N_i)P(N_i)}{P(T/N_1)P(N_1) + P(T/N_2)P(N_2) + \cdots + P(T/N_s)P(N_s)} \tag{12.9}$$

We shall illustrate the applicability of Bayes' theorem with our oil drilling problem.

Computational Procedure

The Bayesian method is a systematic approach for establishing the optimal strategy. This decision-making tool makes use of the decision maker's judgment and the economic consequences of the states of nature and the possible actions.

The values of $P(T/N_i)P(N_i)$ and $P(T)$ for each observation are required as seen in expression 12.9. For our oil drilling problem, Table 12.5 shows the computations of $P(T/N_i)P(N_i)$ and $P(T)$ values. Subsequently, these values are substituted in expression 12.9 to compute the $P(N_i/T)$ values as shown in Table 12.6. The decision tree in Figure 12.1 displays the actions and the outcomes incorporating the given data and the computed

TABLE 12.5   Calculation of $P(T)$ Values

		Observation		
		T_1	T_2	T_3
$P(T/N_i)P(N_i)$	N_1	$0.1 \times 0.6 = 0.06$	$0.1 \times 0.6 = 0.06$	$0.8 \times 0.6 = 0.48$
	N_2	$0.1 \times 0.1 = 0.01$	$0.3 \times 0.1 = 0.03$	$0.6 \times 0.1 = 0.06$
	N_3	$0.2 \times 0.2 = 0.04$	$0.5 \times 0.2 = 0.10$	$0.3 \times 0.2 = 0.06$
	N_4	$0.7 \times 0.1 = 0.07$	$0.2 \times 0.1 = 0.02$	$0.1 \times 0.1 = 0.01$
$P(T)$		0.18	0.21	0.61

TABLE 12.6   Calculation of $[P(N_i/T)]$ Values

	Observation		
State of Nature	T_1	T_2	T_3
N_1	$\dfrac{0.06}{0.18} = 0.333$	$\dfrac{0.06}{0.21} = 0.286$	$\dfrac{0.48}{0.61} = 0.787$
N_2	$\dfrac{0.01}{0.18} = 0.056$	$\dfrac{0.03}{0.21} = 0.143$	$\dfrac{0.06}{0.61} = 0.098$
N_3	$\dfrac{0.04}{0.18} = 0.222$	$\dfrac{0.10}{0.21} = 0.476$	$\dfrac{0.06}{0.61} = 0.098$
N_4	$\dfrac{0.07}{0.18} = 0.389$	$\dfrac{0.02}{0.21} = 0.095$	$\dfrac{0.01}{0.61} = 0.016$

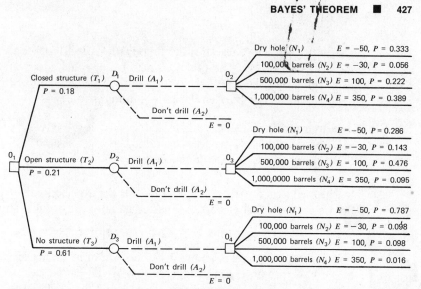

FIGURE 12.1 Decision tree for the oil drilling problem.

values. Next, the expected payoff of each action for different observations are calculated.

$$\text{Expected payoff of } O_2 = 0.333(-50) + 0.056(-30) + 0.222(100)$$
$$+ 0.389(350) = 140.0$$

Likewise, the payoffs of other outcome points are calculated. Table 12.7 presents the expected payoffs for various observation-action combinations. At the decision point D_1, two alternatives, A_1 and A_2, are available. The expected payoff of action A_1 is \$140,000, and that of action A_2 is \$0. We select action A_1 because it yields the largest expected payoff. We can choose the preferred actions for each observation from Table 12.7. The column indexed T_1 gives the expected payoffs for various actions

TABLE 12.7 Expected Payoffs for Different Observation-Action Combinations

	Observation		
Action	T_1	T_2	T_3
A_1	140.0	62.4	−26.7
A_2	0	0	0
Selected action	A_1	A_1	A_2

available at D_1. The largest value among them is \$140.0, and it corresponds to action A_1. The preferred actions for other observations are chosen similarly. The set of selected actions becomes the decision rule for the optimal strategy. Therefore, the optimal strategy is

If observation is T_1, select action A_1,
If observation is T_2, select action A_1,
If observation is T_3, select action A_2.

The expected gain for the optimal strategy is the expected payoff at point O_1 in Figure 12.1. Therefore, the expected gain for the optimal strategy is

$$0.18(140.0) + 0.21(62.4) + 0.61(0) = 38.3$$

The optimal strategy is shown by double lines in Figure 12.2. Recall that we have obtained the same solution by the enumeration method. The optimal strategy is known as the *optimal Bayesian Strategy*. The Bayes' theorem is used to sharpen our judgment on the occurrence of the states of nature by utilizing the experimental information together with the prior probabilities. The $P(N_i/T)$ values are called the *posterior probabilities*.

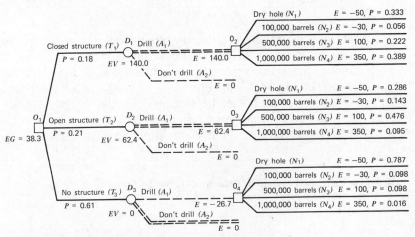

FIGURE 12.2 Decision tree for the oil drilling problem.

The computational procedure can be divided into four steps:

STEP 1. Calculate the $P(T/N_i)P(N_i)$ values for all combinations of states of nature and observations.

STEP 2. Calculate the posterior probabilities employing the Bayes' theorem.

STEP 3. Determine the expected payoff values for all observation-action combinations.

STEP 4. For each observation select the action that has the maximum payoff among the possible alternatives. Calculate the expected gain for the selected strategy.

A Computer Model

We shall develop a computer program for determining the Bayesian strategy in this section. The following variables are employed in the program:

NAT	= number of possible states of nature (s)
LACT	= number of actions available (m)
IOBVS	= number of observations (k)
U(K, L)	= estimated payoff value for state of nature K and action L
CONDTN(K, M)	= conditional probability of observation M occurring given that state of nature K has occurred.
	= $P(T_j/N_i)$ for j = M and i = K
STATP(K)	= prior probability for state of nature K
TABLE1(K, M)	= value of $P(T/N_i)P(N_i)$ for state of nature i = K and observation M
TABLE2(K, M)	= value of $P(N_i/T)$ for state of nature i = K and observation M
POBVS(M)	= value of $P(T)$ for observation M
MAXACT(M)	= value that specifies the action with the largest expected payoff for observation M
EFFECT(M)	= the largest expected payoff for observation M
VALUE	= value of $P(T)$ for the specified observation
XVAL	= expected payoff for the specified observation-action combination
XLRGST	= the largest expected payoff for the specified observation
YVAL	= expected gain for the optimal Bayesian strategy

The Variable STATP( ) is a one-dimensional array with NAT elements, while the variables POBVS( ), MAXACT( ), and EFFECT( ) are one-dimensional arrays of size IOBVS elements. Matrix U(,) is the payoff table with NAT rows and LACT columns. The matrices CONDTN(,), TABLE1(,), and TABLE2(,) are of size NAT by IOBVS. Matrix TABLE3 (,) has LACT rows and IOBVS columns.

We might employ the following statements to transmit the input data into the computer:

```
        READ 10, NAT, LACT, IOBVS
        DO 100 I = 1, NAT
        READ 30, (U(I, J), J = 1, LACT)
100     CONTINUE
        DO 200 I = 1, NAT
        READ 40, (CONDTN(I, J), J = 1, IOBVS)
200     CONTINUE
        READ 20, (STATP(I), I = 1, NAT)
 10     FORMAT ( )
 30     FORMAT ( )
 40     FORMAT ( )
 20     FORMAT ( )
```

The following set of instructions would cause the computer to calculate the $P(T/N_i)P(N_i)$ and $P(T)$ values for different observations:

```
        DO 400 J = 1, IOBVS
        VALUE = 0.0
        DO 500 I = 1, NAT
        TABLE1(I, J) = CONDTN(I, J) * STATP(I)
        VALUE = VALUE + TABLE1(I, J)
500     CONTINUE
        POVBS(J) = VALUE
400     CONTINUE
```

The $P(N_i/T)$ values are calculated by utilizing the instructions:

```
        DO 600 I = 1, NAT
        DO 600 J = 1, IOBVS
        TABLE2(I, J) = TABLE1(I, J)/POBVS(J)
600     CONTINUE
```

For determing the values of expected payoffs for different observation-action combinations, we might employ the statements:

```
      DO 700 J = 1, IOBVS
      DO 710 I = 1, LACT
      XVAL = 0.00
      DO 720 MX = 1, NAT
      XVAL = XVAL + TABLE2(MX, J) * U(MX, I)
  720 CONTINUE
      TABLE3(I, J) = XVAL
  710 CONTINUE
  700 CONTINUE
```

The MAXACT and EFFECT values for each observation are determined by the following instructions:

```
      DO 730 J = 1, IOBVS
      XLRGST = -9999999.9
      DO 740 I = 1, LACT
      IF (TABLE3(I, J).LT.XLRGST) GO TO 740
      MAXACT(J) = I
      XLRGST = TABLE3(I, J)
  740 CONTINUE
      IC = MAXACT(J)
      EFFECT(J) = TABLE3(IC, J)
  730 CONTINUE
```

The expected gain for the optimal Bayesian strategy is calculated by including the statements:

```
      YVAL = 0.00
      DO 750 J = 1, IOBVS
      YVAL = YVAL + (POBVS(J) * EFFECT(J))
  750 CONTINUE
```

Then, the input data and the computed values are printed out and the program is terminated. Figure 12.3 exhibits the complete computer program composed of our previously designed program segment. The resulting computer printout for our oil drilling problem is shown in Figure 12.4.

```
                DIMENSION U(4,2), CONDTN(4,3), STATP(4)
                DIMENSION TABLE1(4,3), TABLE2(4,3), TABLE3(2,3)
                DIMENSION POBVS(3), MAXACT(3), EFFECT(3)
        C
        C   READ THE INPUT DATA
                READ 10, NAT, LACT, IOBVS
                DO 100 I = 1,NAT
                READ 30, (U(I,J), J = 1,LACT)
          100 CONTINUE
                DO 200 I = 1,NAT
                READ 40, ( CONDTN(I,J), J = 1,IOBVS)
          200 CONTINUE
                READ 20,(STATP(I), I = 1,NAT)
        C
        C   COMPUTE TABLE1 AND POBVS VALUES
                DO 400 J = 1,IOBVS
                VALUE = 0.0
        C
                DO 500 I = 1,NAT
                TABLE1(I,J) = CONDTN(I,J) * STATP(I)
                VALUE = VALUE + TABLE1(I,J)
          500 CONTINUE
        C
                POBVS(J) = VALUE
          400 CONTINUE
        C
        C   COMPUTE TABLE2 VALUES
                DO 600 I = 1,NAT
                DO 600 J = 1,IOBVS
                TABLE2(I,J) = TABLE1(I,J)/POBVS(J)
          600 CONTINUE
        C
        C   COMPUTE TABLE3 VALUES
                DO 700 J = 1,IOBVS
                DO 710 I = 1,LACT
                XVAL = 0.00
        C
                DO 720 MX = 1,NAT
                XVAL = XVAL + TABLE2(MX,J) * U(MX,I)
          720 CONTINUE
        C
                TABLE3(I,J) = XVAL
          710 CONTINUE
          700 CONTINUE
        C
        C   COMPUTE EFFECT VALUES
                DO 730 J = 1,IOBVS
                XLRGST = -9999999.9
        C
                DO 740 I = 1,LACT
                IF (TABLE3(I,J).LT.XLRGST) GO TO 740
                MAXACT(J) = I
                XLRGST = TABLE3(I,J)
          740 CONTINUE
        C
                IC = MAXACT(J)
                EFFECT(J) = TABLE3(IC,J)
          730 CONTINUE
```

FIGURE 12.3   Computer program for the Bayesian method (illustrated for the oil drilling problem).

```
C
C   DETERMINE YVAL
      YVAL = 0.00
      DO 750 J = 1,IOBVS
      YVAL = YVAL + (POBVS(J) * EFFECT(J))
  750 CONTINUE
C
C   PRINTOUT THE INPUT DATA
      PRINT 8
      PRINT 805
      PRINT 806
C
      DO 810 I=1,NAT
      PRINT 820, I, ( U(I,J), J= 1,2 ), STATP(I)
  810 CONTINUE
C
      PRINT 815
      PRINT 817
C
      DO 825 I=1,NAT
      PRINT 830, I, ( CONDTN(I,J), J= 1,3)
  825 CONTINUE
C
      PRINT 850
C
C
C   PRINTOUT THE COMPUTED VALUES
      DO 860 I=1,NAT
      PRINT 855, I, ( TABLE1 (I,J), J= 1,3 )
  860 CONTINUE
      PRINT 845, (POBVS(K),K=1,3)
C
      PRINT 880
C
      DO 885 I=1,NAT
      PRINT 890,I,(TABLE2(I,J), J=1,3)
  885 CONTINUE
C
      PRINT 895
C
      DO 900 I=1,LACT
      PRINT 905,I,(TABLE3(I,J), J=1,3)
  900 CONTINUE
C
      PRINT 910
C
      DO 920 I=1,IOBVS
      PRINT 925,I,MAXACT(I)
  920 CONTINUE
C
      PRINT 945, YVAL
C
C
C
    8 FORMAT (1H1)
   10 FORMAT (3I4)
   20 FORMAT (3F6.3)
   30 FORMAT (3F6.2)
   40 FORMAT (3F6.4)
  805 FORMAT (5X,12HPAYOFF TABLE)
  806 FORMAT (13X,1H1,5X,1H2,2X,9HP(NATURE))
  815 FORMAT (/,5X,36HCONDITIONAL PROBABILITY TABLE,P(T/N))
```

FIGURE 12.3 (Continued)

433

```
817 FORMAT (13X,1H1,5X,1H2,5X,1H3)
820 FORMAT (8X, I1, 1X, 2(F5.0,1X), 3X, F3.2)
830 FORMAT ( 9X, I1, 3(1X,F4.3,1X)))
845 FORMAT ( 5X, 5HSUM =, 3(1X,F4.3,1X))
850 FORMAT ( /,5X, 13HTABLE1 VALUES )
855 FORMAT ( 9X, I1, 3(1X,F4.3,1X))
880 FORMAT ( /,5X, 13HTABLE2 VALUES  )
890 FORMAT ( 9X, I1, 3(1X,F4.3,1X))
895 FORMAT ( / 5X, 13HTABLE3 VALUES )
905 FORMAT ( 9X, I1, 3(1X,F5.1 ) )
910 FORMAT (/,10X,18HMAXIMIZING ACTIONS,/)
925 FORMAT (5X,22HIF OBSERVATION IS NO. ,I4,
    219H SELECT ACTION NO.,I3)
945 FORMAT (/,5X,41HEXPECTED GAIN FOR THE SELECTED STRATEGY =,
    1 F9.3)
    STOP
    END
```

FIGURE 12.3 (*Continued*)

```
PAYOFF TABLE
         1     2   P(NATURE)
    1  -50.   -0     .60
    2  -30.   -0     .10
    3  100.   -0     .20
    4  350.   -0     .10

CONDITIONAL PROBABILITY TABLE,P(T/N)
         1     2     3
    1  .100  .100  .800
    2  .100  .300  .600
    3  .200  .500  .300
    4  .700  .200  .100

TABLE1 VALUES
    1  .060  .060  .480
    2  .010  .030  .060
    3  .040  .100  .060
    4  .070  .020  .010
SUM = .180  .210  .610

TABLE2 VALUES
    1  .333  .286  .787
    2  .056  .143  .098
    3  .222  .476  .098
    4  .389  .095  .016

TABLE3 VALUES
    1 140.0  62.4 -26.7
    2    0     0     0

      MAXIMIZING ACTIONS

IF OBSERVATION IS NO.   1  SELECT ACTION NO.   1
IF OBSERVATION IS NO.   2  SELECT ACTION NO.   1
IF OBSERVATION IS NO.   3  SELECT ACTION NO.   2

EXPECTED GAIN FOR THE SELECTED STRATEGY =   38.300
```

FIGURE 12.4   Computer printout for the oil drilling problem.

VALUE OF EXPERIMENT

The Bayesian approach enables us to determine the optimum strategy with the least possible computational effort. We can also use this technique to estimate the value of the experiment. How much can we pay for the seismic soundings? If the experimental data were not available, we would have used, as shown earlier, the prior probabilities to obtain the decision rule: Drill the oil well. It yields an expected gain of $22,000. The experimental data was utilized in establishing the optimal Bayesian strategy, and it yields an expected gain of $38,300. The expected gain has been increased by $16,300 (= $38,300 − $22,000) by using the experimental data. Hence, the value of experimentation is $16,300.

For a more complete discussion of the oil drilling problem, the reader should consult Grayson's book (1960). The interested reader is referred to the text by Halter and Dean (1971) for the application of Bayesian analysis in various fields such as agriculture, forest management, and climatology.

PROBLEMS

12.1 A small import-export company is planning to import slide projectors and market them through its discount stores. The manager estimates from previous experience that demand can be approximately represented to occur with these probabilities:

State of Nature (Demand)	Prior Probability
N_1 = 10,000 units	0.2
N_2 = 15,000 units	0.3
N_3 = 20,000 units	0.5
	$\overline{1.0}$

Management considers the following actions:

$$A_1 = \text{Order 10,000 units}$$
$$A_2 = \text{Order 15,000 units}$$
$$A_3 = \text{Order 20,000 units}$$

If the demand is larger than the order size, the company might suffer damage to its reputation. However, if the demand is less than the order size, the excess units can be sold for a lower unit price. The payoff table given below presents the consequences (in $10,000) of various order sizes for the various demands.

State of Nature	Order Sizes		
	A_1 10,000 units	A_2 15,000 units	A_3 20,000 units
N_1 = 10,000 units	65	60	50
N_2 = 15,000 units	50	80	75
N_3 = 20,000 units	30	60	90

The studies conducted by a market research team indicate whether there would be large demand (T_1), average demand (T_2), or limited demand (T_3). The relationship, $P(T_j/N_i)$, between the findings of the team and the demands are

State of Nature	T_1	T_2	T_3
N_1	0.2	0.4	0.4
N_2	0.1	0.6	0.3
N_3	0.8	0.2	0.0

(a) Calculate the expected gains for the strategies:

Strategy	T_1	T_2	T_3
1	A_1	A_2	A_2
2	A_1	A_2	A_3
3	A_3	A_3	A_1
4	A_3	A_2	A_2
5	A_3	A_2	A_1

   (b) Write a FORTRAN program to cause the computer to calculate and print out the expected gains for all strategies.

12.2 Consider Problem 12.1. Determine the optimal strategy by the Bayes' theorem.

12.3 Consider Problem 12.1. Formulate a FORTRAN program to enable the computer to determine and print out the optimal Bayes' strategy using Bayes' theorem.

12.4 Consider Problem 12.1.

   (a) Calculate the optimal strategy and the corresponding expected gain when the market research data is not available.

   (b) Determine the value of the experiment.

──────────── SELECTED REFERENCES ────────────

Chernoff, H., and L. Moses, *Elementary Decision Theory*, Wiley, New York, 1959.

Giffin, W. C., *Introduction to Operations Engineering*, Richard D. Irwin, Homewood, Ill., 1971.

Grayson, C. J., Jr., *Decisions under Uncertainty; Drilling Decisions by Oil and Gas Operators*, Harvard Business School, Plimpton Press, Boston, 1960.

Halter, A. N., and G. W. Dean, *Decisions under Uncertainty with Research Applications*, South-Western Publishing, Cincinnati, Ohio, 1971.

Morris, W. T., *Management Science—a Bayesian Introduction*, Prentice-Hall, Englewood Cliffs, N.J., 1968.

Raiffa, H., *Decision Analysis—Introductory Lectures on Choices under Uncertainty*, Addison-Wesley, Reading, Mass., 1968.

Schlaifer, R., *Analysis of Decisions under Uncertainty*, McGraw-Hill, New York, 1969.

inventory planning - probabilistic models

THE INVENTORY METHODS discussed in Chapter 9 assume that the requirements of the products as well as the lead time are accurately known. These two assumptions restrict the application of the deterministic models to many real-life inventory systems. The probabilistic inventory methods attempt to take into account the variations in the demands and lead times and thus are employed to a wide variety of situations.

NATURE OF PROBABILISTIC INVENTORY MODELS

The probabilistic inventory methods are classified into three groups:

    a. Variations in demand, while the lead time is a constant.
    b. Variations in lead time, while the demand is a constant.
    c. Variations in both the lead time and demand.

We cannot forecast the exact quantity of units required during the lead-time period because of the range of values that the demand and lead time may take. If an order is placed whenever the inventory level is equal to the expected usage during the lead time, the customers are supplied with their requirements until they total the expected usage. Thereafter, the customers demands are not satisfied. Such a drawback can be avoided by keeping a few extra units as buffer stock. The *buffer stock* is the reserve inventory that is used to fulfill the requirements during the lead time whenever the total demand exceeds the expected demand. The function of buffer stock is illustrated in Figure 13.1.

In Figure 13.1, Case 1 displays the inventory model under certainty. It indicates that the *DDLT* value is a constant for each cycle and hence no buffer stock is required. However, Figure 13.1, Case 2 displays an inventory model under uncertainty, denoting that the *DDLT* for cycle 1 is larger than the expected demand and thus buffer stock is applied. In cycles 2 and 3, the total actual demand values do not exceed the expected

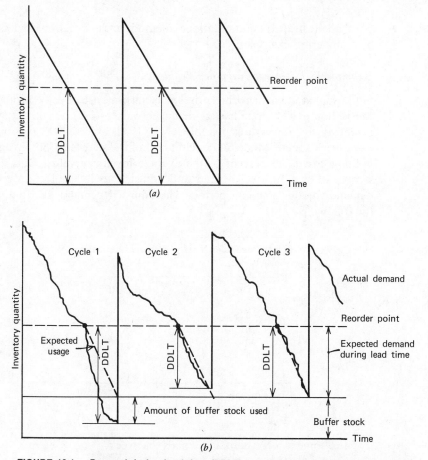

FIGURE 13.1   Demand during lead time (DDLT) values for the two types of inventory models. (*a*) **Case 1. Deterministic model; DDLT is a constant. Buffer stock = 0.** (*b*) **Case 2. Probabilistic model; DDLT is not a constant. Buffer stock is required in cycle 1.**

demand and, therefore, buffer stocks are not used. By storing a larger buffer stock, the service level is increased because the risk of stock runout is reduced. The decision to increase the buffer stock imposes a greater capital investment and inventory cost. We shall develop computational procedures and related computer models for some important inventory systems under risk.

VARIATIONS IN DEMAND

We shall illustrate the computational method for the variations in demand by a numerical example.

Example 13.1   Stereo Inventory Problem

Suppose that the Miller Enterprises sells musical instruments through its music houses. The purchasing manager wishes to know the expected inventory cost for various possible service levels exceeding 80 percent for eight-track stereos Model PQR. Unit cost of the stereos is $50. Inventory holding cost is 20 percent of the average inventory value. Data for a sample of size 50 are gathered and adjusted to reflect the forecasted demand for the planning period. The adjusted demand distribution is the following:

Demand per Week, (in 100 Units), D	Number of Observations F(D)
6	5
7	10
8	20
9	7
10	4
11	3
12	1
Total number of observations	50

The lead time is a constant and equals one week.

STEP 1. Calculate the probability for each demand possibility applying expression 13.1:

$$P(D) = \frac{F(D)}{\text{Total number of observations}} \qquad (13.1)$$

where

$P(D)$ = Probability that the demand during lead
time is D units

$F(D)$ = frequency or the number of observations
that the demand during lead time is D units

D	$P(D)$
600	$5/50 = 0.10$
700	$10/50 = 0.20$
800	$20/50 = 0.40$
900	$7/50 = 0.14$
1000	$4/50 = 0.08$
1100	$3/50 = 0.06$
1200	$1/50 = 0.02$
Total	$= 1.00$

Figure 13.2 displays the probability values.

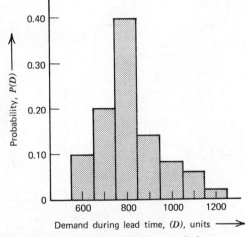

FIGURE 13.2 **Distribution of demand during lead time period.**

STEP 2. Determine the probability [PGTL(D)] for each demand possibility (D), which indicates the chance that the actual demand during lead time may exceed D units.

$$PGTL(D) = PGTL(E) + P(E) \tag{13.2}$$

Where E = the next larger demand possibility than D and PGTL(G) = 0 provided that G denotes the demand possibility with the largest units.

We start the calculations at the bottom for 1200 and then move step by step to 600 units.

D	P(D)	PGTL(D)
600	0.10	0.90
700	0.20	0.70
800	0.40	0.30
900	0.14	0.16
1000	0.08	0.08
1100	0.06	0.02
1200	0.02	0.00
	Total = 1.00	

PGTL(D) also refers to the probability of stock runout for the related D value. The probability of stock runout is shown in Figure 13.3.

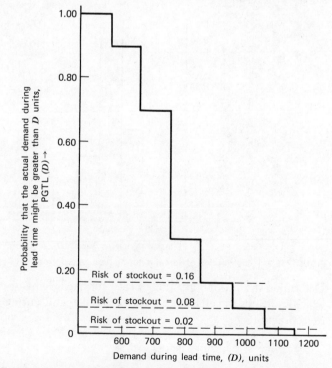

FIGURE 13.3   Cumulative distribution of the actual demand during lead time indicating the chances of exceeding the planned D units.

STEP 3. Compute the expected demand during lead time by the expression 13.3

$$\text{DVMEAN} = \frac{\sum\{(D) \cdot F(D)\}}{\sum F(D)} \qquad (13.3)$$

where DVMEAN is the expected demand during lead time.
Thus,

$$(600) \cdot (5) + (700) \cdot (10) + (800) \cdot (20) + (900) \cdot (7)$$
$$+ (1000) \cdot (4) + (1100) \cdot (3) + (1200) \cdot (1) = 40,800$$

and

$$5 + 10 + 20 + 7 + 4 + 3 + 1 = 50$$

$$\text{DVMEAN} = \frac{40,800}{50} = 816 \text{ units}$$

STEP 4. Calculate the buffer stock and average inventory for the desired D values. The manager wants the inventory cost information for service levels larger than 80 percent. This corresponds to less than 0.20 (= 1 − 0.80) of PGTL(D) value. Thus, we note that only for the D values of 900, 1000, 1100, and 1200 units the buffer stock and average inventory are computed.

For any given D value,

$$\text{BUFFER} = D - \text{DVMEAN} \qquad (13.4)$$

$$\text{AI} = \frac{D - \text{BUFFER}}{2} + \text{BUFFER} \qquad (13.5)$$

where AI = average inventory

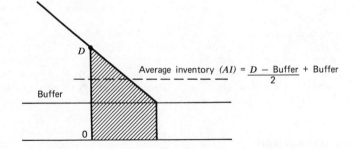

D	DVMEAN	BUFFER	AI, dollars
900	816	900 − 816 = 84	(900 − 84)/2 + 84 = 492.00
1000	816	1000 − 816 = 184	(1000 − 184)/2 + 184 = 592.00
1100	816	1100 − 816 = 284	(1100 − 284)/2 + 284 = 692.00
1200	816	1200 − 816 = 384	(1200 − 384)/2 + 384 = 792.00

STEP 5. Establish the service level, the value of the average inventory and the inventory cost.

$$\text{Service level, } Y = 1 - \text{PGTL(D)} \tag{13.6}$$

$$\text{Value of average inventory, COSTAI} = (\text{AI}) \cdot \text{UCOST} \tag{13.7}$$

$$\text{Inventory cost, COSTI} = (\text{COSTAI}) \cdot (\text{FACTOR}) \tag{13.8}$$

where

UCOST = unit cost of the product
FACTOR = inventory holding cost represented as a fraction of the average inventory value

The computed values are shown in Table 13.1. The purchasing manager can analyze the effects of various service levels and the related inventory costs and select the one that fulfills the company's requirements. If the service level is changed from 84 to 92 percent, the inventory cost would be increased by $1,000.00 (= 5920.00 − 4920.00). Furthermore, if the service level is to be incremented to 98 percent, an additional $1000.00 (= 6920.00 − 5920.00) must be expended as inventory cost. If no stockouts are desired, the service level is 100 percent requiring a supplemental inventory cost of $1000.00 (= 7920.00 − 6920.00). The manager can use one of these alternatives as their operative plan.

TABLE 13.1   Calculation of Inventory Cost Values for Various Service Levels

D	Service Level	Value of Average Inventory ($)	Inventory Cost ($)
900	1 − 0.16 = 0.84	(492)(50) = 24,600.00	(24,600)(0.20) = 4920.00
1000	1 − 0.08 = 0.92	(592)(50) = 29,600.00	(29,600)(0.20) = 5920.00
1100	1 − 0.02 = 0.98	(692)(50) = 34,600.00	(34,600)(0.20) = 6920.00
1200	1 − 0.00 = 1.00	(792)(50) = 39,600.00	(39,600)(0.20) = 7920.00

A Computer Program

We shall develop a general computer program in this section. We might initialize the program with the input instruction

```
        READ 700, UCOST, FACTOR
    700 FORMAT ( )
```

Let

SUM = total number of observations

WTOTAL = $\sum(D) \cdot F(D)$ applied in expression 13.3.

DV(K) = demand value, in units, arranged in the Kth position. The values are arranged in the ascending order. For instance, $DV(3) = 800$ in our example.

JSIZE = total number of demand possibilities.

The program can be applied to problems where the total number of demand possibilities cannot exceed 19. In the data deck, the last card indicates a negative value for frequency, and it signifies to the computer system that it is the last data card. The following set of statements would cause the computer to read DV( ) and associated F( ) values, and compute JSIZE, SUM, and WTOTAL.

```
        SUM = 0
        WTOTAL = 0
        DO 100 K = 1, 20
        READ 705, DV(K), F(K)
        DV(K) = DV(K) * 100.0
        IF (F(K).LT.0.0) GO TO 102
        SUM = SUM + F(K)
        WTOTAL = WTOTAL + DV(K) * F(K)
100     CONTINUE
102     JSIZE = K − 1
705     FORMAT ( )
```

The values of P and PGTL for each demand might be computed by the following statements:

```
        DO 110 J = 1, JSIZE
        P(J) = F(J)/SUM
110     CONTINUE
        K = JSIZE
        PGTL(K) = 0
        DO 120 L = 1, JSIZE
        J = K
        K = K − 1
        IF (K.LE.O) GO TO 122
        PGTL(K) = PGTL(J) + P(J)
120     CONTINUE
```

The input and computed values as well as DVMEAN can be printed out by the following instructions:

```
122   CONTINUE
      PRINT 300, UCOST, FACTOR
      DVMEAN = WTOTAL/SUM
      DO 125 N = 1, JSIZE
      PRINT 320, DV(N), F(N), P(N), PGTL(N)
125   CONTINUE
      PRINT 330, DVMEAN
300   FORMAT ( )
320   FORMAT ( )
330   FORMAT ( )
```

The buffer stock, average inventory, value of the average inventory, and the inventory cost might be determined and printed out by these statements:

```
      DO 130 M = 1, JSIZE
      IF (PGTL(M).GT.0.20) GO TO 130
      BUFFER = DV(M) − DVMEAN
      AI = (DV(M) − BUFFER)/2.0 + BUFFER
      COSTAI = AI * UCOST
      COSTI = COSTAI * FACTOR
      Y = (1.00 − PGTL(M) ) * 100.00
      PRINT 340, Y, DV(M), BUFFER, AI, COSTAI, COSTI
130   CONTINUE
340   FORMAT ( )
```

The complete program is presented in Figure 13.4. The input data deck for our example has the following appearance:

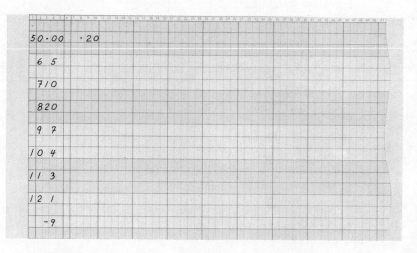

```
      DIMENSION DV(20),F(20),P(20),PGTL(20)
C
C  READ THE DATA
      READ 700, UCOST,FACTOR
C
C  COMPUTE DVMEAN
      SUM=0
      WTOTAL=0
      DO 100 K=1,20
      READ 705, DV(K),F(K)
      DV(K)=DV(K)*100.0
      IF (F(K).LT.0.0) GO TO 102
      SUM=SUM+F(K)
      WTOTAL=WTOTAL+DV(K)*F(K)
  100 CONTINUE
  102 JSIZE=K-1
C
C  FIND OUT P VALUES
      DO 110 J=1,JSIZE
      P(J)=F(J)/SUM
  110 CONTINUE
C
C  CALCULATE PGTL VALUES
      K=JSIZE
      PGTL(K)=0
      DO 120 L=1,JSIZE
      J=K
      K=K-1
      IF(K.LE.0) GO TO 122
      PGTL(K)=PGTL(J)+P(J)
  120 CONTINUE
  122 CONTINUE
      DVMEAN=WTOTAL/SUM
C
C  PRINT OUT THE GIVEN AND CALCULATED VALUES
      PRINT 800
      PRINT 300, UCOST,FACTOR
      DO 125 N=1,JSIZE
      PRINT 320, DV(N),F(N),P(N),PGTL(N)
  125 CONTINUE
      PRINT 330, DVMEAN
C
C  COMPUTE BUFFER,AI,COSTAI,
C  COSTI AND Y VALUES
C  PRINT OUT THE COMPUTED VALUES
      PRINT 341
      DO 130 M=1,JSIZE
      IF(PGTL(M).GT.0.20) GO TO 130
      BUFFER=DV(M)-DVMEAN
      AI=(DV(M)-BUFFER)/2.0+BUFFER
```

FIGURE 13.4 Computer program for the inventory problems with variations in demand (illustrated for the stereo inventory problem).

```
      COSTAI=AI*UCOST
      COSTI=COSTAI*FACTOR
      Y=(1.00-PGTL(M))*100.00
      PRINT 340, Y,DV(M),BUFFER,AI,COSTAI,COSTI
  130 CONTINUE
C
  300 FORMAT (2F10.2)
  320 FORMAT (4F10.2)
  330 FORMAT( / ,5X,26HEXPECTED DEMAND PER WEEK =,F7.2, / )
  340 FORMAT (/,6F10.2)
  341 FORMAT (7X,1HY,6X,5HDV(M),5X,6HBUFFER,6X,2HAI,5X,
     16HCOSTAI,5X,5HCOSTI,/,5X,
      256H-------------------------------------------------------,)
  700 FORMAT (2F5.2)
  705 FORMAT (2F2.0)
  800 FORMAT (1H1)
      STOP
      END
```

FIGURE 13.4 (*Continued*)

Note that the cards are arranged so that DV values are provided in ascending order. Recall that the last card having the negative *F* value denotes that it is the last card. The computer printout of the results is displayed in Figure 13.5.

The same approach can be applied to the inventory problem in which variability in lead time prevails, while the daily demand is a constant. The main difference is that of interchanging the terms "lead time" and "demand" in the procedure presented in this section.

```
    50.00        .20
   600.00       5.00        .10         .90
   700.00      10.00        .20         .70
   800.00      20.00        .40         .30
   900.00       7.00        .14         .16
  1000.00       4.00        .08         .08
  1100.00       3.00        .06         .02
  1200.00       1.00        .02          0
```

EXPECTED DEMAND PER WEEK = 816.00

Y	DV(M)	BUFFER	AI	COSTAI	COSTI
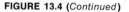
84.00	900.00	84.00	492.00	24600.00	4920.00
92.00	1000.00	184.00	592.00	29600.00	5920.00
98.00	1100.00	284.00	692.00	34600.00	6920.00
100.00	1200.00	384.00	792.00	39600.00	7920.00

FIGURE 13.5 Computer printout for the stereo inventory problem.

INVENTORY MODEL FOR PERISHABLE PRODUCTS

Perishable products such as fresh fruits, vegetables, and hot lunches are usually stored for a limited time because they are liable to spoil or deteriorate. Perishable goods may not be bought and stored in large quantities if they cannot be expended or sold by their short-storage life-time. The unused units can be disposed of, if at all possible, for a lower price incurring a loss. More often than not, the demand for the perishable products during their storage periods are subject to variability and is represented by a demand distribution. The decision makers are often faced with the problem of establishing the amount of units to be procured so that the total expected profit is maximized. To illustrate the computa-tional method, let us consider the following numerical example.

Example 13.2 Tomato Inventory Problem

A grocery store manager buys tomatoes at a cost of $4 per box and sells them for $7 per box. If the tomatoes are not sold in a week, they are sold at the beginning of the following week for $1 per box (salvage value). The demand distribution is given below.

Demand per Week (Boxes)	Probability
1	0.05
2	0.10
3	0.20
4	0.40
5	0.20
6	0.05
Total = 1.00	

The manager wants to implement the alternative that maximizes the total expected profit. In other words, how many boxes or units of tomatoes must be ordered each week to maximize the total expected profit.

STEP 1. Compute the unit profit and unit loss values.

$$UG = UR - UC \tag{13.9}$$

$$UL = UC - US \tag{13.10}$$

where

UG = gain or profit per unit
UL = loss per unit if not sold in a week
UR = revenue or sales price per unit
UC = cost per unit
US = salvage value per unit

Thus

UG = 7 − 4 = $3
UL = 4 − 1 = $3

(Note that UG and UL do not have to be the same value.)

STEP 2. Find out the alternatives available for the number of units to be ordered. For each alternative, determine the conditional profits for the given demand distribution values. We observe from the demand distribution that the demand cannot be less than one unit and also cannot exceed six units in any week. Therefore, the order possibilities are the same as those of the demand, namely, 1, 2, 3, 4, 5, and 6 units. The conditional profit for each order-demand combination is computed by applying the expressions 13.11 and 13.12.

$$P_{Q,D} = [UG \cdot D] - [UL \cdot (Q - D)] \text{ IF } Q > D \qquad (13.11)$$

$$P_{Q,D} = UG \cdot Q \qquad\qquad \text{IF } Q \leqq D \qquad (13.12)$$

where

$P_{Q,D}$ = profit obtained when Q units are ordered
and D units are sold
Q = number of units ordered
D = number of units sold

Note that a negative $P_{Q,D}$ value denotes a loss. For Q > D, more units were ordered, while a few of them are sold and (Q − D) units are unsold by the end of the week. On the other hand, for Q ≦ D, the demanded units at least equals the ordered quantity and thus no units are left over by the end of the week. The computed $P_{Q,D}$ values are shown in Table 13.2. For instance, if Q = 3 units

$$P_{3,1} = (3) \cdot (1) - 3 \cdot (3 - 1) = -3$$
$$P_{3,2} = (3) \cdot (2) - 3 \cdot (3 - 2) = \quad 3$$
$$P_{3,3} = (3) \cdot (3) = 9$$
$$P_{3,4} = (3) \cdot (3) = 9$$
$$P_{3,5} = (3) \cdot (3) = 9$$
$$P_{3,6} = (3) \cdot (3) = 9$$

TABLE 13.2   Calculation of Expected Profit Values for the Alternatives

Order, Units (Q)	Demand (D), units						Expected Gain E(G) ($)
	1	2	3	4	5	6	
1	3.00	3.00	3.00	3.00	3.00	3.00	3.00
2	0	6.00	6.00	6.00	6.00	6.00	5.70
3	− 3.00	3.00	9.00	9.00	9.00	9.00	7.80
4	− 6.00	0	6.00	12.00	12.00	12.00	8.70 ←Maximum
5	− 9.00	− 3.00	3.00	9.00	15.00	15.00	7.20
6	− 12.00	− 6.00	0	6.00	12.00	18.00	4.50
P(D) =	0.05	0.10	0.20	0.40	0.20	0.05	

STEP 3. Determine the expected profit (or gain) for each Q value. The expression applied to find out the expected profit is

$$\bar{E}(G) = \sum_{D=0}^{D=\infty} [P_{Q,D} \cdot P(D)] \qquad (13.13)$$

where

$E(G) =$ expected total profit for the order quantity of Q units

$P(D) =$ probability that there will be a demand for D units in a week

For our inventory problem, the range of D is 1 through 6. Hence,

$$E(G) = \sum_{D=1}^{D=6} [P_{Q,D} \cdot P(D)] \qquad (13.14)$$

The calculated E(G) values are shown in Table 13.2. For example, when Q = 3,

$$E(G) = -3(0.05) + 3(0.10) + 9(0.20) + 9(0.4) + 9(0.2) + 9(0.05)$$
$$= \$7.80$$

STEP 4. Select the largest value among the computed E(G) values. The selected profit value and the related order size, Q_0, refer to the optimum solution. On the basis of these calculations, our conclusion is to order four boxes of tomatoes each time.

A Computer Model

We develop a computer model in this section for the inventory algorithm for the perishable products. Let us define the following list of variables:

COSTU	= cost per unit
SELLU	= sales price per unit
SVALUE	= salvage value per unit
KDSIZE	= total number of demand possibilities
UGAIN	= gain or profit per unit
ULOSS	= loss per unit if not sold in a week
SVAL	= maximum expected total profit
DUNITS(J)	= the Jth demand possibility, denotes number of units
P(J)	= probability that there will be a demand for DUNITS(J) units
QUNITS(J)	= the Jth order-size possibility indicates number of units
EGAIN(J)	= expected total profit for the Jth order-size possibility
PTABLE(J, K)	= profit attained when J units are ordered and K units are demanded

The variables DUNITS( ), P( ), QUNITS( ), and EGAIN( ) are one-dimensional arrays with KDSIZE elements. The two-dimensional sub-scripted variable PTABLE(,) is of the size KDSIZE by KDSIZE.
We might input the data by the instructions:

```
      READ 100, COSTU, SELLU, SVALUE, KDSIZE
      DO 250 J = 1, KDSIZE
      READ 105, DUNITS(J), P(J)
  250 CONTINUE
  100 FORMAT ( )
  105 FORMAT ( )
```

The profit per unit as well as the loss per unit are calculated by the statements:

```
      UGAIN = SELLU − COSTU
      ULOSS = COSTU − SVALUE
```

The order-size possibilities and the profit values for order demand combinations might be determined by the instructions:

```
      DO 255 J = 1, KDSIZE
 255  QUNITS(J) = DUNITS(J)
      DO 300 K = 1, KDSIZE
      DO 300 L = 1, KDSIZE
      IF (QUNITS(K).GT.DUNITS(L)) GO TO 303
      GAIN = QUNITS(K) * UGAIN
      PTABLE(K, L) = GAIN
      GO TO 300
 303  X = UGAIN * DUNITS(L)
      Y = ULOSS * (QUNITS(K) − DUNITS(L))
      PTABLE(K, L) = X − Y
 300  CONTINUE
```

The following statements might be used to compute the expected total profit values for the order-size possibilities:

```
      DO 320 M = 1, KDSIZE
      SUM = 0
      DO 310 N = 1, KDSIZE
      SUM = SUM + PTABLE(M, N) * P(N)
 310  CONTINUE
 320  EGAIN(M) = SUM
```

The largest expected total profit and the related order-size value, INDEX, are established by the instructions:

```
      SVAL = −999999.99
      DO 330 K = 1, KDSIZE
      IF (SVAL.GT.EGAIN(K)) GO TO 330
      SVAL = EGAIN(K)
      INDEX = K
 330  CONTINUE
```

Then, the input values and the computed values are printed out. Our complete computer program for our example problem is shown in Figure

13.6. The input data deck has the following appearance:

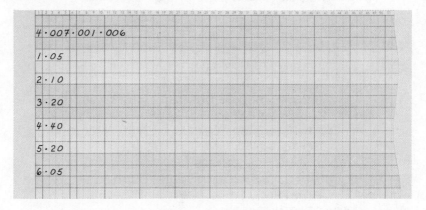

The printout is shown in Figure 13.7. The perishable products problem can also be solved by applying the cost minimization or the regret minimization approaches. These approaches are presented by Groff and Muth (1972).

```
          DIMENSION DUNITS(6),P(6)
          DIMENSION PTABLE(6,6),QUNITS(6),EGAIN(6)
   C
   C   READ THE GIVEN VALUES
          READ 100,COSTU,SELLU,SVALUE,KDSIZE
          DO 250 J=1,KDSIZE
          READ 105,DUNITS(J),P(J)
      250 CONTINUE
   C
   C   COMPUTE UGAIN AND ULOSS
          UGAIN=SELLU-COSTU
          ULOSS=COSTU-SVALUE
   C
   C   FIND OUT POSSIBLE QUNITS VALUES
          DO 255 J=1,KDSIZE
      255 QUNITS(J)=DUNITS(J)
   C
   C   COMPUTE PTABLE VALUES
          DO 300 K=1,KDSIZE
          DO 300 L=1,KDSIZE
          IF(QUNITS(K).GT.DUNITS(L))GO TO 303
   C
          GAIN=QUNITS(K)*UGAIN
          PTABLE(K,L)=GAIN
          GO TO 300
      303 X=UGAIN*DUNITS(L)
          Y=ULOSS*(QUNITS(K)-DUNITS(L))
          PTABLE(K,L)=X-Y
      300 CONTINUE
```

FIGURE 13.6   Computer program for perishable products inventory problems (illustrated for the tomato inventory problem).

```
C
C   FIND OUT EGAIN VALUES
      DO 320 M=1,KDSIZE
      SUM=0
      DO 310 N=1,KDSIZE
      SUM=SUM+PTABLE(M,N)*P(N)
C
  310 CONTINUE
C
  320 EGAIN(M)=SUM
C
C   DETERMINE THE LARGEST EGAIN VALUE
      SVAL=-999999.99
      DO 330 K=1,KDSIZE
      IF(SVAL.GT.EGAIN(K))GO TO 330
      SVAL=EGAIN(K)
      INDEX=K
  330 CONTINUE
C
C   PRINT OUT THE GIVEN VALUES
C   PRINT OUT THE COMPUTED VALUES
      PRINT 700
      PRINT 200,COSTU
      PRINT 201,SELLU
      PRINT 202,SVALUE
      PRINT 701
      PRINT 260
      PRINT 265
      PRINT 270
      PRINT 207,(DUNITS(K),K=1,KDSIZE)
      PRINT 275
      PRINT 265
      DO 230 J=1,KDSIZE
      PRINT 234,QUNITS(J),(PTABLE(J,N),N=1,KDSIZE),EGAIN(J)
  230 CONTINUE
      PRINT 265
      PRINT 215,(P(L),L=1,KDSIZE)
      PRINT 265
      PRINT 701
      K = QUNITS(INDEX)
      PRINT 237, K
      PRINT 239,EGAIN(INDEX)
C
C
  100 FORMAT (3F4.2,I1)
  105 FORMAT (F1.0,F3.2)
  200 FORMAT (5X,13HCOST PER UNIT,10X,3H= $, F6.2)
  201 FORMAT (5X,16HREVENUE PER UNIT, 7X, 3H= $,F6.2 )
  202 FORMAT (5X,26HSALVAGE VALUE PER UNIT = $, F6.2 )
  207 FORMAT (1X,5HUNITS,6(F5.0,2X),2X,6HPROFIT)
  215 FORMAT (1X,5HP(D)=, 6(2X,F4.2,1X))
  234 FORMAT (2X,F3.0,1X,7F7.2)
  237 FORMAT (5X,18HOPTIMUM ORDER SIZE,9X,1H=,I3,6H UNITS)
  239 FORMAT (5X,30HEXPECTED PROFIT AT OPTIMUM = $, F6.2)
  260 FORMAT (16X, 24HCONDITIONAL PROFIT TABLE )
  265 FORMAT (1X,26H--------------------------,
     530H------------------------------ )
  270 FORMAT (1X,5HORDER,13X,15HDEMAND(D),UNITS,
     715X,8HEXPECTED)
  275 FORMAT (51X, 3H($) )
  700 FORMAT (1H1)
  701 FORMAT (//)
      STOP
      END
```

FIGURE 13.6 (Continued)

```
COST PER UNIT          = $  4.00
REVENUE PER UNIT       = $  7.00
SALVAGE VALUE PER UNIT = $  1.00
```

CONDITIONAL PROFIT TABLE

ORDER UNITS	\	DEMAND (D) \UNITS	\	\	\	\	EXPECTED PROFIT ($)
	1.	2.	3.	4.	5.	6.	
1.	3.00	3.00	3.00	3.00	3.00	3.00	3.00
2.	0	6.00	6.00	6.00	6.00	6.00	5.70
3.	-3.00	3.00	9.00	9.00	9.00	9.00	7.80
4.	-6.00	0	6.00	12.00	12.00	12.00	8.70
5.	-9.00	-3.00	3.00	9.00	15.00	15.00	7.20
6.	-12.00	-6.00	0	6.00	12.00	18.00	4.50
P(D)=	.05	.10	.20	.40	.20	.05	

```
OPTIMUM ORDER SIZE          =  4 UNITS
EXPECTED PROFIT AT OPTIMUM = $  8.70
```
FIGURE 13.7   Computer printout for the tomato inventory problem.

VARIATIONS IN LEAD TIME WITH OPPORTUNITY COSTS

The inventory problem in which the lead time is subject to variability and the demand is a constant can be solved by the approach described in the first model. Recall that the effect of stockout was not included directly in the procedure. If orders are received whenever the products are out of stock, the customers' orders cannot be fulfilled and thus sales are lost. Consequently, the customers' goodwill may be lost. If such a situation occurs many times, they may change their loyalty to the competitors' products. In the long run, this may cost more than just the profit from the unsatisfied order. This cost is called the *opportunity cost*. We shall discuss the inventory model in which variations in lead time prevail and opportunity costs for out-of-stock situations are considered. Let us consider the following numerical example:

Example 13.3   Bicycle Inventory Problem

A retailer procures and sells 10-speed bicycles. Here are data for model PXZ:

Demand per year = 1000
Number of working days per year = 250
Ordering cost per order = $75
Holding cost based on the average inventory = $20 per unit per year
Opportunity cost per unit = $30
Daily demand is a constant
The estimated variability in lead time is given below.

Lead Time (LT)	Probability P(LT)
4	0.05
5	0.10
6	0.30
7	0.25
8	0.20
9	0.06
10	0.03
11	0.01
Total = 1.00	

The retailer wishes to determine the optimum inventory plan. The optimum inventory plan is the plan that minimizes the total inventory cost.

The total cost for any plan consists of the four components: ordering cost, holding cost for the ordered quantity, expected holding cost for the excess units available when the shipment arrives, and the expected opportunity cost. If we order the goods providing a small lead time, the consequence is that a fewer number of units would be left out as excess units and, thus, the expected holding cost would be a small expenditure. Also large orders received are left unsatisfied and impose huge opportunity costs. On the other hand, if orders are placed providing a large lead time, the effect is that many excess units and fewer demands during stockouts would occur, resulting in large holding costs and small opportunity costs. Moreover, as we also know, Figure 9.2 represents the effect of order size on ordering cost and holding cost. The total cost curves for the lead-time values are shown in Figure 13.8. We note that the TIC curve for any given lead time is U-shaped. The optimum value for each lead time corresponds to that point on the related curve having the smallest total cost similar to the optimum value shown in Figure 9.2. The optimum for the problem can be established by drawing the cost curves for each one of the lead times, finding their optimum costs, and then selecting the smallest among the optimum values. This enumerative procedure requires a prohibitively large

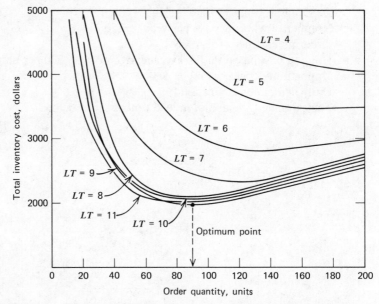

FIGURE 13.8   Total inventory cost curves for the bicycle inventory problem.

if not impossible, computational effort, especially when we are to determine the optimum policies for numerous products. We discuss an efficient technique in this section.

Figure 13.9 shows the major steps for determining the optimum policy. First, the excess or shortage units are determined for each possible scheduled lead time and required lead time combination. Since the daily demand is a constant, we have

$$\text{Daily demand} = \frac{\text{Demand per year}}{\text{Number of working days per year}} \qquad (13.15)$$

Thus,

$$\text{Daily demand} = \frac{1000}{250} = 4 \text{ units}$$

The excess or shortage units are calculated by the expression

$$U_{m,n} = (m - n)DD \qquad (13.16)$$

where

$U_{m,n}$ = excess or shortage units when the allotted lead time is m days and the required lead time is n days

m = number of days of allotted lead time

STEP 1

Calculate the excess or shortage units for each lead time allotted and lead time required combination.

↓

STEP 2

Calculate the values of excess or shortage costs for the lead time allotted and lead time required combinations.

↓

STEP 3

Determine the expected holding costs and opportunity costs for each lead time.

↓

STEP 4

Compute the economic order sizes for each lead time value.

↓

STEP 5

Calculate the minimum total inventory costs for the lead time values.

↓

STEP 6

Select the smallest total cost. This and the associated values refer to the optimum inventory plan.

FIGURE 13.9 Steps for establishing the optimum inventory plan for the inventory system with variations in lead time and opportunity costs.

n = number of days of lead time actually required

DD = daily demand, units

The computed values are given in Table 13.3. A positive quantity indicates an excess, a negative value denotes a shortage, and a zero value implies the case when the lead time provided exactly equals the required lead time. For example when $m = 5$,

$$U_{5,4} = (5 - 4)4 = 4$$

$$U_{5,5} = (5 - 5)4 = 0$$

$$U_{5,6} = (5 - 6)4 = -4$$

$$U_{5,7} = (5 - 7)4 = -8$$

$$U_{5,8} = (5 - 8)4 = -12$$

$$U_{5,9} = (5 - 9)4 = -16$$

$$U_{5,10} = (5 - 10)4 = -20$$

$$U_{5,11} = (5 - 11)4 = -24$$

TABLE 13.3   Calculation of Excess or Shortage Bicycles

Lead Time Allotted (m days)	Lead Time Required, n Days							
	4	5	6	7	8	9	10	11
4	0	−4	−8	−12	−16	−20	−24	−28
5	4	0	−4	−8	−12	−16	−20	−24
6	8	4	0	−4	−8	−12	−16	−20
7	12	8	4	0	−4	−8	−12	−16
8	16	12	8	4	0	−4	−8	−12
9	20	16	12	8	4	0	−4	−8
10	24	20	16	12	8	4	0	−4
11	28	24	20	16	12	8	4	0

In the second step, the excess and shortage costs are calculated, applying the expressions

$$C_{m,n} = U_{m,n}UHC \qquad \text{if } U_{m,n} > 0 \qquad (13.17)$$

$$C_{m,n} = U_{m,n}(-UOC) \qquad \text{if } U_{m,n} \leqq 0 \qquad (13.18)$$

where

$C_{m,n}$ = excess or shortage cost when the allotted lead time is m days and the required lead time is n days

UHC = unit holding cost

UOC = unit opportunity cost

Table 13.4 displays the calculated $C_{m,n}$ values. We illustrate again the calculations for $m = 5$,

$$C_{5,4} = 4(20) = \$80$$

$$C_{5,5} = 0(-30) = \$0$$

$$C_{5,6} = -4(-30) = \$120$$

$$C_{5,7} = -8(-30) = \$240$$

$$C_{5,8} = -12(-30) = \$360$$

$$C_{5,9} = -16(-30) = \$480$$

$$C_{5,10} = -20(-30) = \$600$$

$$C_{5,11} = -24(-30) = \$720$$

The $C_{m,n}$ values in Table 13.4 denote the costs due to excess storage and stockout as follows:

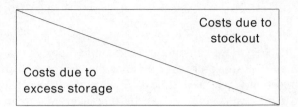

The expected holding costs as well as the opportunity costs for the allotted lead time periods are calculated in Step 3. For any allotted lead time period, m, the costs due excess storage are given in the respective row of Table 13.4. The expected holding cost for that particular lead time, EHC_m, is then obtained by multiplying these cost values by their respective probabilities and adding them up.

TABLE 13.4   Calculation of Expected Holding Costs and Opportunity Costs

Lead Time Allotted (m_{days})	Lead Time Required, n Days								Expected Cost	
	4	5	6	7	8	9	10	11	EHC_m	EOC_m
4	0	120	240	360	480	600	720	840	0	336.00
5	80	0	120	240	360	480	600	720	4.00	222.00
6	160	80	0	120	240	360	480	600	16.00	120.00
7	240	160	80	0	120	240	360	480	52.00	54.00
8	320	240	160	80	0	120	240	360	108.00	18.00
9	400	320	240	160	80	0	120	240	180.00	6.00
10	480	400	320	240	160	80	0	120	256.80	1.20
11	560	480	400	320	240	160	80	0	336.00	0
$P(LT)$	0.05	0.10	0.30	0.25	0.20	0.06	0.03	0.01		

For $m = 7$, we have

$$EHC_7 = 240(0.05) + 160(0.10) + 80(0.3)$$
$$= \$52$$

Similarly, the costs due to stockout for any allotted lead time, m, are presented in the associated row of Table 13.4. The expected opportunity cost for that specific lead time, EOC_m, is then attained by multiplying these cost values by their related probabilities and adding them up. Thus, for $m = 7$,

$$EOC_7 = 120(0.2) + 240(0.06) + 360(0.03) + 480(0.01)$$
$$= \$54$$

In Step 4, we compute the economic order sizes for the various lead times. The stockouts may arise at the end of each cycle. The expected opportunity cost is the measure of stockout effect for each cycle. The ordering cost denotes the expenditure incurred once in each cycle. Therefore, the expected opportunity cost may be added with the ordering cost in establishing the economic order quantity formula. Expression 9.9 is thus modified as

$$EOQ_m = \sqrt{\frac{2(OC + EOC_m)DYU}{UHC}} \qquad (13.19)$$

where

EOQ_m = economic order quantity for the allotted lead time m

OC = ordering cost

DYU = demand per year, units

The EOQ values for the various lead times are shown in Table 13.5. Applying for $m = 9$, we obtain

$$EOQ_9 = \sqrt{\frac{2(75 + 6)1000}{20}}$$
$$= 90 \text{ bicycles}$$

TABLE 13.5 Economic Order Quantities and Total Inventory Costs for the Lead Times

Lead Time Allotted, (m days)	Economic Order Quantity (EOQ_m)	Total Inventory Cost (TIC_m)	
4	202.73	4054.63	
5	172.34	3450.74	
6	139.64	2808.85	
7	113.58	2323.56	
8	96.44	2036.73	
9	90.00	1980.00	← Optimum
10	87.29	2002.65	
11	86.60	2068.05	

The *EOQ* values refer to the minimum total cost points on the respective cost curves, shown in Figure 13.8. The total inventory costs (TIC_m) for the *EOQ* values are determined in Step 5.

TIC_m = Total ordering cost + Expected total opportunity cost + Total holding cost for the lot size + Expected total holding cost for the reserve stock provided for the delivery delays

$$= OC \cdot \left(\frac{DYU}{EOQ_m}\right) + EOC_m \cdot \left(\frac{DYU}{EOQ_m}\right) + \frac{EOQ_m}{2} \cdot UHC + EHC_m$$

Hence,

$$TIC_m = (OC + EOC_m)\left(\frac{DYU}{EOQ_m}\right) + \left(\frac{EOQ_m}{2}\right) \cdot UHC + EHC_m \quad (13.20)$$

The total inventory costs are presented in Table 13.5. Applying expression 13.20 for $m = 9$, we obtain

$$TIC_9 = (75 + 6)\left(\frac{1000}{90}\right) + \left(\frac{90}{2}\right)20 + 180$$

$$= \$1980.00$$

In the last step, we select the smallest total cost. The smallest TIC_m is $1980.00, and it corresponds to $m = 9$. Thus, the optimum inventory plan is to place an order for 90 ($= EOQ_9$) bicycles whenever the inventory reaches $9 \times 4 = 36$ bicycles.

A Computer Program

Let us establish the following list of variables that are used in the formulation of the computer model:

LTIMES	= number of lead time possibilities
DEMAND	= demand per year, units
DAYS	= number of working days per year
UORC	= ordering cost per order
UHOC	= unit holding cost per year
UOPC	= opportunity cost per unit
DUSE	= daily demand, units
LTOPT	= number of the lead time possibility that corresponds to the optimum inventory plan
Y	= optimum lead time value
JROPT	= reorder point for the optimum inventory plan

JQ	= order quantity for the optimum inventory plan
LEADT(K)	= Kth lead time possibility
PROB(K)	= probability for LEADT(K)
THOC(K)	= expected total holding cost for LEADT(K)
TOPC(K)	= expected total opportunity cost for LEADT(K)
QOPT(K)	= economic order quantity for LEADT(K)
TOTALC(K)	= total inventory cost for QOPT(K)
EORSU(J, L)	= excess or shortage units when the allotted lead time is LEADT(J) and the required lead time is LEADT(L)
EORSC(J, L)	= excess or shortage cost when the allotted lead time is LEADT(J) and the required lead time is LEADT(L)

The variables LEADT( ), PROB( ), THOC( ), TOPC( ), QOPT( ), and TOTALC( ) are one-dimensional arrays of size LTIMES elements. The subscripted variables EORSU(,) and EORSC(,) are two-dimensional arrays of size LTIMES by LTIMES.

With the following instructions we might initialize the program:

```
      READ 301, LTIMES
      READ 302, DEMAND, DAYS, UROC, UHOC, UOPC
      READ 303, (LEADT(I), I = 1, LTIMES)
      READ 304, (PROB(I), I = 1, LTIMES)
  301 FORMAT ( )
  302 FORMAT ( )
  303 FORMAT ( )
  304 FORMAT ( )
```

The excess or shortage units for various combinations of allotted lead times and required lead times are computed by the statements:

```
      DUSE = DEMAND/DAYS
      DO 10 I = 1, LTIMES
      DO 10 J = 1, LTIMES
      DIFFLT = LEADT(I) − LEADT(J)
      EORSU(I, J) = DIFFLT * DUSE
   10 CONTINUE
```

The following set of statements might be used to determine the excess or shortage costs for the possible combinations of lead times provided and lead time required:

```
      DO 20 I = 1, LTIMES
      DO 20 J = 1, LTIMES
      IF (EORSU(I, J).GT.0) GO TO 24
      X = EORSU(I, J) * (−UOPC)
      GO TO 26
   24 X = EORSU(I, J) * UHOC
   26 EORSC(I, J) = X
   20 CONTINUE
```

We might calculate the expected total holding cost values for the lead time possibilities by including the instructions:

```
      DO 30 I = 1, LTIMES
      THOC(I) = 0
      DO 30 J = 1, LTIMES
      IF (LEADT(I).LT.LEADT(J)) GO TO 30
      THOC(I) = THOC(I) + EORSC(I, J) * PROB(J)
   30 CONTINUE
```

Similarly, the following statements could be used to calculate the expected total opportunity cost values for the lead time possibilities:

```
      DO 40 I = 1, LTIMES
      TOPC(I) = 0
      DO 40 J = 1, LTIMES
      IF (LEADT(I).GE.LEADT(J)) GO TO 40
      TOPC(I) = TOPC(I) + EORSC(I, J) * PROB(J)
   40 CONTINUE
```

For calculating the economic order quantity for each lead time possibility, we can employ the instructions:

```
      DO 50 I = 1, LTIMES
      E = UORC + TOPC(I)
      QOPT(I) = SQRT(2.0 * E * DEMAND/UHOC)
   50 CONTINUE
```

The following statements would cause the computer to compute the total inventory costs for the economic order quantity values:

```
       DO 60 I = 1, LTIMES
       F = (UORC + TOPC(I)) * DEMAND/QOPT(I)
       G = UHOC * QOPT(I)/2.00
       TOTALC(I) = F + G + THOC(I)
  60   CONTINUE
```

We might select the smallest total cost and calculate values pertaining to the optimum inventory plan by the instructions:

```
       XLARGE = 999999.99
       DO 70 I = 1, LTIMES
       IF (TOTALC(1).GE.XLARGE) GO TO 70
       LTOPT = I
       XLARGE = TOTALC(I)
  70   CONTINUE
       Y = LEADT(LTOPT)
       JROPT = DEMAND/DAYS * Y
       JQ = QOPT(LTOPT)
```

Then, the given values as well as the calculated values are printed out. Our complete program for the bicycles inventory problem is shown in Figure 13.10.

There are four cards in the data deck and they have the following appearance:

The computer printout is displayed in Figure 13.11.

```
      DIMENSION LEADT(8),PROB(8),EORSU(8,8),EORSC(8,8)
      DIMENSION THOC(8),TOPC(8),QOPT(8),TOTALC(8)
C
C
C   READ THE GIVEN VALUES
      READ 301,LTIMES
      READ 302,DEMAND,DAYS,UORC,UHOC,UOPC
      READ 303,(LEADT(I),I=1,LTIMES)
      READ 304,(PROB(I),I=1,LTIMES)
C
C
C   PRINT OUT THE GIVEN VALUES
      PRINT 404
      PRINT 800
      PRINT 800
      PRINT 106, LTIMES
      PRINT 101, DEMAND
      K = DAYS
      PRINT 105, K
      PRINT 102, UORC
      PRINT 103, UHOC
      PRINT 104, UOPC
C
C   COMPUTE EORSU VALUES
      DUSE=DEMAND/DAYS
      DO 10 I=1,LTIMES
      DO 10 J=1,LTIMES
      DIFFLT=LEADT(I)-LEADT(J)
      EORSU(I,J)=DIFFLT*DUSE
   10 CONTINUE
C
C   CALCULATE EORSC VALUES
      DO 20 I=1,LTIMES
      DO 20 J=1,LTIMES
      IF(EORSU(I,J).GT.0)GO TO 24
      X=EORSU(I,J)*(-UOPC)
      GO TO 26
   24 X=EORSU(I,J)*UHOC
   26 EORSC(I,J)=X
   20 CONTINUE
C
C
C   CALCULATE THOC VALUES
      DO 30 I=1,LTIMES
      THOC(I)=0
C
      DO 30 J=1,LTIMES
      IF(LEADT(I).LT.LEADT(J))GO TO 30
      THOC(I)=THOC(I)+EORSC(I,J)*PROB(J)
   30 CONTINUE
```

FIGURE 13.10 Computer program for inventory systems with variations in lead time together with opportunity costs (illustrated for the bicycle inventory problem).

```
C
C   FIND OUT TOPC VALUES
      DO 40 I=1,LTIMES
      TOPC(I)=0
C

      DO 40 J=1,LTIMES
      IF(LEADT(I).GE.LEADT(J))GO TO 40
      TOPC(I)=TOPC(I)+EORSC(I,J)*PROB(J)
   40 CONTINUE
C
C   DETERMINE QOPT VALUES
      DO 50 I=1,LTIMES
      E=UORC+TOPC(I)
      QOPT(I) = SQRT(2.0*E*DEMAND/UHOC)
   50 CONTINUE
C
C   COMPUTE TOTALC VALUES
      DO 60 I=1,LTIMES
      F=(UORC+TOPC(I))*DEMAND/QOPT(I)
      G=UHOC*QOPT(I)/2.00
      TOTALC(I)=F+G+THOC(I)
   60 CONTINUE
C
C   ESTABLISH THE SMALLEST TOTALC
C   DETERMINE LTOPTVALUE
      XLARGE=999999.99
      DO 70 I=1,LTIMES
      IF (TOTALC(I).GE.XLARGE) GO TO 70
      LTOPT = I
      XLARGE=TOTALC(I)
   70 CONTINUE
C
C
C   PRINT OUT THE CALCULATED VALUES
      PRINT 200
      PRINT 205
      PRINT 210
      PRINT 215
      PRINT 220, (LEADT(K),K=1,LTIMES )
      PRINT 205
      DO 80 I=1,LTIMES
      PRINT 225,LEADT(I),(EORSU(I,J),J=1,LTIMES)
   80 CONTINUE
      PRINT 205
C
C
      PRINT 230
      PRINT 205
      PRINT 210
      PRINT 215
      PRINT 220, (LEADT(K),K=1,LTIMES )
      PRINT 205
      DO 90 I=1,LTIMES
      PRINT 225,LEADT(I),(EORSC(I,J),J=1,LTIMES)
   90 CONTINUE
```

FIGURE 13.10 (Continued)

```
      PRINT 205
      PRINT 235, (PROB(K),K=1,LTIMES )
      PRINT 205
      PRINT 240
      PRINT 245
      PRINT 250
      PRINT 255
      PRINT 240
      DO 100 I =1,LTIMES
      PRINT 260,LEADT(I),QOPT(I),THOC(I),TOPC(I),TOTALC(I)
  100 CONTINUE
      PRINT 240
      PRINT 265
      PRINT 270, LEADT(LTOPT)
      Y=LEADT(LTOPT)
      JROPT = DEMAND/DAYS*Y
      PRINT 275,JROPT
      JQ = QOPT(LTOPT)
      PRINT 280, JQ
      PRINT 285,TOTALC(LTOPT)
C
  101 FORMAT (5X,19HANNUAL DEMAND,UNITS,11X,1H=,F8.2)
  102 FORMAT (5X,23HORDERING COST PER ORDER,7X,3H= $,F6.2)
  103 FORMAT (5X,26HUNIT HOLDING COST PER YEAR,4X,3H= $,F6.2)
  104 FORMAT (5X,25HOPPORTUNITY COST PER UNIT,5X,3H= $,F6.2)
  105 FORMAT (5X,31HNO. OF WORKING DAYS PER YEAR  =,I4)
  106 FORMAT (5X,31HNO. OF LEAD TIME POSSIBILITIES=,I2)
  200 FORMAT (15X,24HEXCESS OR SHORTAGE UNITS )
  205 FORMAT (1X,25H-------------------------,
     530H------------------------------  )
  210 FORMAT (2X,4HLEAD,15X,23HLEAD TIME REQUIRED,(LT) )
  215 FORMAT (2X,4HTIME )
  220 FORMAT (1X,7HALLOTED, 8(I2,4X) )
  225 FORMAT (I4,2X,8F6.1)
  230 FORMAT (15X,24HEXCESS OR SHORTAGE COSTS )
  235 FORMAT (1X,6HP(LT)=,8(F4.3,2X))
  240 FORMAT (6X,20H--------------------,
     424H----------------------  )
  245 FORMAT (7X,4HLEAD,3X,7HOPTIMUM,5X,
     520HEXPECTED COST VALUES  )
  250 FORMAT (7X,4HTIME,4X,5HORDER )
  255 FORMAT (6X,7HALLOTED,2X,4HSIZE,3X,7HHOLDING,
     22X,17HOPPORTUNITY TOTAL  )
  260 FORMAT (9X,I2,3X,4(F7.2,2X) )
  265 FORMAT (5X,22HOPTIMUM INVENTORY PLAN )
  270 FORMAT (10X,9HLEAD TIME,11X,1H=,I4)
  275 FORMAT (10X,13HREORDER POINT,7X,1H=,I3,6H UNITS )
  280 FORMAT (10X,14HORDER QUANTITY,6X,1H=,I3,6H UNITS )
  285 FORMAT (10X,21HTOTAL INVENTORY COST=,2H $,F8.2 )
  301 FORMAT (I1)
  302 FORMAT (5F7.2)
  303 FORMAT (8I2)
  304 FORMAT (8F3.2)
  404 FORMAT(1H1)
  800 FORMAT(2(/))
      END
```

FIGURE 13.10 (Continued)

```
NO. OF LEAD TIME POSSIBILITIES= 8
ANNUAL DEMAND,UNITS            = 1000.00
NO. OF WORKING DAYS PER YEAR   = 250
ORDERING COST PER ORDER        = $ 75.00
UNIT HOLDING COST PER YEAR     = $ 20.00
OPPORTUNITY COST PER UNIT      = $ 30.00
          EXCESS OR SHORTAGE UNITS
```

LEAD TIME ALLOTED	LEAD TIME REQUIRED,(LT)						
	4	5	6	7	8	9	10	11
4	0	-4.0	-8.0	-12.0	-16.0	-20.0	-24.0	-28.0
5	4.0	0	-4.0	-8.0	-12.0	-16.0	-20.0	-24.0
6	8.0	4.0	0	-4.0	-8.0	-12.0	-16.0	-20.0
7	12.0	8.0	4.0	0	-4.0	-8.0	-12.0	-16.0
8	16.0	12.0	8.0	4.0	0	-4.0	-8.0	-12.0
9	20.0	16.0	12.0	8.0	4.0	0	-4.0	-8.0
10	24.0	20.0	16.0	12.0	8.0	4.0	0	-4.0
11	28.0	24.0	20.0	16.0	12.0	8.0	4.0	0

```
          EXCESS OR SHORTAGE COSTS
```

LEAD TIME ALLOTED	LEAD TIME REQUIRED,(LT)						
	4	5	6	7	8	9	10	11
4	0	120.0	240.0	360.0	480.0	600.0	720.0	840.0
5	80.0	0	120.0	240.0	360.0	480.0	600.0	720.0
6	160.0	80.0	0	120.0	240.0	360.0	480.0	600.0
7	240.0	160.0	80.0	0	120.0	240.0	360.0	480.0
8	320.0	240.0	160.0	80.0	0	120.0	240.0	360.0
9	400.0	320.0	240.0	160.0	80.0	0	120.0	240.0
10	480.0	400.0	320.0	240.0	160.0	80.0	0	120.0
11	560.0	480.0	400.0	320.0	240.0	160.0	80.0	0

```
P(LT)=.050   .100   .300   .250   .200   .060   .030   .010
```

LEAD TIME ALLOTED	OPTIMUM ORDER SIZE	EXPECTED COST VALUES		
		HOLDING	OPPORTUNITY	TOTAL
4	202.73	0	336.00	4054.63
5	172.34	4.00	222.00	3450.74
6	139.64	16.00	120.00	2808.85
7	113.58	52.00	54.00	2323.56
8	96.44	108.00	18.00	2036.73
9	90.00	180.00	6.00	1980.00
10	87.29	256.80	1.20	2002.65
11	86.60	336.00	0	2068.05

```
OPTIMUM INVENTORY PLAN
     LEAD TIME             =  9
     REORDER POINT         = 36 UNITS
     ORDER QUANTITY        = 90 UNITS
     TOTAL INVENTORY COST= $ 1980.00
```

FIGURE 13.11 Computer printout for the bicycle inventory problem.

Sensitivity of TIC at Optimum to UOC Changes

We have established the optimum inventory policy for the bicycle problem by hand computations and also by applying the computer model. This optimum policy is valid insofar as the parameters such as the demand per year, unit ordering cost, unit opportunity cost, and unit holding cost have really the same assigned values as in Example 13.3. It is apparent that all of the parameters values can be obtained with sufficient accuracy except the unit opportunity cost. It is indeed a difficult task to estimate the opportunity cost. Hence, we wish to know, "Is it important that the unit opportunity cost is determined accurately?" Namely, "How sensitive is the total inventory cost for changes in the opportunity cost?" If the changes in opportunity cost shifts the total inventory cost significantly, care should be taken to estimate accurately the unit stockout cost. On the other hand, if the changes in the stockout cost do not alter the total cost significantly, we conclude that a reasonable guess on the unit stockout cost is adequate in the inventory problem formulation.

Assigning the same original values for the parameters of our bicycle problem but only substituting the UOC values of \$10 through \$100 in steps of \$10, the problems have been solved. The resulting optimal inventory policies are displayed in Table 13.6. We note that although the

TABLE 13.6 Optimum Inventory Plans for Various Unit Opportunity Cost Values

| Unit Opportunity Cost, *UOC* $ | Optimum Inventory Plan | | |
	Lead Time, Days	Reorder Point Units	Ordering Quantity, Units	Total Inventory Cost $
10	8	32	90	1908
20	9	36	89	1958
30	9	36	90	1980
40	9	36	91	2002
50	10	40	88	2012
60	10	40	88	2016
70	10	40	88	2021
80	10	40	88	2025
90	10	40	89	2030
100	10	40	89	2034

unit stockout costs have been changed from $10 to $100, the related total costs at optimum are not affected seriously. Thus, we conclude that the total inventory cost is not highly sensitive to changes in the unit opportunity cost. Hence, if the unit stockout cost is not available with required accuracy, a reasonable guess on the unit opportunity cost would be adequate.

VARIATIONS IN DEMAND AND LEAD TIME

In this inventory model, both the demand and lead time are considered to vary. Most inventory systems can be formulated according to this model. We shall illustrate the computational procedure by a simple example.

Example 13.4   Refrigerator Inventory Problem

The P. R. Enterprises sells refrigerators Model R460 through its hardware store. Variability in demand and lead time prevails and it is represented as follows:

Demand per Day	Probability	Lead Time	Probability
1 refrigerator	0.30	1 day	0.60
2 refrigerators	0.50	2 days	0.40
3 refrigerators	0.20		1.00
	1.00		

It is apparent that the total demand during lead time, $DDLT$, cannot be stated as a deterministic quantity. The shortest lead time is one day and the smallest demand per day is one refrigerator. Hence, $DDLT$ will be at least one refrigerator.

The longest lead time is two days and the largest demand per day is three units. Thus, the largest value for $DDLT$ is three units during the first day plus three units the second day, resulting in six units. The possible $DDLT$ values are one through six units.

The probability of occurrence for each possible $DDLT$ is shown in Table 13.7. For any given $DDLT$, all possibilities of the event happening are listed. Then, the joint probability for each one of them is calculated. The joint probability is obtained by multiplying the related probability of lead time required by the associated demand probabilities. For instance, $DDLT = 4$ can occur in the following three ways:

Possible Way	Joint Probability of Occurrence
First day 2 units Second day 2 units	$(0.4)(0.5)(0.5) = 0.100$
First day 1 unit Second day 3 units	$(0.4)(0.3)(0.2) = 0.024$
First day 3 units Second day 1 unit	$(0.4)(0.2)(0.3) = 0.024$
Probability for $DDLT = 4$	$= 0.148$

Subsequently, for each $DDLT$ value, the probability is determined by adding the joint probabilities of the related possible ways. Thus, for $DDLT = 4$, we have,

$$P(DDLT = 4) = 0.100 + 0.024 + 0.024 = 0.148$$

TABLE 13.7 **Probability Calculations for** *DDLT* **Values**

Demand during Lead Time, DDLT	Lead Time Required,	Demand during First Day	Demand during Second Day	Probability Calculations	Probability P(DDLT)
1	1	1		$(0.6)(0.3)$ = 0.180	0.180
2	1	2		$(0.6)(0.5)$ = 0.300	0.336
	2	1	1	$(0.4)(0.3)(0.3)$ = 0.036	
3	1	3		$(0.6)(0.2)$ = 0.120	0.240
	2	1	2	$(0.4)(0.3)(0.5)$ = 0.060	
	2	2	1	$(0.4)(0.5)(0.3)$ = 0.060	
4	2	2	2	$(0.4)(0.5)(0.5)$ = 0.100	0.148
	2	1	3	$(0.4)(0.3)(0.2)$ = 0.024	
	2	3	1	$(0.4)(0.2)(0.3)$ = 0.024	
5	2	2	3	$(0.4)(0.5)(0.2)$ = 0.040	0.080
	2	3	2	$(0.4)(0.2)(0.5)$ = 0.040	
6	2	3	3	$(0.4)(0.2)(0.2)$ = 0.016	0.016
				Total	1.000

The sum of all the probabilities for the *DDLT* values must equal one. If not, care should be taken to see that all possibilities are included for each one of the *DDLT* values.

Next, the probability of stockout for each *DDLT* value is determined. The probability of stockout for any *DDLT* value equals the probability that the actual demand during lead time exceeds the particular *DDLT* value. The desired probability for any *DDLT* value is computed by expression 13.2:

$$PGTL(DDLT) = PGTL(E) + P(E)$$

and

$$PGTL(G) = 0, \text{ where } G \text{ denotes the largest } DDLT$$

The computations are shown in Table 13.8. The computational method of determining *PGTL(DDLT)* values are discussed under Model 1, Step 2. The *PGTL*(6) equals zero, indicating that if *DDLT* = 6, no stockouts would occur. Thus, the service level is 100 percent. Similarly, we can interpret the value of *PGTL*(5). *PGTL*(5) equals 0.016, denoting that the expected stockout is 16 times in 1000 lead times and the service level is 0.984 or 98.4 percent (= 1.00 − 0.016). If an order is placed when 5 units are available in the inventory, stockouts may happen only when the *DDLT* exceeds 5, namely, 6 units. *P*(6) in this case is thus equal to *PGTL*(5). Likewise, the stockout for *DDLT* = 4 is *PGTL*(4) which equals 0.08 plus 0.016 (=0.096). The related service level is 0.904 or 90.4 percent (= 1.00 − 0.096). The service levels are given in Table 13.8. The company may select one of the *DDLT* value as the reorder point that would yield the desired service level.

TABLE 13.8   Service Level Calculations

Demand during Lead Time, DDLT	Probability P(DDLT)	Probability that the Actual Demand during Lead Time Exceeds DDLT Value PGTL(DDLT)	Service Level, 1-PGTL(DDLT)
1	0.180	0.820	0.180
2	0.336	0.484	0.516
3	0.240	0.244	0.756
4	0.148	0.096	0.904
5	0.080	0.016	0.984
6	0.016	0.000	1.000

The expected demand during lead time is obtained by multiplying the *DDLT* values by the associated probabilities and adding them up.

$$DVMEAN = (1)(0.180) + (2)(0.336) + (3)(0.240) + (4)(0.148)$$
$$+ (5)(0.08) + (6)(0.016)$$
$$= 2.66 \text{ units}$$

where

$$DVMEAN = \text{expected demand during lead time}$$

The buffer stock is calculated by applying the expression,

$$\text{Buffer stock} = DDLT - DVMEAN \qquad (13.21)$$

Thus,

$$\text{For } DDLT = 6, \text{ buffer stock} = 6 - 2.66 = 3.34 \text{ units}$$
$$\text{For } DDLT = 5, \text{ buffer stock} = 5 - 2.66 = 2.34 \text{ units}$$

The task of establishing the buffer stock for a large-size inventory system with numerous demand and lead time possibilities is considerably more complicated and time consuming. In such situations, the simulation technique discussed in Chapter 15 would be a valuable tool.

PROBLEMS

13.1 The Alan Supply Corporation sells drug D79 by wholesale. The estimated demand distribution is shown below:

Demand per Week (in 100 Boxes)	Probability
3	0.05
4	0.30
5	0.40
6	0.20
7	0.03
8	0.02

The drug is directly supplied to the corporation by the manufacturers. The required lead time is one week.
(a) Calculate the expected demand during lead time.
(b) Calculate the probability of stockout for each one of the demand alternatives.
(c) Determine the buffer stock for the service levels of 0.95 and 0.98.
(d) Find out the service level for the reorder point of 800 boxes.

13.2 The demand for 12-volt batteries is a constant, at the rate of 10 units per day, and the lead time is subject to variability. The lead time distribution is shown below:

Lead Time (Days)	Number of Observations
0	3
1	12
2	14
3	17
4	13
5	11
6	10
7	5
8	3
9	2

(a) Determine the expected lead time and the probability distribution for lead time.

(b) Draw the lead time distribution as well as the cumulative lead time distribution graphs.

(c) Determine the buffer stock for 0.05, 0.10, and 0.20 risk levels.

13.3 The Barnhart Brothers are the sole distributors of the Crown suits on the West Coast. The sales records indicate that the demand for the suits has occurred according to the following distribution:

Demand per Month (in 1000 Units)	Number of Observations
1	8
2	20
3	40
4	25
5	4
6	2
7	1

Estimated lead time = 1 month

Assume that the demand distribution reflects the expected monthly sales for the planning period. The reorder point of the current program is 5000 units. The related costs are

Cost per unit = $65

Inventory holding cost = 15 percent of the average inventory value

The company wants to estimate the cost of increasing its service level. Write a program that will cause the computer to compute and print out the inventory cost values for the reorder points, 6000 units and 7000 units.

13.4 A manufacturer of paint solvents requires 700 gallons of raw material R each week. The variations on the supply of R are as follows:

Lead Time (Weeks)	Number of Observations
2	8
3	15
4	12
5	10
6	2
7	2
8	1

The cost per gallon of R is $4.50 and the inventory holding cost is 30 percent of the average inventory value. Prepare a program that will enable the computer to calculate and print out the following for the various risk levels of stockout up to 20 percent.

Risk of stockout
Reorder point
Buffer stock
Average inventory ($)
Inventory cost ($)

13.5 A small restaurant is planning to sell hot lunches in the forthcoming local concert. Each hot lunch can be prepared at a cost of $0.60 and sold for $1.00. The probable demand of hot lunches is as shown below.

Demand (Number of Lunches)	Probability
50	0.05
60	0.15
70	0.40
80	0.30
90	0.10

Assume that unsold hot lunches have no value.
(a) Calculate the best preparation level (number of lunches to be prepared).
(b) Determine the total expected profit for the optimum preparation level.

13.6 The Rodriguez Newsstand can buy a particular daily newspaper for $0.05 per copy and sell it for $0.15 per copy. The unsold copies, if any are left at the end of a day, have no value. Here is the estimated demand distribution:

Demand (Number of Copies)	Probability
100	0.04
110	0.10
120	0.25
130	0.30
140	0.14
150	0.12
160	0.05

Write a program that will enable the computer to calculate and print out the number of copies to be procured that will maximize the total expected profit. What is the total expected profit for the optimum plan?

13.7 The distribution of lead times pertaining to television sets model M19 is as follows:

Lead Time (Weeks)	Probability
2	0.03
3	0.07
4	0.20
5	0.65
6	0.04
7	0.01

The costs associated with $M19$ are

Ordering cost = $150 per order

Holding cost = $70 per unit per year based on the average inventory

Opportunity cost = $100 per unit

Assume that the weekly demand is 30 units. Prepare a computer program to calculate and print out the optimum inventory plan that minimizes the total cost.

13.8 Consider Problem 13.7. Write a computer program that will calculate and print out the optimum inventory plans for the 20 possible unit opportunity costs, starting with $10 to $200 with the increments of $10. The desired form of output is shown in Table 13.6. Discuss on the sensitivity of the unit opportunity cost over the reorder point, ordering quantity, and total inventory cost.

13.9 The demand distribution and lead-time distribution for dishwashing machine Eazyklean Model 0726 have been as follows:

Demand		Lead Time	
Demand per Week	Probability	Lead Time (Weeks)	Probability
0	0.10	1	0.25
1	0.15	2	0.55
2	0.45	3	0.20
3	0.30		

(a) For the various possible expected demands during the lead time, compute the probability.
(b) What is the expected demand during lead time?
(c) For the maximum demand values of 9, 8, 7, and 6, calculate the buffer stocks and risk levels.

13.10 The demand distribution and the lead-time distribution for sewing machines have been as follows:

Demand		Lead Time	
Demand per Week	Probability	Lead Time (Weeks)	Probability
0	0.10	1	0.35
1	0.20	2	0.55
2	0.25	3	0.10
3	0.30		
4	0.15		

(a) For the different possible expected demands during the lead time, compute the probability.
(b) Find out the expected demand during lead time.
(c) Determine the buffer stocks and risk levels for the maximum demand values of 11, 10, 9, and 8.

------SELECTED REFERENCES------

Buchan, J., and E. Koenigsberg, *Scientific Inventory Management*, Prentice-Hall, Englewood Cliffs, N.J., 1963.
Buffa, E. S., *Operations Management: Problems and Models*, Wiley, New York, Third Edition, 1972.

Buffa, E. S., and W. H. Taubert, *Production—Inventory Systems: Planning and Control*, Richard D. Irwin, Homewood, Ill., Revised Edition, 1972.

Felter, R. B., and W. C. Dalleck, *Decision Models for Inventory Management*, Richard D. Irwin, Homewood, Ill., 1961.

Groff, G. E., and J. F. Muth, *Operations Management: Analysis for Decisions*, Richard D. Irwin, Homewood, Ill., 1972.

McMillan, C., and R. F. Gonzalez, *Systems Analysis—A Computer Approach to Decisions Models*, Richard D. Irwin, Homewood, Ill., Third Edition, 1973.

Moore, F. G., and R. Jablonski, *Production Control*, McGraw-Hill, New York, Third Edition, 1969.

Naddor, E., *Inventory Systems*, Wiley, New York, 1966.

Riggs, J. L., *Economic Decision Models for Engineers and Managers*, McGraw-Hill, New York, 1968.

Schlaifer, R., *Probability and Statistics for Business Decisions*, McGraw-Hill, New York, 1959.

Star, M. K., and D. W. Miller, *Inventory Control: Theory and Practice*, Prentice-Hall, Englewood Cliffs, N.J., 1962.

Wagner, H. M., *Statistical Inventory Management*, Wiley, New York, 1962.

CHAPTER 14

analysis of waiting lines

A WAITING SITUATION basically involves a flow of units (customers) arriving at one or more service facilities. For example, customers wait for teller service at the bank, break down machines wait for a repair crew to be available, and airplanes wait for runways to be clear. The elements of some systems involving waiting lines are given in Table 14.1.

TABLE 14.1   Examples of Systems Involving Waiting Lines

System	Unit Arriving to Be Serviced	Service or Processing Facility	Service or Operation Being Performed
Car Wash	Automobiles	Attendants and car washing facility	Washing automobiles
Medical clinic	Patients	Doctors, their staff, and facilities	Medical care
Supermarket	Customers with groceries	Clerks and check-out counters	Tabulation of items, receipt of payment, and bagging of groceries
Harbor	Ships	Docks	Unloading and loading

Waiting line or queuing theory can be employed for waiting situations to construct models and to analyze existing programs as well as alternative plans to obtain valuable insights.

WAITING LINE CHARACTERISTICS

We can describe the major features of a waiting situation by the following important characteristics:

Population source
Arrivals
Service discipline
Service facility

Figure 14.1 shows the characteristics of a waiting process. The population source of customers that arrive at a queuing system may have a finite or an infinite number of potential customers. A medical clinic, for example, may have a finite number of patients. On the other hand, a car wash in a big city may be considered to have an infinite number of customers. The calculations are much easier for an infinite population source. Therefore, when the size of the calling population is relatively large, the input source is assumed to be infinite.

The important feature of arrivals is the distribution of arrivals. The arrival process can be described by the time elapsed between successive arrivals and determined by the probability of different times between arrivals. Another way is to describe the number of arrivals in some specified length of time and to establish the probability of various number of arrivals during that time. For many real-world queuing situations, the interarrival times can be represented by the exponential distribution. If the arrival rates are expressed as the number of arrivals per some unit of time, the patterns closely resemble the Poisson distribution. Other distributions such as the constant, Erlang, and normal distributions can be found useful to model arrival distributions of different queuing problems.

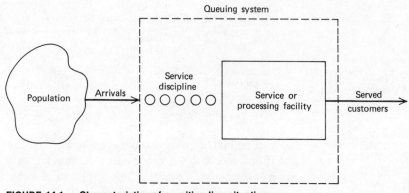

FIGURE 14.1   Characteristics of a waiting line situation.

The service discipline is a priority rule or a set of rules for selecting customers from the waiting line to start service. Perhaps the most common service discipline is first-come first-served (*FCFS*). The priority rule of *FCFS* specifies that customers are admitted to start service in the order of their arrivals. Other priority rules include reservations first, emergencies first, and customer with the shortest expected service time first.

The significant features of any service facility are its structure and the service time. There are four basic structures of service facilities. Figure 14.2 shows the four arrangements of service facilities. The selection of structure

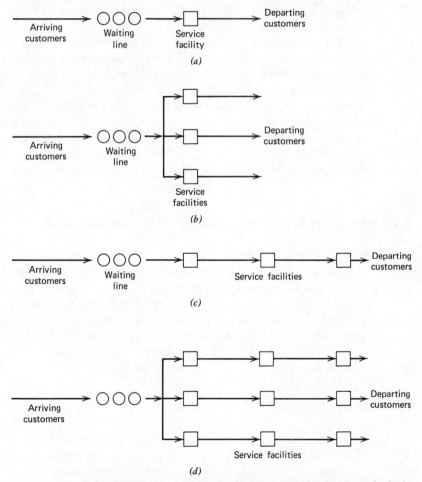

FIGURE 14.2 **Basic structure of waiting line situations.** (*a*) **Single channel–single phase situation,** (*b*) **Multiple channel–single phase situation,** (*c*) **Single channel–multiple phase situation,** (*d*) **Multiple channel–multiple phase situation.**

for a waiting line situation is based upon the nature of the service performed, the volume of customers, and the kinds of operations performed with their technological order according to which operations are to be performed. A single channel–single phase situation is the simplest kind of service structure. A one-man barber shop is an example of the single channel–single phase system. A bank where two or more tellers are servicing their customers is a multiple channel–single phase system. If a supermarket has one checkout counter with one checkout clerk to tabulate the items and receive payment, one person to bag the groceries, and another person to carry the bags to the customer's car, we have a single channel–multiple phase system. If two or more parallel checkout counters of the kind described above are encountered, we have a multiple channel–multiple phase system. The service time or holding time refers to the time elapsed from the commencement of service to its completion for a customer in a service facility. The probability distribution of service times is used to describe the service times. The exponential distribution can frequently be found useful in modeling the distribution of service times. In case that the distribution of service times cannot be approximated according to the exponential distribution, other distributions such as Erlang and hyperexponential distributions may be tried.

ANALYSIS OF ARRIVAL AND SERVICE TIME DISTRIBUTIONS

The waiting line models are most often formulated assuming that the arrival rate for a specified time is a Poisson distribution and that the service time is an exponential distribution. There are several good reasons for assuming Poisson arrivals and exponential service times. First, the mathematical analysis is simple and, therefore, requires less computational effort. Second, the assumptions have been found to be valid for many real world waiting line problems. We shall describe the arrival and service time patterns by a numerical example.

An Illustrative Example

A manufacturer of air conditioners has a tool crib with a single clerk. The tool crib handles all tools required by mechanics. A study was conducted for five days where 12 five-minute periods were chosen at random for each day. The number of arrivals (mechanics) that were observed for the 60 periods are shown in Table 14.2. The service times required to handle the mechanic's request of tools for 80 mechanics selected randomly are summarized as follows:

Service Time (Minutes)	Mean Service Time (Minutes)	Observed Frequency
0 and under 5	2.5	61
5 and under 10	7.5	16
10 and under 15	12.5	1
15 and under 20	17.5	2
20 or above		0
	Total	80

TABLE 14.2   Number of Customers Arrived in the Study Period

					Interval Number in the Day							
Day	1	2	3	4	5	6	7	8	9	10	11	12
1	1	0	1	3	1	0	2	0	2	0	1	0
2	0	1	2	0	2	1	0	0	0	1	0	1
3	1	1	1	0	0	1	0	4	2	0	2	1
4	0	0	2	0	2	0	1	1	0	0	1	0
5	3	0	0	1	0	1	0	0	2	0	0	2

The data shown in Table 14.2 can be summarized as follows:

Number of Arrivals in Five-Minute Period	Observed Frequency
0	29
1	18
2	10
3	2
4	1
5	0
Total	60

Mean arrivals in five-minute period (λ)

$$= \frac{(0)(29) + (1)(18) + (2)(10) + (3)(2) + (4)(1) + (5)(0)}{60}$$

$$= \frac{48}{60} = 0.80$$

The Greek letter lambda (λ) is commonly used to represent the mean arrival rate. The expected frequency of x arrivals $[f(x)]$ for the Poisson distribution with $\lambda = 0.80$ is:

$$f(x) = \frac{\lambda^x e^{-\lambda}}{x!} (n_1) \qquad (14.1)$$

where

$$n_1 = \text{number of observations } (= 60)$$

Therefore, the expected frequency for zero arrivals is

$$\text{when} \quad x = 0 \quad f(0) = \frac{\lambda^0 e^{-0.8}}{0!} (60)$$

$$= 26.958 \text{ arrivals per five-minute period}$$

Likewise, $f(x)$ values for other x values may be computed. The expected frequencies are:

x	$f(x)$
0	26.958
1	21.570
2	8.628
3	2.298
4	0.462
5	0.072
6	0.012
7	0.000
Total	1.000

The observed frequencies as well as the expected frequencies based on Poisson distribution are shown in Figure 14.3. We note that the expected frequencies closely correspond to the observed frequencies. Hence, we conclude that the distribution of arrival rates can be considered a Poisson distribution with a mean rate of 0.80 per five-minute period.

The mean service time is the weighted arithmetic mean of the service times; therefore

$$\text{Mean service time } (m) = \frac{(2.5)(61) + (7.5)(16) + (12.5)(1) + (17.5)(2)}{80}$$

$$= \frac{320}{80} = 4 \text{ minutes per mechanic}$$

The expected frequency of service times less than t minutes for an exponential distribution with a mean of four minutes per mechanic is

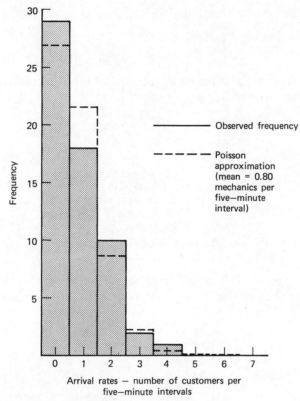

FIGURE 14.3   Arrival rates distribution in tool crib.

calculated using the expression

$$f \text{ (less than } t \text{ minutes)} = n_2\left[1 - e^{-(t/m)}\right] \qquad (14.2)$$

where

$$n_2 = \text{number of observations } (=80)$$

The expected frequency for the interval "t and under T" is determined by the formula

$$f \text{ (t and under T)} = n_2\left[1 - e^{-(T/m)}\right] - n_2\left[1 - e^{-(t/m)}\right]$$
$$= n_2\left[e^{-(t/m)} - e^{-(T/m)}\right] \qquad (14.3)$$

Therefore, the expected frequency for the interval "0 and under 5 minutes" is

$$f \text{ (0 and under 5)} = 80\left[e^{-(0/4)} - e^{-(5/4)}\right]$$
$$= 80(1 - 0.2865) = 57.1 \text{ mechanics}$$

The expected frequency for the interval "5 and under 10 minutes" is

$$f \text{ (5 and under 10)} = 80[e^{-(5/4)} - e^{-(10/4)}]$$
$$= 80(0.2865 - 0.08208) = 16.4 \text{ mechanics}$$

Likewise, the expected frequencies for other intervals may be calculated. The expected frequencies are:

Service Time (Minutes)	Expected Frequency
0 and under 5	57.1
5 and under 10	16.4
10 and under 15	4.7
15 and under 20	1.3
20 or above	0.5
Total	80.0

Figure 14.4 shows the observed frequencies and the expected frequencies plotted on a graph. The service times are approximated closely

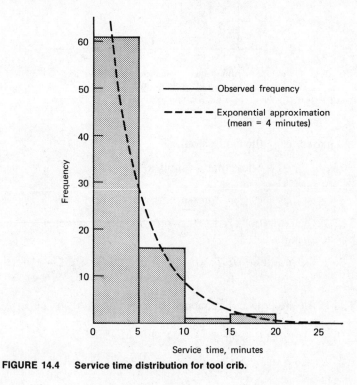

FIGURE 14.4   Service time distribution for tool crib.

by an exponential distribution with the mean service time of four minutes per mechanic.

$$\left[\begin{array}{l} \text{Mean service rate for the specified} \\ \quad \text{time interval of five minutes } (\mu) \end{array}\right] = \frac{5}{4}$$

$$= 1.25 \text{ mechanics}$$
$$\text{per five-minute period}$$

The Greek letter mu (μ) is commonly used to represent the mean service rate. We shall discuss the single channel–single phase queuing system in the following section.

SINGLE CHANNEL–SINGLE PHASE MODEL

In this section we assume that the following conditions are valid:

1. The calling source and the queue length are unlimited.
2. Poisson arrival rates and exponential service times are applicable.
3. The queue discipline is first-come-first-served.
4. The mean service rate μ is greater than the mean arrival rate λ.

We can determine some important characteristics of the single channel–single phase system by employing the equations:

Average utilization of the service facility is ρ (the Greek letter rho) $= \dfrac{\lambda}{\mu}$

$$\text{(14.4)}$$

Mean number of units in the system, including the one unit being served, is

$$L_s = \frac{\lambda}{\mu - \lambda} \tag{14.5}$$

Mean number of units in the queue is

$$L_q = \frac{\lambda^2}{\mu(\mu - \lambda)} \tag{14.6}$$

Mean time in the system, including the service time, is

$$T_s = \frac{1}{\mu - \lambda} \tag{14.7}$$

Mean waiting time is

$$T_q = \frac{\lambda}{\mu(\mu - \lambda)} \tag{14.8}$$

Probability of exactly n units in the system is

$$P_n = \left(1 - \frac{\lambda}{\mu}\right)\left(\frac{\lambda}{\mu}\right)^n \qquad (14.9)$$

Probability of time that the facility is idle would be

$$P_0 = 1 - \frac{\lambda}{\mu} \qquad (14.10)$$

For the tool crib example, the average utilization of the tool crib is

$$\rho = \frac{\lambda}{\mu} = \frac{0.80}{1.25} = 64 \text{ percent}$$

This indicates that the clerk in the tool crib is expected to work 64 percent of the time and the remaining 36 percent of the time he would be idle.
   Mean number of units in the system is

$$L_s = \frac{\lambda}{\mu - \lambda} = \frac{0.80}{1.25 - 0.80} = 1.778 \text{ mechanics}$$

Mean number of units in the queue is

$$L_q = \frac{\lambda^2}{\mu(\mu - \lambda)} = \frac{(0.80)^2}{1.25(1.25 - 0.80)} = 1.138 \text{ mechanics}$$

Mean time in the system is

$$T_s = \frac{1}{\mu - \lambda} = \frac{1}{1.25 - 0.80} = 2.222 \text{ time intervals}$$

Mean waiting time is

$$T_q = \frac{\lambda}{\mu(\mu - \lambda)} = \frac{0.80}{1.25(1.25 - 0.80)} = 1.422 \text{ time intervals}$$

If we are interested to know the probability of two mechanics in the system, we have

$$P_2 = \left(1 - \frac{\lambda}{\mu}\right)\left(\frac{\lambda}{\mu}\right)^2 = \left(1 - \frac{0.80}{1.25}\right)\left(\frac{0.80}{1.25}\right)^2 = 14.7 \text{ percent}$$

It signifies that 14.7 percent of the time, there would be two mechanics in the system, including the one mechanic being served. Likewise

$$P_1 = \left(1 - \frac{0.80}{1.25}\right)\left(\frac{0.80}{1.25}\right)^1 = 23 \text{ percent}$$

$$P_0 = \left(1 - \frac{0.80}{1.25}\right)\left(\frac{0.80}{1.25}\right)^0 = 36 \text{ percent}$$

The probability that two or more mechanics are in the system is

$$P_2 + P_3 + P_4 + \cdots$$

Since $P_0 + P_1 + P_2 + \cdots = 1$, the desired probability is

$$= 1 - (P_0 + P_1) = 1 - (0.36 + 0.23) = 41 \text{ percent}$$

That is, there would be two or more mechanics in the system for 41 percent of the time.

The probability of two mechanics in the waiting line is the same as the probability of three mechanics in the system. Therefore,

Probability of two mechanics in the waiting line $= P_3$

$$= \left(1 - \frac{\lambda}{\mu}\right)\left(\frac{\lambda}{\mu}\right)^3$$

$$= \left(1 - \frac{0.80}{1.25}\right)\left(\frac{0.80}{1.25}\right)^3$$

$$= 9.4 \text{ percent}$$

There will be zero mechanics in the waiting line for the two cases: (1) there is one unit in the system, and (2) there are no units in the system. Hence,

Probability of zero mechanics in the waiting line $= P_0 + P_1$

$$= 0.36 + 0.23$$

$$= 59 \text{ percent}$$

Formula (14.10) is used to find the probability of the time that the facility is idle as follows:

$$P_0 = 1 - \frac{\lambda}{\mu} = 1 - \frac{0.80}{1.25} = 36 \text{ percent}$$

Management can use these attributes in evaluating the waiting line systems.

A Computer Model

We shall develop a computer program for determining various characteristics of the single channel–single phase system in this section. The following variables are employed in the program:

$$
\begin{aligned}
\text{XLAM} \quad &= \text{mean arrival rate } (\lambda) \\
\text{YMU} \quad &= \text{mean service rate } (\mu) \\
\text{RHO} \quad &= \text{average utilization of the service facility } (\rho) \\
\text{XMNUS} &= \text{mean number of units in the system } (L_s) \\
\text{XMNUQ} &= \text{mean number of units in the queue } (L_q) \\
\text{XMTS} \quad &= \text{mean time in the system } (T_s) \\
\text{XMWT} \quad &= \text{mean waiting time } (T_q) \\
\text{PTFI} \quad &= \text{probability of time that the facility is idle } (P_0) \\
\text{PNS} \quad &= \text{probability of exactly } n \text{ units in the system } (P_n) \\
\text{PNQ} \quad &= \text{probability of } n \text{ units in the queue}
\end{aligned}
$$

The input data are the values of XLAM and YMU. These values might be read into the computer and printed out by the instructions:

```
      READ 100, XLAM, YMU
      PRINT 2, XLAM
      PRINT 3, YMU
100   FORMAT ( )
  2   FORMAT ( )
  3   FORMAT ( )
```

Next, the values of RHO, XMNUS, and XMNUQ are calculated and printed out. These operations can be performed by including the statements:

```
      RHO = XLAM/YMU
      PRINT 4, RHO
      XMNUS = XLAM/(YMU − XLAM)
      PRINT 5, XMNUS
      XMNUQ = (XLAM ** 2.)/(YMU * (YMU − XLAM))
      PRINT 6, XMNUQ
  4   FORMAT ( )
  5   FORMAT ( )
  6   FORMAT ( )
```

The following set of instructions would cause the computer to calculate the values of XMTS, XMWT, and PTFI and print out the results:

```
      XMTS = 1.0/(YMU − XLAM)
      PRINT 7, XMTS
      XMWT = (1.0/YMU) * XMNUS
      PRINT 8, XMWT
      PTFI = 1.0 − XLAM/YMU
      PRINT 9, PTFI
   7  FORMAT ( )
   8  FORMAT ( )
   9  FORMAT ( )
```

The values of PNS and PNQ are calculated and printed out for 10 successive units ranging from one unit through 10 units by utilizing the statements:

```
        DO 130 K = 1, 11
        N = K − 1
        IF (N.NE.O) GO TO 135
        PNS = PTFI
        PNQ = PTFI + PTFI * RHO
        GO TO 130
   135  JK = N + 1
        PNS = PTFI * (RHO ** N)
        PNQ = PTFI * (RHO ** JK)
        PRINT 12, N, PNS, PNQ
   130  CONTINUE
    12  FORMAT ( )
        STOP
        END
```

A complete computer program composed of our previously developed segments is shown in Figure 14.5. For solving Example 14.1 by the computer program, the input data is prepared according to FORMAT statement 100. The data card has the following appearance:

```
0·80    1.25
```

Figure 14.6 displays the resulting computer printout for our example.

```
                    PRINT 1
                    READ 100,XLAM,YMU
                    PRINT 2,XLAM
                    PRINT 3,YMU
      C
      C             CALCULATE AND PRINTOUT THE SERVICE
      C             FACILITY UTILIZATION
                    RHO=XLAM/YMU
                    PRINT 4,RHO
      C
      C             CALCULATE AND PRINTOUT THE MEAN
      C             NUMBER OF UNITS IN THE SYSTEM
                    XMNUS=XLAM/(YMU-XLAM)
                    PRINT 5,XMNUS
      C
      C             CALCULATE AND PRINTOUT THE MEAN
      C             NUMBER OF UNITS IN THE QUEUE
                    XMNUQ=(XLAM**2.)/(YMU*(YMU-XLAM))
                    PRINT 6,XMNUQ
      C
      C             CALCULATE AND PRINTOUT THE MEAN
      C             TIME IN THE SYSTEM
                    XMTS=1.0/(YMU-XLAM)
                    PRINT 7,XMTS
      C
      C             CALCULATE AND PRINTOUT THE MEAN
      C             WAITING TIME
                    XMWT=(1.0/YMU)*XMNUS
                    PRINT 8,XMWT
      C
      C             CALCULATE AND PRINTOUT THE PROBABILITY
      C             OF THE FACILITY IDLE TIME
                    PTFI=1.0-XLAM/YMU
                    PRINT 9,PTFI
      C
      C             CALCULATE AND PRINT OUT PNS AND PNQ
                    PRINT 10
                    PRINT 11
                    DO 130 K=1,11
                    N=K-1
                    IF(N.NE.0)    GO TO 135
                    PNS = PTFI
                    PNQ = PTFI+PTFI*RHO
                    GO TO 130
                135 JK = N+1
                    PNS = PTFI*(RHO**N)
                    PNQ = PTFI*(RHO**JK)
                    PRINT 12,N,PNS,PNQ
                130 CONTINUE
```

FIGURE 14.5   Computer program for the single channel−single phase model (illustrated for the tool crib problem.

```
C
   1 FORMAT(/,15X,35HSINGLE CHANNEL - SINGLE PHASE MODEL)
   2 FORMAT(/,10X,17HMEAN ARRIVAL RATE,22X,1H=,F6.3)
   3 FORMAT(/,10X,17HMEAN SERVICE RATE,22X,1H=,F6.3)
   4 FORMAT(/,10X,28HSERVICE FACILITY UTILIZATION,11X,
     11H=,F6.3)
   5 FORMAT(/,10X,34HMEAN NUMBER OF UNITS IN THE SYSTEM,5X,
     21H=,F6.3)
   6 FORMAT(/,10X,33HMEAN NUMBER OF UNITS IN THE QUEUE,6X,
     31H=,F6.3)
   7 FORMAT(/,10X,23HMEAN TIME IN THE SYSTEM,16X,1H=,F6.3)
   8 FORMAT(/,10X,17HMEAN WAITING TIME,22X,1H=,F6.3)
   9 FORMAT(/,10X,37HPROBABILITY OF THE FACILITY IDLE TIME,2X,
     11H=,F6.3)
  10 FORMAT(//,10X,12HNO. OF UNITS,5X,10HPROB. OF N,5X,
     210HPROB. OF N)
  11 FORMAT(14X,3H(N),11X,9HIN SYSTEM,6X,8HIN QUEUE,/)
  12 FORMAT(14X,I3,12X,F6.4,9X,F6.4)
 100 FORMAT(2F6.2)
     STOP
     END
```

FIGURE 14.5 (*Continued*)


```
           SINGLE CHANNEL - SINGLE PHASE MODEL

    MEAN ARRIVAL RATE                          = 0.800

    MEAN SERVICE RATE                          = 1.250

    SERVICE FACILITY UTILIZATION               = 0.640

    MEAN NUMBER OF UNITS IN THE SYSTEM         = 1.778

    MEAN NUMBER OF UNITS IN THE QUEUE          = 1.138

    MEAN TIME IN THE SYSTEM                    = 2.222

    MEAN WAITING TIME                          = 1.422

    PROBABILITY OF THE FACILITY IDLE TIME      = 0.360


    NO. OF UNITS      PROB. OF N        PROB. OF N
       (N)           IN SYSTEM          IN QUEUE

        1             0.2304            0.1475
        2             0.1475            0.0944
        3             0.0944            0.0604
        4             0.0604            0.0387
        5             0.0387            0.0247
        6             0.0247            0.0158
        7             0.0158            0.0101
        8             0.0101            0.0065
        9             0.0065            0.0042
       10             0.0042            0.0027
```

FIGURE 14.6 Computer printout for the tool crib problem.

APPLICATION OF THE WAITING LINE MODEL

In operating the waiting line systems, if a lower level of service rate (μ) is provided, service costs are reduced. However, this causes the formation of a long waiting line. If the cost of waiting can be estimated, it would be a large amount. On the other hand, if a higher level of service rate (μ) is recommended, service costs are increased. But this would result in a shorter waiting line giving a lower cost of waiting. Hence management often faces the problem of deciding the number of servers at a service facility that minimizes the total cost of servicing and waiting. The computations are illustrated by the following numerical example.

Example 14.1

The J. S. Nolan Textile Company manufactures draperies and bedspreads in Evansville. It distributes them throughout the United States using their trucks. The shipping department operates one docking facility for loading purposes. Suppose that the arrival rate is Poisson-distributed with a mean of two arrivals per hour. The service times are modeled by an exponential distribution with a mean of 1.5 trucks per man-hour. Assume that the service rates are linearly related to the size of the dock crew, that is, a two-man crew would service at the rate of 3 trucks per hour, and a three-man crew would service at the rate of 4.5 trucks per hour, and so on. The hourly rate for each man on the crew is $7. The estimated cost of an idle truck and driver is $20 per hour. Determine the optimum crew size that minimizes the total cost of servicing and waiting.

We shall calculate the total expected cost for an 8-hour day for different crew sizes and select that crew for which the total expected cost is the smallest. The cost of idle trucks and drivers for a k-man crew for 8 hours is

$$IC_k = 8 \text{ hours} \times \begin{bmatrix} \text{Mean number of} \\ \text{units in} \\ \text{the system} \end{bmatrix} \times \begin{bmatrix} \text{Cost of} \\ \text{idle trucks and} \\ \text{drivers per hour} \end{bmatrix} \quad (14.11)$$

Mean number of units in the system (L_s) is obtained employing formula (14.5).

Salary for the k-man crew for 8 hours is

$$SC_k = 8 \text{ hours} \times \begin{bmatrix} \text{Size of} \\ \text{crew} \end{bmatrix} \times \begin{bmatrix} \text{Hourly} \\ \text{rate} \end{bmatrix} \quad (14.12)$$

Total expected cost for the k-man crew is

$$TC_k = IC_k + SC_k \quad (14.13)$$

For a one-man crew, $\lambda = 2$ and $\mu = 1.5$. Since the condition $\lambda < \mu$ is not satisfied, a waiting line of infinite length would form. Hence, this is an infeasible program.

For a two-man crew, $\lambda = 2$ and $\mu = 3$. This is a feasible alternative. Cost of idle trucks and drivers is

$$IC_2 = 8 \times \left[\frac{2}{3 - 2} \right] \times 20 = \$320$$

Salary for the crew is

$$SC_2 = 8 \times 2 \times 7 = \$112$$

$$\text{Total expected cost} = 320 + 112 = \$432$$

For a three-man crew, $\lambda = 2$ and $\mu = 4.5$. This is a feasible alternative. Cost of idle trucks and drivers is

$$IC_3 = 8 \times \left[\frac{2}{4.5 - 2} \right] \times 20 = \$128$$

Salary for the crew is

$$SC_3 = 8 \times 3 \times 7 = \$168$$

Total expected cost $= 128 + 168 = \$296$

For a four-man crew, $\lambda = 2$ and $\mu = 6$. This is a feasible alternative. Cost of idle trucks and drivers is

$$IC_4 = 8 \times \left[\frac{2}{6 - 2} \right] \times 20 = \$80$$

Salary for the crew is

$$SC_4 = 8 \times 4 \times 7 = \$224$$

$$\text{Total expected cost} = 80 + 224 = \$304$$

Among the three total costs, the least cost corresponds to the three-man crew. Beginning with the four-man crew, the total costs of successive larger crew sizes would be ever increasing. The appropriate characteristics and the total cost of alternative crew sizes must be considered in selecting the crew size.

Consider a waiting line with Poisson arrivals and exponential service times. If the cost of waiting per unit per period (C_w), the cost of servicing one unit (C_s), as well as the mean arrival rate (λ) are known, management might wish to determine the optimum service rate (μ_{opt}) that minimizes the total cost of waiting and servicing. Expected cost of waiting per period is

$$ECW = \begin{bmatrix} \text{Cost of waiting} \\ \text{per period} \\ \text{for one unit} \end{bmatrix} \times \begin{bmatrix} \text{Mean number of} \\ \text{units in the} \\ \text{system} \end{bmatrix}$$

$$= C_w \frac{\lambda}{\mu - \lambda} \qquad (14.14)$$

Expected cost of servicing per period is

$$ECS = \begin{bmatrix} \text{Cost of servicing} \\ \text{one unit} \end{bmatrix} \times \begin{bmatrix} \text{Mean service} \\ \text{rate per period} \end{bmatrix}$$

$$= C_s \mu \qquad (14.15)$$

Total cost of waiting and servicing per period for the service rate μ is the sum of ECW and ECS.

$$TC = ECW + ECS \qquad (14.16)$$

We might calculate TC values for different μ values and the μ with the least TC value denotes the optimum service rate.

Example 14.2

The parameters of a queuing system are $C_w = \$0.36$, $C_s = \$0.40$, and $\lambda = 0.10$. Determine μ_{opt}.

TABLE 14.3   Computation of *TC* Values for Different Service Rates

μ	ECW	ECS	TC
0.10	∞	\$0.04	∞
0.20	0.36	0.08	0.44
0.30	0.18	0.12	0.30
0.40	0.12	0.16	0.28
0.50	0.09	0.20	0.29
0.60	0.07	0.24	0.31
0.70	0.06	0.28	0.34
0.80	0.05	0.32	0.37
0.90	0.05	0.36	0.41
1.00	0.04	0.40	0.44

Table 14.3 gives TC values for different μ values. Note that μ_{opt} is 0.40 and the associated TC is \$0.28. Because μ is continuous, μ_{opt} can be determined by differentiating expression 14.16 with respect to μ, setting the resulting expression equal to zero, and solving for μ to obtain

$$\mu_{opt} = \lambda + \sqrt{\frac{\lambda C_w}{C_s}} \qquad (14.17)$$

For Example 14.2, we have

$$\mu_{opt} = 0.10 + \sqrt{\frac{(0.10)(0.36)}{0.40}}$$

$$= 0.40$$

Total cost for μ_{opt} is obtained by employing equation 14.16.

$$TC = \frac{(0.36)(0.10)}{0.40 - 0.10} + (0.40)(0.40)$$

$$= \$0.28$$

When the assumptions for Poisson arrivals and exponential service times are found to be invalid, the solution procedure becomes complex. We might then use the simulation technique discussed in the next chapter.

PROBLEMS

14.1 The number of customers at a copy center for a sample of 50 ten-minute intervals arrive according to the following distribution:

Number of Arrivals in 10 minute Interval	Frequency
0	38
1	10
2	1
3	1
4	0
	50

A study of the time required to service a sample of 60 customers yields the following distribution:

Service Time, Minutes	Frequency
0 and under 10	21
10 and under 20	14
20 and under 30	10
30 and under 40	7
40 and under 50	5
50 and under 60	3
60 or above	0
	60

(a) Examine whether or not the distribution of arrival rates can be modeled by the Poisson distribution.

(b) Examine whether or not the distribution of service times can be modeled by the exponential distribution.

(c) Assuming Poisson arrivals and exponential service time, find the average utilization of the clerk in the copy center.

14.2 The Speedy Carwash Company operates a single-channel auto wash facility in Rialto. The arrival rate of cars follows the Poisson distribution and the average arrival rate is 8 per hour. The distribution of service times is modeled by the exponential; mean service rate is five minutes.

(a) Determine the average utilization of the facility.

(b) Determine the probability of time that the facility is idle.

(c) Determine the probability of three cars in the system.

(d) Determine the mean number of cars in the system.

(e) Determine the mean number of cars in the waiting line.

(f) Determine the probability of two cars in the waiting line.

14.3 The arrival rate of ships at a loading dock is 1.5 per day and the mean service rate is 0.5 day. Assuming Poisson arrivals and exponential service time, find the following:

(a) The mean number of ships in the system, including the one being served.

(b) The mean number of ships in the queue.

(c) The mean time in the system, including the service time.

(d) The mean waiting time.

(e) The probability of four ships in the system.

(f) The probability of no ships in the queue.

14.4 The arrival rates of trucks at a docking facility are Poisson-distributed at the rate of 3.6 per 24 hours. The average service time is 4.80 hours. The distribution of service times is adequately modeled by the exponential distribution.

(a) Calculate the probability of time that there will be at least three trucks in the system.

(b) Calculate the probability of time that there will be at least three trucks in the waiting line.

(c) Determine the mean number of units in the queue.

(d) Determine the mean waiting time.

14.5 An electronic company maintains a bank of machines in the assembly division. The distribution of breakdowns is Poisson-distributed at a mean rate of 2 per hour. Among the three available repairmen, one is to be hired. The wages and service rates of the repairmen are the following:

Repairman	A	B	C
Wages per hour	$4	$5	$7
Service rate (machines per hour)	2.5	3	3.8

The service rates may be approximated by the exponential distribution. Nonproductive time of any one machine is estimated to cost the company $6.50 per hour. Determine which repairman should be hired.

14.6 Mechanics arrive at a tool crib of a factory to check out the special tools required to perform the assigned job. The arrival of mechanics is approximately Poisson-distributed with an average rate of 10 per hour. An attendant is to be hired to serve the mechanics in the tool crib. The estimated costs and service rates for two attendants are as follows:

Attendant	A	B
Wages per hour	$3	$4.5
Mean service time	5 minutes	3 minutes

Assume that the distribution of service times can be adequately described by the exponential distribution. If the cost of nonproductive time is $6 per hour per mechanic, determine which attendant should be hired?

14.7 The parameters of a queueing system are $C_w = \$0.75$, $C_s = \$0.60$, and $\lambda = 0.20$.
   (a) Determine the optimal μ that minimizes the total cost (TC) of waiting and servicing per period by calculating TC values for the 10 different μ values from 0.10 to 1.00 in steps of 0.10.
   (b) Determine the optimal μ using expression (14.17).

14.8 The parameters of a queueing system are $C_w = \$3.60$, $C_s = 0.72$, and $\lambda = 0.80$.
   (a) Establish the optimal μ by employing expression 14.17.
   (b) Calculate the total cost (TC) per period for the optimal μ.

───────────────── **SELECTED REFERENCES** ─────────────────

Bhatia, A., and A. Garg, "Basic Structure of Queuing Problems," *The Journal of Industrial Engineering*, Vol. 14, No. 1, January–February 1963.

Clark, C. T., and L. L. Schkade, *Statistical Methods for Business Decisions*, South-Western Publishing, Cincinnatti, 1969.

Fabrycky, W. J., P. M. Ghare, and P. E. Torgersen, *Industrial Operations Research*, Prentice-Hall, Englewood Cliffs, N.J., 1972.

Hillier, F. S., "The Application of Waiting Line Theory to Industrial Problems," *Journal of Industrial Engineering*, Vol. 15, No. 1, January–February 1964.

Hillier, F. S., and G. J. Lieberman, *Introduction to Operations Research*, Holden-Day, San Francisco, Calif., 1967.

Morse, P. M., *Queues, Inventories, and Maintenance*, Wiley, New York, 1958.

Nelson, R. T., "An Empirical Study of Arrival, Service Time, and Waiting Time Distributions of a Job Shop Production Process," Research Report No. 60, *Management Science Research Project*, UCLA, 1959.

Panico, J. A., *Queuing Theory*, Prentice-Hall, Englewood Cliffs, N.J., 1969.

Paul, R. J., and R. E. Stevens, "Staffing Service Activities with Waiting Line Models," *Decision Sciences*, Vol. 2, No. 2, April 1971.

Saaty, T. L., *Elements of Queuing Theory*, McGraw-Hill, New York, 1961.

Taha, H. A., *Operations Research—An Introduction*, Macmillan, New York, 1971.

CHAPTER 15

simulation

THE TECHNIQUE OF SIMULATION is an important tool for analyzing models of systems. Simulation models have been extensively employed for assisting management in the decision-making process. In the previous chapters, we have formulated mathematical models for the systems under study and solved them through analytical approaches. However, mathematical analysis cannot be utilized for several real systems because the analytical approach is too complex, costly, or impractical. Simulation frequently provides the only approach for analyzing complex problems. For many years, engineers have used scale models of machines in simulating plant layouts and scale models of airplanes in wind tunnels in simulating airplane flights. Simulation enables managers to perform simulated experiments on the mathematical model of some real system and observe the behavior of the model to make inferences about the real system.

Simulation basically involves the construction of some kind of mathematical model that describes the system, performing the simulation on the model, and evaluating the results. In constructing a simulation model, we represent the elements of the system and the interrelationships between the elements by arithmetic, analogic, or logical processes so that the simulated sampling can be executed on the model. The solution of a mathematical model through the analytical method gives a direct and an overall solution to the problem. On the other hand, in the simulation approach, we observe the system's behavior over time and accumulate the relevant data as the simulation progresses in time to obtain valuable insight into the aggregate behavior of the system. Simulation can be used to try out various policies, compare their performance measures, understand the interrelationships between the components, and select the most promising policy.

ILLUSTRATIVE EXAMPLES

Example 15.1 Simulation of Inventory Requirements during Lead Time

We have analyzed in Chapter 13 the inventory problem in which both daily demand and lead time are subject to variations. By enumerating all possible demands during a lead-time period, we calculated the buffer stock. The same problem can be solved using a simulation model.

Figure 15.1 and 15.2 show the probability distributions of lead time and daily demand for the refrigerators of Model R460. We wish to determine the buffer stocks for various service levels. Recall that the expected demand during lead time is required in calculating the buffer stocks, as shown in expression 13.21. We can compute the desired values by performing simulated sampling of lead time and observing the total demand simulated in each of these simulated lead times.

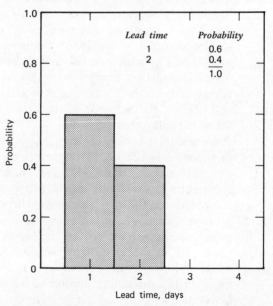

FIGURE 15.1   Probability distribution of lead time for refrigerators model R460.

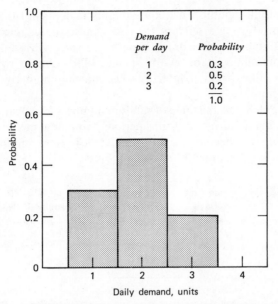

Demand per day	Probability
1	0.3
2	0.5
3	0.2
	1.0

FIGURE 15.2   Probability distribution of daily demand for refrigerators model R460.

The simulated sampling can be performed by obtaining random observations from probability distributions of the system's events. One way of generating random observations is to use a game of roulette. The ball in a roulette wheel, with 100 slots, which are numbered from 00 through 99, has an equally likely chance of stopping at any of the 100 numbers in each trial. Because the probability of one-day lead time is .6, 60 percent of the slots, namely 00 through 59, are allotted for one-day lead time. Likewise, slots 60 through 99 are assigned for a two-day lead time. The daily demand can be obtained in a similar way. Slots 00 to 29 are allotted for one unit per day, slots 30 to 79 for two units, and 80 to 99 for three units. If the outcome of a trial is slot 79, it denotes a two-day lead time. Suppose that the outcomes of the next two trials are slot 22 and 83. They indicate that the daily demands are one unit and three units for the first and the second day of the lead time period, respectively. Total demand during lead time for the trial is 4 units ($= 1 + 3$). The simulated sampling is often called *Monte Carlo method*. The disadvantages of generating random observations using physical devices such as a roulette wheel are that the process is slow and the observations generated are nonreproducible. Tables of random numbers such as Appendix III are available to perform random sampling. A random number is a number in a group of numbers arranged in no particular order with each number having just as equal a chance to occur as any other number. The RAND

Corporation published 1,000,000 random digits in 1955; these digits were produced using a random frequency pulse source (electric device). Random numbers can be drawn according to any consistent pattern from the random number table in any manner—for example, by columns or rows, up or down, and diagonally. The starting point on the table is unimportant.

We shall use the table of random numbers in Appendix III to simulate the demand during lead-time periods by manual method. When two-digit random numbers are used to simulate lead times and daily demands, the following assignments can be employed:

Lead Time, Days	Probability	Cumulative Probability	Random Numbers
1	0.6	0.6	00–59
2	0.4	1.0	60–99
	1.0		

Daily Demand Units	Probability	Cumulative Probability	Random Numbers
1	0.3	0.3	00–29
2	0.5	0.8	30–79
3	0.2	1.0	80–99
	1.0		

The following procedure can be used to determine the demand during lead time for 25 trials.

STEP 1. Draw a random number from Appendix III. We shall start with column 1 at the top and read down the first two digits in each entry, then column 2, column 3, etc.

STEP 2. Determine the lead time (LT) corresponding to the random number.

STEP 3. Draw LT random numbers from the table of random numbers.

STEP 4. For each of the LT random numbers, find the corresponding daily demands $(D_1, D_2, \ldots, D_{LT})$.

STEP 5. Calculate the total demand during lead time.

$$DDLT = D_1 + D_2 + \cdots + D_{LT}$$

STEP 6. Repeat the steps 1 to 5 until 25 lead-time periods are simulated.

Table 15.1 shows a summary of results obtained for 25 trials. The frequency distribution of $DDLT$ values is determined in order to calculate

TABLE 15.1   Hand Simulation of Demands during Lead Times

Trial Number	Random Number for Lead Time	Lead Time (LT), Days	Random Number for Daily Demand during LT Days		Daily Demand during LT Days		Total Demand during Lead Time (DDLT)
			Day 1	Day 2	Day 1	Day 2	
1	21	1	28	—	1	—	1
2	58	1	49	—	2	—	2
3	39	1	64	—	2	—	2
4	63	2	73	32	2	2	4
5	76	2	44	73	2	2	4
6	00	1	63	—	2	—	2
7	99	2	47	43	2	2	4
8	64	2	30	96	2	3	5
9	61	2	61	35	2	2	4
10	20	1	54	—	2	—	2
11	73	2	16	19	1	1	2
12	22	1	60	—	2	—	2
13	71	2	94	65	3	2	5
14	21	1	02	—	1	—	1
15	43	1	82	—	3	—	3
16	58	1	07	—	1	—	1
17	67	2	75	70	2	2	4
18	88	2	66	43	2	2	4
19	57	1	43	—	2	—	2
20	40	1	86	—	3	—	3
21	18	1	73	—	2	—	2
22	90	2	04	62	1	2	3
23	25	1	62	—	2	—	2
24	09	1	05	—	1	—	1
25	87	2	92	58	3	2	5

the probability distribution of *DDLT* and the service level estimates. Table 15.2 displays the computed values for Example 15.1.

Expected demand during lead time is

$$DVMEAN = 1(0.16) + 2(0.36) + 3(0.12) + 4(0.24) + 5(0.12)$$
$$= 2.8 \text{ units}$$

Therefore, the buffer stocks for various possible demands during lead times are

DDLT	*Buffer Stock*	*Service Level*
4	$4 - 2.8 = 1.2$ units	0.88
5	$5 - 2.8 = 2.2$ units	1.00

TABLE 15.2   Service Level Calculations

DDLT	Frequency of DDLT, F	Probability of DDLT, P (DDLT) = F/SF	Probability that the Actual Demand during Lead Time Exceeds DDLT Value PGTL(DDLT)	Service Level, 1 − PGTL(DDLT)
1	4	0.16	0.84	0.16
2	9	0.36	0.48	0.52
3	3	0.12	0.36	0.64
4	6	0.24	0.12	0.88
5	3	0.12	0.00	1.00
	SF = 25	1.00		

A computer program for establishing the buffer stocks by Monte Carlo simulation is discussed later in this chapter.

Example 15.2   Simulation of Machine Repair Operations

Ken's Sportswear operates over 70 knitting machines in its sport shirts division at Southfield. An analysis of the past records indicates the following probability distribution of time between breakdown of knitting machines:

Time between Breakdowns, Hours	Probability
2	0.05
2.5	0.30
3	0.25
3.5	0.20
4	0.15
4.5	0.05
	1.00

Nonproductive time on any one knitting machine is estimated to cost the company $12 per hour. Two alternative plans are available to management on making the repair mechanics assignment for the breakdowns. It can either hire a skilled mechanic at a wage rate of $10 per hour or hire a skilled mechanic and a helper at a total wage rate of $17 per hour. The estimated service time distributions for the plans are as follows:

| Plan 1 | | Plan 2 | |
| One Skilled Mechanic Only | | One Skilled Mechanic and a Helper | |
Service Time, Hours	Probability	Service Time, Hours	Probability
2	0.14	1.5	0.35
2.5	0.20	2	0.35
3	0.40	2.5	0.25
3.5	0.10	3	0.05
4	0.10		1.00
4.5	0.06		
	1.00		

Which of the two plans should be selected? Assume in plan 2 that both the skilled mechanic and helper do the repair work on one machine at any time. The policy of the maintenance section is to service according to the first-come, first-serve basis.

The customers that arrive at the maintenance section requiring service are the breakdown machines. If the repair mechanics are busy, the arrivals have to wait until the customers who arrived earlier are served. We have a single channel–single phase waiting line situation for the repair operation. Figures 15.3 and 15.4 show that the time between arrivals and the service

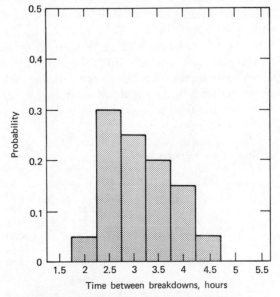

FIGURE 15.3   **Probability distribution of time between breakdowns.**

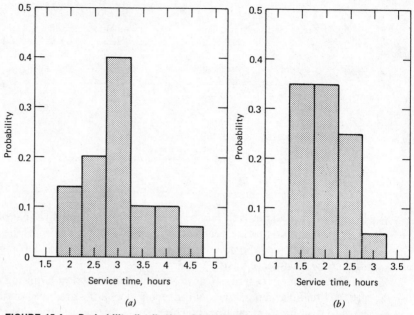

FIGURE 15.4   Probability distribution of the machine repair time. (*a*) **Plan 1.** (*b*) **Plan 2.**

times do not follow the exponential distribution. Therefore, we cannot model the repair operation according to the standard waiting line situation discussed in Chapter 14.

Monte Carlo simulation can be employed to evaluate the plans and select the plan that is beneficial for the company. We shall illustrate the hand simulation procedure for 25 breakdowns. When two-digit random numbers are used in the simulated sampling, we can assign the random numbers for the time between breakdowns and service time as follows:

Time between Breakdowns, Hours	Probability	Cumulative Probability	Random numbers
2	0.05	0.05	00–04
2.5	0.30	0.35	05–34
3	0.25	0.60	35–59
3.5	0.20	0.80	60–79
4	0.15	0.95	80–94
4.5	0.05	1.00	95–99
	1.00		

Plan 1—One skilled mechanic only:

Service Time, Hours	Probability	Cumulative Probability	Random numbers
2	0.14	0.14	00–13
2.5	0.20	0.34	14–33
3	0.40	0.74	34–73
3.5	0.10	0.84	74–83
4	0.10	0.94	84–93
4.5	0.06	1.00	94–99
	1.00		

Plan 2—One skilled mechanic and a helper:

Service Time, Hours	Probability	Cumulative Probability	Random Numbers
1.5	0.35	0.35	00–34
2	0.35	0.70	35–69
2.5	0.25	0.95	70–94
3	0.05	1.00	95–99

The starting conditions of the system must be determined before beginning simulation. For our repair operations example, how many breakdowns are waiting to be repaired at the beginning of simulation? And when does breakdown 1 arrive? The starting conditions of a simulation model significantly influence the resulting outcome while the model is "warming up." Therefore, it might be necessary to record the aggregate behavior of the system when it is functioning in steady state. Mize and Cox (1968, p. 157) suggest that "... we will wish to attain equilibrium before recording system performance, in which case two alternatives are open to us: (1) ignore data generated during some initial period; or (2) choose starting conditions that approximate the steady-state condition of the system. Owing to the unfortunate fact that we cannot generally specify even reasonably accurate starting conditions for all variables, a combination of (1) and (2) is usually necessary." We shall assume that the system is idle and empty at the beginning of simulation. The first machine is assumed to arrive x hours after the starting time of the system (0 hour), and x is the arrival time measured for the first machine from previous arrival. The value of x is obtained from Table 15.3.

We shall draw random numbers from Appendix III starting with row 1 at the left end and read to the right the first two digits in each entry,

TABLE 15.3 Hand Simulations of Time between Arrivals and Service Time for 25 Machine Breakdowns (Plan 1)

Arrival	Random Number for Time between Arrivals	Arrival Time Measured from Previous Arrival	Random Number for Service Time	Service Time
1	21	2.5	62	3
2	22	2.5	98	4.5
3	03	2	95	4.5
4	00	2	40	3
5	52	3	21	2.5
6	28	2.5	25	2.5
7	71	3.5	41	3
8	12	2.5	55	3
9	58	3	54	3
10	46	3	28	2.5
11	58	3	62	3
12	99	4.5	44	3
13	38	3	63	3
14	70	3.5	05	2
15	32	2.5	58	3
16	49	3	09	2
17	94	4	47	3
18	69	3.5	57	3
19	25	2.5	29	2.5
20	97	4.5	49	3
21	39	3	05	2
22	44	3	85	4
23	93	4	75	3.5
24	09	2.5	64	3
25	30	2.5	39	3
			Total service time	74.5

then row 2, row 3, etc. We require two random numbers for each breakdown—the first number for time between arrivals and the second for service time. Table 15.3 shows the random numbers and the corresponding arrival time measured from previous arrival together with the the service time for 25 breakdowns.

Table 15.4 summarizes the arrival times, machine waiting times, and repair mechanics idle time for the 25 arrivals. At the starting of the system, no breakdowns are waiting for service. The arrival time from previous arrival and service time for breakdown 1 are 2.5 hours and 3 hours, respectively. Hence, breakdown 1 arrives 2.5 hours after the operations

TABLE 15.4   Hand Simulation of Machine Waiting Time and Mechanics Idle Time (Plant)

Arrival	Arrival Time	Time Repair Begins	Time Repair Ends	Machine Waiting Time	Repair Mechanics Idle Time
1	2.5	2.5	5.5	0	2.5
2	5	5.5	10	0.5	0
3	7	10	14.5	3	0
4	9	14.5	17.5	5.5	0
5	12	17.5	20	5.5	0
6	14.5	20	22.5	5.5	0
7	18	22.5	25.5	4.5	0
8	20.5	25.5	28.5	5	0
9	23.5	28.5	31.5	5	0
10	26.5	31.5	34	5	0
11	29.5	34	37	4.5	0
12	34	37	40	3	0
13	37	40	43	3	0
14	40.5	43	45	2.5	0
15	43	45	48	2	0
16	46	48	50	2	0
17	50	50	53	0	0
18	53.5	53.5	56.5	0	0.5
19	56	56.5	59	0.5	0
20	60.5	60.5	63.5	0	1.5
21	63.5	63.5	65.5	0	0
22	66.5	66.5	70.5	0	1
23	70.5	70.5	74	0	0
24	73	74	77	1	0
25	75.5	77	80	1.5	0
			Total	59.5	5.5

have started. At that time, the repair mechanic is idle. Therefore, he begins the repair operation immediately. The repair work lasts for the next 3 hours. The repair work ends 5.5 hours after starting the system. Breakdown 1 did not wait for service. The mechanic was not working from the beginning until the arrival of breakdown 1. Thus, the idle time for the mechanic is 2.5 hours. The arrival time measured from previous arrival for breakdown 2 is 2.5 hours. Therefore, the arrival time for breakdown 2 is 5 ($= 2.5 + 2.5$). Since the mechanic is working on machine 1 until 5.5 hours, the waiting time for breakdown 2 is 0.5 hour, that is, from 5 to 5.5. Likewise, the simulated sampling for 25 breakdowns shown in Table 15.4 is calculated based on the data in Table 15.3. Note in

Table 15.4 that the repair work on breakdown 17 ends at 53 while the arrival time of breakdown 18 is 53.5. Therefore, the idle time of the mechanic is 0.5 hour.

We notice in Table 15.4 that the mechanic was idle for 5.5 hours during the 80 hour operation simulation. The total machine waiting time and

TABLE 15.5   Current Real-World Applications of Simulation Methods

Air traffic control queuing	Industry models
Aircraft maintenance scheduling	Textile
Airport design	Petroleum (financial aspects)
Ambulance location and dispatching	Information system design
Assembly line scheduling	Intergroup communication (sociological studies)
Bank teller scheduling	Inventory reorder rule design
Bus (city) scheduling	Aerospace
Circuit design	Manufacturing
Clerical processing system design	Military logistics
Communication system design	Hospitals
Computer time sharing	Job shop scheduling
Telephone traffic routing	Aircraft parts
Message system	Metals forming
Mobile communications	Work-in-progress control
Computer memory-fabrication test-facility design	Shipyard
Consumer behavior prediction	Library operations design
Brand selection	Maintenance scheduling
Promotion decisions	Airlines
Advertising allocation	Glass furnaces
Court system resource allocation	Steel furnaces
Distribution system design	Computer field service
Warehouse location	National manpower adjustment system
Mail (post office)	Natural resource (mine) scheduling
Soft drink bottling	Iron ore
Bank courier	Strip mining
Intrahospital material flow	Parking facility design
Enterprise models	Numerically controlled production facility design
Steel production	Personnel scheduling
Hospital	Inspection department
Shipping line	Spacecraft trips
Railroad operations	Petrochemical process design
School district	Solvent recovery
Equipment scheduling	Police response system design
Aircraft	Political voting prediction
Facility layout	Rail freight car dispatching
Pharmaceutical center	Railroad traffic scheduling
Financial forecasting	Steel mill scheduling
Insurance	Taxi dispatching
Schools	Traffic light timing
Computer leasing	Truck dispatching and loading
Insurance manpower hiring decisions	University financial and operational forecasting
Grain terminal operation	Urban traffic system design
Harbor design	Water resources development

Source: Emshoff, J. R., and R. L. Sisson, Design and Use of Computer Simulation Models, Macmillan, New York, 1970, p. 264.

machine repair time were 59.5 hours and 74.5 hours, respectively. This information would be useful in determining the most promising alternative. However, the results would be more reliable when the performance of repair operation is simulated for a long time, say 1000 hours of repair operation. Although we may perform a hand simulation for a large simulated experiment, Monte Carlo simulation is usually executed on a computer so that the calculations are performed faster. Moreover, most simulation models are of practical value only when the time required to solve large-size problems is greatly reduced. Therefore, a computer-oriented simulation model is often the only feasible approach to executing many real-world simulation studies. We shall discuss a computer model for the repair operations situation and establish the preferred plan later in this chapter. Simulation methods have been employed for analyzing various situations. Table 15.5 summarizes the situations in which simulation methodology has been used.

RANDOM NUMBER GENERATION

Random numbers are used to obtain random observations in Monte Carlo simulations, as seen in the illustrative examples discussed earlier in this chapter. The generation of random numbers is an essential aspect of simulation. The simulation calculations may have to be repeated to check calculations. The same sequence of random numbers employed in the initial calculations are used in such situations. Hence, the process that generates random numbers must be reproducible. Long simulation runs are executed to obtain a realistic picture of the system's behavior. Long simulation runs involve the use of a great many random numbers. Therefore, methods that generate random numbers at a low cost and high speed are preferred.

Random numbers can be generated by manual methods involving the use of such devices as a disc or a roulette wheel. The manual approach is slow and nonreproducible. Using tables of random numbers in the computer simulation is clumsy because the numbers are either stored in the computer memory which requires a large storage space, or read from tape or punched cards when the numbers are required. The most popular way of obtaining random numbers for computer simulation is to have the computer itself generate the numbers through a computer program. There are several numerical methods available to produce a series of random numbers. The Lehmer Multiplicative Congruential Method is a widely used procedure for generating random numbers.

The following mathematical formula is applied over and over again in the multiplicative congruential method to produce random numbers:

$$X_{n+1} = AX_n \text{ (modulo } m) \tag{15.1}$$

where

X_n = the nth random number generated by the process

A and m are known positive integers $A < m$. The $(n + 1)$th random number is generated from the nth random number. Equation 15.1 means that X_{n+1} is the remainder of dividing AX_n by m.

X_0 = the initial value of the sequence and it is any odd number

$m = r^b$

r is the radix of the computing machine
$r = 2$ for binary computer
$r = 10$ for decimal computer

Hence $m = 2^b$ for a binary computer. We shall design the mathematical formula assuming $b = 4$. Now, $m = 2^b = 2^4 = 16$. The value of A is established from the two expressions:

$$A \simeq 2^{(b/2)} \tag{15.2}$$

$$A = 8t \pm 3 \qquad \text{and} \qquad t \text{ is an integer} \tag{15.3}$$

Equation 15.2 gives $A \simeq 2^2 \simeq 4$. Substituting $t = 1$ into (15.3), we obtain

$$A = 11 \text{ or } 5$$

Therefore, we choose $A = 5$.

Let $X_0 = 9$. Our mathematical formula can be represented as follows:

$$X_{n+1} = 5X_n \text{ (modulo 16)} \qquad \text{and} \qquad X_0 = 9 \tag{15.4}$$

The first random number X_1 is obtained from X_0.

$$X_1 = 5X_0 \text{ (modulo 16)}$$
$$5X_0 = 5 \times 9 = 45$$
$$45 = (2 \times 16) + 13$$

Thus $X_1 = 13$. Likewise, X_2 is generated from X_1; X_3 is generated from X_2; and so on. Table 15.6 summarizes the calculations for five iterations. Note that the sequence will repeat itself with a cycle of only four numbers. The length of a cycle of the congruential method can be predetermined by the expression:

$$\text{Length of cycle} = 2^{b-2} \text{ numbers} \tag{15.5}$$

TABLE 15.6   Random Number Generation by the Congruential Method

			$X_{n+1} = 5X_n$ (Modulo 16) and $X_0 = 9$	
n_n	X_n	$5X_n$	Modulo 16	X_{n+1}
0	9	45	13	First number = 13
1	13	65	1	Second number = 1
2	1	5	5	Third number = 5
3	5	25	9	Fourth number = 9
4	9	45	13	

By assigning suitable values for A, b, and m, we can generate a great many numbers before the numbers are repeated. The sequence of numbers that resulted from the congruential method will range from 1 to m. In order to have uniformly distributed numbers (U_i) from 0 to 1, we can divide the resulting random numbers by m. If we wish to obtain three-digit numbers in the range of 000 to 999, we can multiply each U_i value by 1000. SUBROUTINE RANDOM presented in Figure 15.5 can produce over 1000 random numbers with three-digits each. Observe in the main program of Figure 15.5 that the initial value of the sequence is 15.

The random numbers produced by the congruential method has passed many statistical tests for randomness. With a computer these numbers can be generated at high speed and at low cost. Since the numbers generated are predictable and reproducible, they cannot really be called random numbers. The numbers are sufficiently random in appearance and are usable in simulation to obtain random observations. Hence, they are often called pseudorandom numbers.

COMPUTER MODELS FOR THE ILLUSTRATIVE EXAMPLES

In this section we shall develop computer programs to analyze the illustrative examples by Monte Carlo simulation. We shall also establish the preferred alternative for the examples.

Simulation of Inventory Requirements during Lead Time

We shall design a computer model for Example 15.1 to determine the buffer stock and service level for various demands during lead time. The following variables are employed in the computer program.

LODAY	= the smallest possible lead time, days
LRGDAY	= the largest possible lead time, days
LOUSE	= the smallest possible demand in a day
LRGUSE	= the largest possible demand in a day
LODMD	= the smallest possible total demand during a lead-time period
LRGDMD	= the largest possible total demand during a lead-time period
ITIMES	= number of lead-time periods to be simulated
IGEN	= random number
JCYCLE	= counter that indicates the trial number being simulated
IDAY	= number of days for the particular lead-time period
IRAND	= random number
QANDMD(K)	= frequency of K-units demand occurred during lead-time periods
PRBDMD(K)	= probability of K-units demand occurring during ITIMES lead-time trials
STKPRB(K)	= probability of stockout for DDLT = K
IQUANT	= total demand during the particular lead-time period
KCYCLE	= counter that indicates the number of days chosen for the lead time
IUSE	= demand during the particular day
EXPDMD	= expected demand during lead time
RANDOM	= this subprogram generates a three-digit random number
SLEVEL	= service level
BUFSK	= buffer stock, units

The variables QUANDMD( ), PRBDMD( ), and STKPRB( ) are one-dimensional arrays with LRGDMD elements. We might transmit the input data and initialize the program by the instructions:

```
     READ 904, LODAY, LRGDAY, LOUSE, LRGUSE, ITIMES
     LODMD = LODAY * LOUSE
     LRGDMD = LRGDAY * LRGUSE
     DO 460 K = LODMD, LRGDMD, 1
460  QANDMD(K) = 0
     IGEN = 15
904  FORMAT ( )
```

The following statements will cause the computer to calculate the total demand during a lead-time period with IDAY days.

```
      IQUANT = 0
      DO 600 KCYCLE = 1, IDAY
      CALL RANDOM (IGEN, IRAND)
      IUSE = 2
      IF (IRAND.LE.299) IUSE = 1
      IF (IRAND.GE.800) IUSE = 3
      IQUANT = IQUANT + IUSE
600   CONTINUE
```

We can perform ITIMES trials and determine the frequency distribution of DDLT values by placing the statements developed above in the space left blank in the following instructions:

```
      DO 650 JCYCLE = 1, ITIMES
      CALL RANDOM (IGEN, IRAND)
      IDAY = 2
      IF (IRAND.LE.599) IDAY = 1
        ⋮
      QANDMD(IQUANT) = QANDMD(IQUANT) + 1.0
650   CONTINUE
```

For each DDLT value, the probability of occurrence and the probability of stockout can be calculated and printed out by the following set of statements:

```
      TIMES = ITIMES
      BEFORE = 0
      FORE = 0
      JQUANT = LRGDMD + 1
      DO 630 KQUANT = LODMD, LRGDMD, 1
      JQUANT = JQUANT − 1
      PRBDMD(JQUANT) = QANDMD(JQUANT)/TIMES
      STKPRB(JQUANT) = FORE + BEFORE
      FORE = PRBDMD(JQUANT)
      BEFORE = STKPRB(JQUANT)
630   CONTINUE
      DO 700 K = LODMD, LRGDMD, 1
      PRINT 925, K, PRBDMD(K), STKPRB(K)
700   CONTINUE
925   FORMAT ( )
```

The following set of instructions would enable the computer to calculate and print out the average demand during lead time.

```
        EXPDMD = 0
        DO 800 INDEX = LODMD, LRGDMD
        DMD = INDEX
        EXP = DMD * PRBDMD(INDEX)
        EXPDMD = EXPDMD + EXP
   800  CONTINUE
        PRINT 930, EXPDMD
   930  FORMAT ( )
```

Next, the buffer stock for various DDLT values is calculated. The calculated values for the DDLT values with 10 percent or fewer stockout probabilities are printed out.

```
        DO 810 L = LODMD, LRGDMD
        IF (STKPRB(L).GT.0.10) GO TO 810
        SLEVEL = 1.0 − STKPRB(L)
        YL = L
        BUFSK = YL − EXPDMD
        PRINT 935, L, BUFSK, SLEVEL
   810  CONTINUE
   935  FORMAT ( )
        STOP
        END
```

SUBROUTINE RANDOM is designed to generate uniformly distributed numbers in the range of 000 to 999 using the multiplicative congruential method.

```
        SUBROUTINE RANDOM (IRAND, IGEN)
        GEN = IGEN
        A = 61.
        ZMODUL = 4096.
        GGEN = (A * GEN)/ZMODUL
        IGGEN = GGEN
        ZGEN = IGGEN
        GEN = (GGEN − ZGEN) * ZMODUL
        IGEN = GEN
        GEN = IGEN
        RANNUB = (GEN/ZMODUL) * 1000.
        IRAND = RANNUB
        RETURN
        END
```

A complete computer program composed of our previously designed program segments is shown in Figure 15.5. The computer program has

```
      DIMENSION QANDMD(6), PRBDMD(6), STKPRB(6)
C
C  PRINT PAGE HEADINGS.
      PRINT 902
C
      READ 904, LODAY, LRGDAY, LOUSE, LRGUSE, ITIMES
C  DETERMINE LODMD AND LRGDMD.
      LODMD = LODAY*LOUSE
      LRGDMD = LRGDAY*LRGUSE
C
      DO 460 K= LODMD,LRGDMD,1
  460 QANDMD(K) = 0
C
      IGEN = 15
C
C  SET UP A LOOP TO FIND QANDMD VALUES.
      DO 650 JCYCLE = 1, ITIMES
C
C  CALL THE RANDOM NUMBER (IRAND), AND DETERMINE DAY.
      CALL RANDOM (IGEN,IRAND)
      IDAY = 2
      IF (IRAND.LE.599) IDAY = 1
      IQUANT = 0
C
C  USE THE VALUE IDAY TO LOOP IUSE.
      DO 600 KCYCLE = 1, IDAY
C  CALL IRAND AND FIND USE.
      CALL RANDOM (IGEN,IRAND)
      IUSE = 2
      IF (IRAND.LE.299) IUSE = 1
      IF (IRAND.GE.800) IUSE = 3
C
C  SUM THE DAILY USAGES.
      IQUANT = IQUANT + IUSE
  600 CONTINUE
C
C  SUM THE NUMBER OF TIMES A DEMAND HAS OCCURED.
      QANDMD(IQUANT) = QANDMD(IQUANT) + 1.0
C
  650 CONTINUE
C
C  DETERMINE PRBDMD AND STKPRB.
      TIMES = ITIMES
      BEFORE = 0
      FORE = 0
      JQUANT = LRGDMD + 1
      DO 630 KQUANT = LODMD, LRGDMD, 1
      JQUANT = JQUANT - 1
      PRBDMD(JQUANT) = QANDMD(JQUANT)/TIMES
      STKPRB(JQUANT) =  FORE + BEFORE
      FORE = PRBDMD(JQUANT)
      BEFORE = STKPRB(JQUANT)
  630 CONTINUE
```

FIGURE 15.5 Computer program for simulating the demands during lead time periods (illustrated for Example 15.1).

```
C   PRINT PRBDMD AND STKPRB VALUES.
        PRINT 920
        PRINT 921
        DO 700 K = LODMD, LRGDMD, 1
        PRINT 925, K, PRBDMD(K), STKPRB(K)
  700 CONTINUE
C
C   FIND EXPDMD.
        EXPDMD = 0
        DO 800 INDEX = LODMD, LRGDMD
        DMD = INDEX
        EXP = DMD*PRBDMD(INDEX)
        EXPDMD = EXPDMD + EXP
  800 CONTINUE
C
C   PRINT EXPDMD.
        PRINT 930, EXPDMD
C
C       FIND AND PRINT BUFSK, AND SLEVEL
        PRINT 933
        DO 810 L = LODMD, LRGDMD
        IF (STKPRB(L).GT.0.10) GO TO 810
        SLEVEL = 1.0 - STKPRB(L)
        YL = L
        BUFSK = YL - EXPDMD
        PRINT 935, L, BUFSK, SLEVEL
  810 CONTINUE
C
C
  902 FORMAT (1H1,///,24X,20HINVENTORY SIMULATION)
  904 FORMAT (5(2X,I5))
  921 FORMAT (19X,6HDEMAND,8X,1HP,6X,10HSTOCKOUT P,/)
  920 FORMAT (////,17X,38HPROBABILITIES (P) FOR POSSIBLE DEMANDS,//)
  925 FORMAT (20X, I3, 2(5X,F7.4))
  930 FORMAT (//20X,16HEXPECTED DULT = ,F6.2,6H UNITS)
  933 FORMAT (///,22X,4HDDLT,8X,6HBUFFER,7X,7HSERVICE,
     1/,34X,5HSTOCK,8X,5HLEVEL,/)
  935 FORMAT (22X,I2,9X,F5.3,9X,F6.4)
        STOP
        END



        SUBROUTINE RANDOM (IGEN, IRAND)
C   THIS SUBPROGRAM IS USED TO
C   GENERATE RANDOM NUMBERS.
        GEN = IGEN
        A = 61.
        ZMODUL = 4096.
        GGEN = (A*GEN)/ZMODUL
        IGGEN = GGEN
        ZGEN = IGGEN
        GEN = (GGEN - ZGEN)* ZMODUL
        IGEN = GEN
        GEN = IGEN
        RANNUB = (GEN/ZMODUL)*1000.
        IRAND = RANNUB
        RETURN
        END
```

FIGURE 15.5 (*Continued*)

```
                        INVENTORY SIMULATION


           PROBABILITIES (P) FOR POSSIBLE DEMANDS

               DEMAND          P         STOCKOUT P

                  1          .1600         .8400
                  2          .3600         .4800
                  3          .2200         .2600
                  4          .1450         .1150
                  5          .0900         .0250
                  6          .0250           0


               EXPECTED DDLT =    2.72 UNITS


              DDLT          BUFFER          SERVICE
                            STOCK           LEVEL

               5           2.280           .9750
               6           3.280          1.0000
```

FIGURE 15.6   Computer printout for Example 15.1.

been utilized to simulate 500 lead-time periods. Figure 15.6 shows the resulting computer printout. Management can use the service level and the related buffer stock for the DDLT values given in Figure 15.6 in selecting the appropriate reorder point.

Simulation of Machine Repair Operations

We shall structure a computer model to simulate 500 hours of repair operations for plan 1 and obtain the aggregate behavior of the system. Then, we shall utilize this information to select the most appropriate alternative. The following symbols are employed in the computer model.

TIMES   = number of hours to be simulated
TSERVT = total service time
TCWT   = total waiting time for the customers
TSWT   = total idle time for the service facility
TPARR   = arrival time of previous customer
TSENP   = service ending time of previous customer
UNITS   = number of machines repaired
IGEN   = random number in the range of 0 to 4096
IRAND   = random number in the range of 0 to 999
JR1   = random number for determining the time between arrivals
JR2   = random number for determining the service time

TBARR   = time between arrivals in hours
TARR   = arrival time of the particular customer
TSBN   = beginning time of service for the particular customer
TSEN   = ending time of service for the particular customer
SEVT   = service time
CWTIME = waiting time for the customer
SWTIME = idle time for the service facility

We might employ the following statements to initialize the program:

```
TIMES = 500
TSEVT = 0
TCWT = 0
TSWT = 0
TPARR = 0
TSENP = 0
UNITS = 0
IGEN = 15
```

Now, we can generate a random number and find the beginning time of service for the breakdown machine. These statements should perform the operations:

```
CALL RANDOM (JR1, IGEN)
TBARR = 2.0
IF (JR1.GE. 50) TBARR = 2.5
IF (JR1.GE.350) TBARR = 3.0
IF (JR1.GE.600) TBARR = 3.5
IF (JR1.GE.800) TBARR = 4.0
IF (JR1.GE.950) TBARR = 4.5
TARR = TPARR + TBARR
IF (TSENP.GE.TARR) TSBN = TSEN
IF (TSENP.LT.TARR) TSBN = TARR
```

Next, we can generate a random number and establish the service time and the ending time of service for the particular customer by including the following instructions:

```
CALL RANDOM (JR2, IGEN)
SEVT = 2.0
IF (JR2.GE.140) SEVT = 2.5
IF (JR2.GE.340) SEVT = 3.0
IF (JR2.GE.740) SEVT = 3.5
IF (JR2.GE.840) SEVT = 4.0
IF (JR2.GE.940) SEVT = 4.5
TSEN = SEVT + TSBN
```

We could calculate the total service time, total waiting time, and other time parameters by the statements:

```
TSERVT = TSERVT + SEVT
CWTIME = TSBN − TARR
TCWT = TCWT + CWTIME
SWTIME = TSBN − TSENP
TSWT = TSWT + SWTIME
TPARR = TARR
TSENP = TSEN
```

In order to repeat the calculations for 500 hours of the service facility operation, we might place the three sets of statements developed above inside the loop in the space left blank:

```
100   UNITS = UNITS + 1.0
      _____
      _____

      IF (TSEN.GE.TIMES) GO TO 200
      GO TO 100
200   J TIMES = TIMES
```

The following statements will cause the computer to print out the number of hours simulated, the number of units served, the total service time, the total waiting time for the customers, and the total idle time for the service facility:

```
      PRINT 300, JTIMES
      PRINT 310, UNITS
      PRINT 320, TSERVT
      PRINT 330, TCWT
      PRINT 340, TSWT
300   FORMAT ( )
310   FORMAT ( )
320   FORMAT ( )
330   FORMAT ( )
340   FORMAT ( )
      STOP
      END
```

SUBROUTINE RANDOM developed for Example 15.1 can be utilized to generate three-digit random numbers. Figure 15.7 shows a complete program for repair operations according to plan 1. The resulting computer printout is displayed in Figure 15.8. We can calculate the

```
C     PRINT TITLES AND HEADINGS
          PRINT 950
          PRINT 951
C
C     INITIALIZE REQUIRED VALUES
          TIMES = 500
          TSEVT = 0
          TCWT = 0
          TSWT = 0
          TPARR = 0
          TSENP = 0
          UNITS = 0
          IGEN = 15
C
C     START SIMULATION LOOP
  100 UNITS = UNITS + 1.0
C
C     DETERMINE TBARR FROM A RANDOM NUMBER
          CALL RANDOM (JR1, IGEN)
          TBARR = 2.0
          IF (JR1.GE. 50)   TBARR = 2.5
          IF (JR1.GE.350)   TBARR = 3.0
          IF (JR1.GE.600)   TBARR = 3.5
          IF (JR1.GE.800)   TBARR = 4.0
          IF (JR1.GE.950)   TBARR = 4.5
          TARR = TPARR + TBARR
C
C     CHECK TO DETERMINE TSBN
          IF (TSENP.GE.TARR) TSBN = TSEN
          IF (TSENP.LT.TARR) TSBN = TARR
C
C     USE A RANDOM NUMBER TO DETERMINE SEVT
          CALL RANDOM (JR2, IGEN)
          SEVT = 2.0
          IF (JR2.GE.140) SEVT = 2.5
          IF (JR2.GE.340) SEVT = 3.0
          IF (JR2.GE.740) SEVT = 3.5
          IF (JR2.GE.840) SEVT = 4.0
          IF (JR2.GE.940) SEVT = 4.5
C
C     DETERMINE WAITING LINE PARAMETERS
          TSEN = SEVT + TSBN
          TSERVT = TSERVT + SEVT
          CWTIME = TSBN - TARR
          TCWT = TCWT + CWTIME
          SWTIME = TSBN - TSENP
          TSWT = TSWT + SWTIME
          TPARR = TARR
          TSENP = TSEN
C
```

FIGURE 15.7   Computer program for simulating the machine repair operations (illustrated for Example 15.2).

```
C    TEST TO SEE IF TSEN IS EQUALL TO TIMES
C   IF YES TERMINATE SIMULATION, IF NO CONTINUE
        IF (TSEN.GE.TIMES) GO TO 200
        GO TO 100
C
C    PRINT OUTPUT VALUES
  200 JTIMES = TIMES
        PRINT 300, JTIMES
        PRINT 310, UNITS
        PRINT 320, TSERVT
        PRINT 330, TCWT
        PRINT 340, TSWT
C
C
  300 FORMAT (//,10X,19HSIMULATION TIME IS ,I3,6H HOURS)
  310 FORMAT (10X,33HTHE NUMBER OF UNITS SIMULATED IS ,F4.0)
  320 FORMAT (10X,45HTHE TOTAL SERVICE TIME FOR THE SIMULATION IS ,
     1F5.1)
  330 FORMAT (10X,35HTHE TOTAL CUSTOMER WAITING TIME IS ,F5.1)
  340 FORMAT (10X,43HTHE TOTAL SERVICE FACILITY WAITING TIME IS ,F5.1)
  950 FORMAT (1H1,///,27X,18HQUEUING SIMULATION)
  951 FORMAT (/,25X,21HMONTE CARLO TECHNIQUE)
        STOP
        END




        SUBROUTINE RANDOM (IRAND, IGEN)
        GEN = IGEN
        A = 61.
        ZMODUL = 4096.
        GGEN = (A*GEN)/ZMODUL
        IGGEN = GGEN
        ZGEN = IGGEN
        GEN = (GGEN - ZGEN)* ZMODUL
        IGEN = GEN
        GEN = IGEN
        RANNUB = (GEN/ZMODUL)*1000.
        IRAND = RANNUB
        RETURN
        END
```

FIGURE 15.7 (*Continued*)

```
                QUEUING SIMULATION

              MONTE CARLO TECHNIQUE


        SIMULATION TIME IS 500 HOURS
        THE NUMBER OF UNITS SIMULATED IS 154.
        THE TOTAL SERVICE TIME FOR THE SIMULATION IS 457.5
        THE TOTAL CUSTOMER WAITING TIME IS 224.5
        THE TOTAL SERVICE FACILITY WAITING TIME IS  42.5
```

FIGURE 15.8   Computer printout for Example 15.2.

following characteristics for plan 1:

$$\text{Probability of time that the service facility is busy} = \frac{\text{Total service time in hours}}{\text{Number of hours simulated}}$$

$$= \frac{457.5}{500} = 0.915$$

$$\text{Probability of time that the service facility is idle} = 1 - \left[\begin{array}{l}\text{Probability of time that the} \\ \text{service facility is busy}\end{array}\right]$$

$$= 1 - 0.915 = 0.085$$

$$\begin{array}{l}\text{Average customer waiting} \\ \text{time } (AWT)\end{array} = \frac{\text{Total waiting time}}{\text{Number of customers served}}$$

$$= \frac{224.5}{154} = 1.46 \text{ hours per customer}$$

$$\text{Average service time } (AST) = \frac{\text{Total service time}}{\text{Number of customers served}}$$

$$= \frac{457.5}{154} = 2.97 \text{ hours per customer}$$

$$\begin{array}{l}\text{Average time (including} \\ \text{service time) per customer} \\ \text{in the system}\end{array} = AWT + AST = 4.43 \text{ hours}$$

$$\begin{array}{l}\text{Total cost for 500 hours of} \\ \text{operation}\end{array} = \left[\begin{array}{l}\text{Total cost of} \\ \text{nonproductive time} \\ \text{for the breakdown} \\ \text{machines}\end{array}\right] + \left[\begin{array}{l}\text{Total} \\ \text{wages} \\ \text{for the} \\ \text{mechanics}\end{array}\right]$$

$$\begin{array}{l}\text{Total cost of nonproductive} \\ \text{time for the breakdown} \\ \text{machines}\end{array} = \left[\begin{array}{l}\text{Total} \\ \text{nonproductive} \\ \text{time, hours}\end{array}\right] \times \$12 \text{ per hour}$$

$$= (224.5 + 457.5) \times 12$$

$$= \$8184$$

$$\begin{array}{l}\text{Total wages for the} \\ \text{mechanics}\end{array} = \left[\begin{array}{l}\text{Number of hours} \\ \text{simulated}\end{array}\right] \times \$10 \text{ per hour}$$

$$= 500 \times 10 = \$5000$$

$$\text{Hence, total cost} = 8184 + 5000$$

$$= \$13,184$$

A computer model was designed for plan 2 and it was simulated for 500 hours of repair operations. The following results were obtained:

Simulation time = 500 hours

Number of units simulated = 154

Total service time for the simulation = 305.5 hours

Total customer waiting time = 6.0 hours

Total service facility waiting time = 194.5 hours

The characteristics for plan 2 are

Probability of time that the service facility is busy = 0.611

Probability of time that the service facility is idle = 0.389

Average customer waiting time (AWT) = 0.04 hours

Average service time (AST) = 1.98 hours

Average time per customer in the system
(including service time) = 2.02 hours

Total cost for 500 hours of operation = $12,238

The total cost of plan 2 is smaller than that of plan 1. The mechanics are idle for 38.9 percent of the time under plan 2. Management can assign "fill-in" jobs to the repair crew to be performed during the idle periods. Plan 2 is the most promising alternative.

SIMULATION LANGUAGES

Computer simulation is a powerful tool for analyzing complete systems. The task of designing and programming a computer model for simulating a complete system can be very time-consuming for the programmers. Moreover, debugging the program and verifying the accuracy of the logic can also become a very difficult problem for systems analysts and programmers. A number of special purpose computer languages have been developed to facilitate the learning and using them for simulation studies. The special purpose languages include GPSS, DYNAMO, SIMSCRIPT, GASP, SIMPAC, and SIMULATE. The special purpose languages provide a generalized structure for developing simulation models. These languages are highly helpful for designing computer programs from the simulation models in considerably less time. Each special purpose language is suitable for certain types of problems. GPSS, DYNAMO, and SIMSCRIPT are perhaps the most widely used special purpose languages. The General Purpose Systems Simulator (GPSS) might be preferred for situations in which the items pass through a series of processes or storage functions such as queues. DYNAMO can be useful in simulating large-size economic systems involving information feedbacks and delays. It uses a difference equation structure. SIMSCRIPT is more flexible and can be applied for a wide variety of situations.

Operational gaming is another popular form of simulation. Situations involving some conflict of interest among players or decision makers are studied through operational gaming. The two widely used applications of operational gaming are the business management games and the military operations. The players make decisions at various stages of the dynamic environment simulated by the game model. The effects of the decisions are simulated; afterwards the players make their decisions and so on. Operational gaming is a valuable tool for training managers.

PROBLEMS

15.1   Consider Problem 13.9. Using a simulated sample of 20, solve the problem.

15.2   Consider Problem 13.10. Simulate the activities for 25 lead-time periods. Solve the problem using the information obtained from the simulation.

15.3   The multiplicative congruential method is used to generate 10,000 random numbers of four-digits each. Establish a mathematical formula for the congruential method.

15.4   Consider Problem 13.9. Prepare a computer program that will enable the computer to simulate 500 samples and print out the solutions.

15.5   Consider Problem 13.10. Design a computer model to simulate 1500 samples and print out the solutions. Compare the solutions with that of Problem 13.10. State your conclusions.

15.6   A printing company is planning to sell its copy machine and rent either one of copier A or one of copier B. The company has estimated the probability distributions of times between arrivals of orders requiring copy machine service and the service times for the customer orders. The firm has given you the following data:

Time between Arrivals, Minutes	Probability
30	0.01
35	0.11
40	0.15
45	0.43
50	0.20
55	0.10
	1.00

Copier A		Copier B	
Service Time, Minutes	Probability	Service Time, Minutes	Probability
25	0.20	20	0.21
30	0.60	25	0.70
35	0.20	35	0.06
	1.00	40	0.03
			1.00

Cost of renting copier A = $1.25 per hour.
Cost of renting copier B = $1.50 per hour.
Cost of customer waiting in the company = $1.00 per hour (order waiting time plus the service time).

The objective is to minimize the combined cost of copier rental cost and the customer waiting cost. Using a simulated sampling of 10 for each alternative, determine which copier must be selected.

15.7 Consider Problem 15.6. Design a program to simulate 10,000 minutes of copier operations on a computer and print out the aggregate performance of the system for both the models. Evaluate the two alternative policies and select the most promising copier model.

───────────── **SELECTED REFERENCES** ─────────────

Basil, D. C., P. R. Cone, and J. A. Fleming, *Executive Decision Making Through Simulation*, Charles E. Merrill, Columbus, Ohio, 1965.

Buffa, E. S., *Operations Management: Problems and Models*, Wiley, New York, 1972.

Chase, R. B., and N. J. Aquilano, *Production and Operations Management*, Richard D. Irwin, Homewood, Ill., 1973.

Emshoff, J. R. and R. L. Sisson, *Design and Use of Computer Simulation Models*, Macmillan, New York, 1970.

Evans, G. W., G. F. Wallace, and G. L. Sutherland, *Simulation Using Digital Computers*, Prentice-Hall, Englewood Cliffs, N.J., 1967.

Forrester, J., *Industrial Dynamics*, MIT Press, Cambridge, Mass., 1961.

Greenlaw, P. S., L. W. Herron, and R. H. Rawden, *Business Simulation: In Industrial and University Education*, Prentice-Hall, Englewood Cliffs, N.J., 1962.

Groff, G. E., and J. F. Muth, *Operations Management: Analysis for Decisions*, Richard D. Irwin, Homewood, Ill., 1972.

Hammersley, J. M., and D. C. Handscomb, *Monte Carlo Methods*, Methuen, London, 1964.

Hillier, F. S., and G. J. Lieberman, *Introduction to Operations Research*, Holden-Day, San Francisco, Calif., 1967.

Malcolm, D. G., "System Simulation—A Fundamental Tool for Industrial Engineering," *Journal of Industrial Engineering*, Vol. 9, No. 3, May–June 1958.

McMillan, C., and R. F. Gonzales, *Systems Analysis*, Richard D. Irwin, Homewood, Ill., Third Edition, 1973.

Meir, R., W. T. Newell, and H. L. Pazer, *Simulation in Business and Economics*, Prentice-Hall, Englewood Cliffs, N.J., 1969.

Mize, J. H., and J. G. Cox, *Essentials of Simulation*, Prentice-Hall, Englewood Cliffs, N.J., 1968.

Naylor, T. H., J. L. Balintfy, D. S. Burdick, and K. Chu, *Computer Simulation Techniques*, Wiley, New York, 1966.

Rand Corporation, *A Million Random Digits with 100,000 Normal Deviates*, Free Press, New York, 1955.

Smith, J., *Computer Simulation Models*, Hafner Publishing, New York, 1968.

Tocher, K. D., *The Art of Simulation*, The English Universities Press, London, 1963.

decision making under uncertainty

WE HAVE DISCUSSED in Chapter 1 that based on the information available to the decision maker, the decision-making situation may be classified into one of the three stages: certainty, risk, or uncertainty. In all of the decision-making situations that we have discussed so far, knowledge on the state of outcomes for each course of action was believed to be certainty or risk. Certainty refers to situations in which each course of action will result in only one specified outcome. On the other hand, risk is characterized by situations in which each action leads to any one of a set of possible outcomes, the probabilities of which are known or can be estimated. In this chapter we shall discuss situations under uncertainty and also different criteria for making decisions under uncertainty. Uncertainty refers to situations in which each action leads to any one of a set of possible outcomes, the probabilities of which, however, are not known. A problem under uncertainty is usually represented by a payoff matrix, in which each row represents a course of action or alternative (A_i) and each column a possible outcome or state of nature (S_j) (Table 16.1). The element that lies at the intersection of the ith row and the jth column (a_{ij}) specifies the outcome that results from the occurrence of A_i and S_j.

When the outcomes of an uncertainty situation are due to chance, the situation is called a noncompetitive situation. The decision maker looks at the environmental nature as a passive opponent. Nature does not try to ruin his expectations. In fact, no active opponent controlling the occurrence of

TABLE 16.1   Structure of a Payoff Matrix

		Outcome or State of Nature				
		S_1	S_2	$\cdots\ S_j$	$\cdots$	S_n
	A_1	a_{11}	a_{12}	a_{1j}		a_{1n}
	A_2	a_{21}	a_{22}	a_{2j}		a_{2n}
Course of action or alternative	$\vdots$ A_i	a_{i1}	a_{i2}	a_{ij}		a_{in}
	$\vdots$ A_m	a_{m1}	a_{m2}	a_{mj}		a_{mn}

outcomes exists for the noncompetitive situations. These decision-making situations are termed games against nature. On the other hand, an opposing force that intends to do everything it can to ruin the expectations of the decision maker exists in some situations. The opposing force may be in the form of an enemy, business enterprise, or army. Such situations are called competitive situations. The competitive problems can be analyzed utilizing game theory, and it is discussed later in this chapter.

ANALYSIS OF NONCOMPETITIVE SITUATIONS

To select a course of action under uncertain environment, several criteria are available to the decision maker. When we choose a criterion, financial circumstances, the attitude toward taking chances, and the psychological makeup of the decision maker must be considered. The following criteria may be employed in selecting a course of action:

1. Maximin criterion
2. Maximax criterion
3. Hurwicz criterion
4. Minimax regret criterion
5. Laplace criterion

SOLUTION PROCEDURES FOR NONCOMPETITIVE SITUATIONS

We shall illustrate the technique of selecting the optimum action for different decision criteria with the following example.

Example 16.1

A construction company has allotted one million dollars to invest in any one of the four projects: office buildings, land development, homes, or apartments. The amount of profit from each project depends on the national economy for the planned period. The possible states of nature are a reduced rate of growth, the same rate of growth, or an increased rate of growth of national economy. Table 16.2 shows the payoff matrix, indicating the estimated profit in $10,000 units. Determine the optimum action for each of the five decision criteria.

TABLE 16.2   Payoff Matrix for Example 16.1 ($10,000 Units)

| | State of Nature | | |
| | --- | --- | --- |
Course of Action	Reduced Growth	Same Growth	Increased Growth
Office buildings (A_1)	− 5	9	30
Land development (A_2)	− 30	20	40
Homes (A_3)	0	7	10
Apartments (A_4)	− 6	5	17

Note. A negative value implies a loss.

Dominance   After obtaining the payoff matrix, the dominated courses of action are identified and eliminated from consideration. When the outcome of a course of action (A_i) is preferable to or equal to that of another course of action (A_k) for each state of nature, then A_i dominates A_k. Comparing the payoff values of alternatives A_1 and A_4 in Table 16.2, we note that the outcome for A_4 is smaller than that of A_1 for each state of nature. Therefore, the company can gain a large outcome by selecting A_1 rather than A_4. Under every circumstance alternative A_4 will not be selected. Alternative A_1 dominates A_4. Alternative A_1 is termed the dominant course of action and A_4 the dominated course of action. Since A_4 will not be preferred, it is removed from the payoff matrix. Table 16.3 shows the revised payoff matrix for our example. The revised payoff matrix is employed to determine the alternative to be selected for the various decision rules.

Maximin Criterion   A decision maker, who is pessimistic about the outlook of a decision situation, looks at the worst thing that might occur and is inherently prepared for the worst rather than the best. The maximin

TABLE 16.3   Revised Payoff Matrix for Example 16.1 ($10,000 Units)

	State of Nature		
Course of Action	Reduced Growth	Same Growth	Increased Growth
Office buildings (A_1)	− 5	9	30
Land development (A_2)	−30	20	40
Homes (A_3)	0	7	10

decision rule was developed by Abraham Wald, and it can be used when the decision maker is pessimistic. The decision rule is to select the alternative that maximizes the minimum gain. The minimum gain for each alternative is recorded, and the alternative with the largest gain is selected for the decision problem. In Table 16.4 note that the selected alternative is A_3 and the maximin value is 0. This criterion considers only the worst outcome for each course of action in making the decision. All other information in the payoff matrix is ignored.

TABLE 16.4   Application of Maximin Criterion for Example 16.1

	State of Nature			
Alternative	Reduced Growth	Same Growth	Increased Growth	Minimum
Office buildings (A_1)	− 5	9	30	− 5
Land development (A_2)	−30	20	40	−30
Homes (A_3)	0	7	10	0 ←Maximin

Maximax Criteria   A decision maker, who is optimistic on the outlook of a decision situation, considers the best of all possible gains and expects the state of nature having the best consequence to occur. The maximax criterion is a suitable decision rule for a decision maker with an optimistic temperament. The decision rule is to select the alternative that maximizes the maximum gain. In Table 16.5 we see that A_2 is the selected alternative and the maximax is 40. This optimistic approach considers only the best outcome for each alternative in selecting a course of action. The other outcomes presented in the payoff table indicate the gains and

TABLE 16.5   Application of Maximax Criterion for Example 16.1

| | State of Nature | | | |
Alternative	Reduced Growth	Same Growth	Increased Growth	Maximum
Office buildings (A_1)	− 5	9	30	30
Land development (A_2)	− 30	20	40	40 ←Maximax
Homes (A_3)	0	7	10	10

penalties for various alternative state of nature combinations. This information is ignored in the decision-making process.

Hurwicz Criterion   The maximin criterion is the ultraconservative approach, and it looks at the dark side of consequences. On the other hand, the maximax criterion is the optimistic approach and it assumes the bright side of consequences. The degree of optimism of a decision maker may be within the two extremes of pessimism and optimism. The criterion is that a weighted combination of the best and worst consequences is calculated for each alternative. The alternative having the highest value is selected.

In calculating the weighted combination, an optimism index (α) is employed and it indicates the decision maker's attitude toward taking chances. The value of α can be between 0 and 1. When the value is closer to 0, it denotes a pessimistic approach as seen in Figure 16.1. If the value is nearer to 1, it signifies an optimistic approach. The weighted combination value for alternative A_i is called the Hurwicz payoff H_i.

$$H_i = \alpha M_i + (1 - \alpha)m_i$$

where

M_i = the maximum gain possible for alternative A_i

m_i = the minimum gain possible for alternative A_i

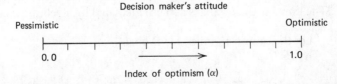

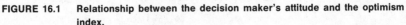

FIGURE 16.1   Relationship between the decision maker's attitude and the optimism index.

**TABLE 16.6   Application of the Hurwicz Criterion for
Example 16.1 ($\alpha = 0.6$)**

Alternative	Maximum Gain (M_i)	Minimum Gain (m_i)	Hurwicz payoff (H_i) $\alpha M_i + (1 - \alpha)m_i$	
Office buildings (A_1)	30	− 5	$0.6(30) + 0.4(-5)$ = 16	←Maximum
Land development (A_2)	40	−30	$0.6(40) + 0.4(-30)$ = 12	
Homes (A_3)	10	0	$0.6(10) + 0.4(0)$ = 6	

For $\alpha = 0.6$, the calculations are displayed in Table 16.6. Action A_1 has the maximum Hurwicz payoff. Therefore, the decision maker should select A_1 if his optimism index is 0.6. Likewise, the optimum alternative can be determined for any desired optimism index.

We can establish the optimum action and the corresponding Hurwicz payoff for any α value using the graphical method. Figure 16.2 shows the graphical approach for Example 16.1. The horizontal axis represents the optimism index. The vertical lines are for outcomes. The vertical line at the left end of the figure denotes the m_i values and that at the right end of the figure shows the M_i values. For any action A_i, the Hurwicz payoff is m_i when $\alpha = 0$ and $H_i = M_i$ when $\alpha = 1$. For instance,

$$\text{when} \quad \alpha = 0, H_i = m_i = -5$$

$$\text{when} \quad \alpha = 1, H_i = M_i = 30$$

These two values are plotted on the two vertical lines, joined by a straight line, and written A_1 along the line to identify it. The other actions are also displayed on the figure in the same way. To determine the Hurwicz payoff of any action, a vertical line through the given α value is drawn until it intersects the corresponding line. A horizontal line is drawn through the point of intersection and the Hurwicz payoff is read on the vertical line. Recall that the action having the highest Hurwicz payoff is the optimal action. Note in Figure 16.2 that the optimal action for $\alpha = 0.6$ is A_1 and it yields a Hurwicz payoff of 16. Figure 16.2 portrays that A_3, A_1, and A_2 are the optimal alternatives for different ranges of α values. The limits of these ranges can be calculated by algebraic method as shown below:

Let the optimism index for B be α_b.

At B, $H_3 = H_1$

$$\alpha_b(10) + (1 - \alpha_b)0 = \alpha_b(30) + (1 - \alpha_b)(-5)$$

$$\alpha_b = 0.2$$

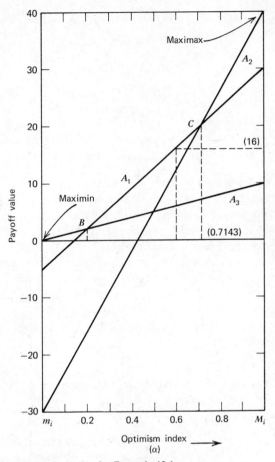

FIGURE 16.2 **Graphical solution for Example 16.1.**

For $\alpha = 0.20$, the highest Hurwicz payoff is obtained by selecting A_3 or A_1. The decision maker, therefore, can select either A_3 or A_1. Let the optimism index for C be α_c. At C, $H_1 = H_2$

$$\alpha_c(30) + (1 - \alpha_c)(-5) = \alpha_c(40) + (1 - \alpha_c)(-30)$$

$$\alpha_c = 0.7143$$

For $\alpha = 0.7143$, the highest Hurwicz payoff is obtained by selecting A_1 or A_2. The decision maker, therefore, can choose either A_1 or A_2. Now, the optimal actions for the various degrees of optimism can be stated as follows:

Decision Maker's Philosophy	Optimum Action
$0 \leqq \alpha \leqq 0.2$	A_3
$0.2 \leqq \alpha \leqq 0.7143$	A_1
$0.7143 \leqq \alpha \leqq 1$	A_2

Observe that the maximin criterion corresponds to the Hurwicz criterion at $\alpha = 0$ and the maximax criterion for $\alpha = 1$.

Minimax Regret Criterion   This criterion was proposed by Savage; it emphasizes the opportunity loss resulting from an incorrect decision. If the decision maker has selected A_2 and nature is S_3, he receives the maximum possible payoff for S_3, namely 40. However, if he had employed A_3 and nature is S_3, the cost of mistake or regret is $30 (= 40 - 10)$. Likewise, when the selected action is A_1, and the prevailing state of nature is S_3, the resulting regret is $10 (= 40 - 30)$. The regret for the payoff of action A_i and state of nature S_i is calculated by subtracting the corresponding return (a_{ij}) from the maximum possible return (MS_j) for the given state of nature (S_j). The regret matrix shown in Table 16.7 displays the regret values for Example 16.1.

TABLE 16.7   Application of the Minimax Regret Criterion for Example 16.1

Regret Matrix

Alternative	Reduced Growth (S_1)	Same Growth (S_2)	Increased Growth (S_3)	Maximum Regret
Office buildings (A_1)	5	11	10	11 ←Minimax regret
Land development (A_2)	30	0	0	30
Homes (A_3)	0	13	30	30
Maximum return (MS_j)	0	20	40	

The decision rule is to select the alternative that minimizes the maximum regret. In Table 16.7 note that the decision maker following minimax regret criterion would employ action A_1. THe minimax regret criterion considers only the largest regret value for each action, and other information is ignored in the analysis.

Laplace Criterion   The Laplace criterion asserts that if the probabilities of future states of nature are not known, equal probability must be assigned to them. Then, the expected payoff is calculated for each action and the action that maximizes the expected payoff is adopted.

Consider that the decision maker is completely ignorant about the occurrences of the states of nature, S_1, S_2, and S_3 in Example 16.1. The three states of nature are equally likely according to the Laplace criterion and, hence, we assign a probability of $\frac{1}{3}$ to each of them. The expected values (EV_i) for the actions are as follows:

$$\text{For } A_1, EV_1 = -5(\tfrac{1}{3}) + 30(\tfrac{1}{3}) = 11\tfrac{1}{3} \leftarrow \text{Maximum}$$

$$\text{For } A_2, EV_2 = -30(\tfrac{1}{3}) + 40(\tfrac{1}{3}) = 10$$

$$\text{For } A_3, EV_3 = 0(\tfrac{1}{3}) + 10(\tfrac{1}{3}) = 5\tfrac{2}{3}$$

Since action A_1 has the largest expected payoff, it is selected. The Laplace criterion takes into account all payoff values for the given states of nature. Nevertheless, the introduction of some new states of nature as well as the elimination of a few contemplated states of nature can result in the selection of a different action.

Each of the five decision criteria that we have discussed do not often result in the same alternative, as observed in the capital investment problem.

Maximin criterion	: Select A_3
Maximax criterion	: Select A_2
Hurwicz criterion	: Select A_3 or A_1 or A_2, depending on the decision maker's philosopy
Minimax regret criterion	: Select A_1
Laplace criterion	: Select A_1

The systems analyst might wish to have a decision rule for selecting the appropriate criterion for any specific uncertain situation. No such decision rule has yet been proposed. Therefore, factors such as the financial circumstances, attitude toward taking chances, and the psychological make-up of the decision maker must be considered in choosing the criterion.

Development of a Computer Model

We shall design a computer model in this section for establishing the optimum action for various decision criteria that we have discussed for the noncompetitive situations. Let us define the following list of variables that are used in the formulation of the computer program:

M	= number of alternatives proposed for the uncertain situation
N	= number of states of nature
YMIN	= minimum profit possible for the given alternative
YMAX	= maximum profit possible for the given alternative
KPLAN	= selected action for the given criterion
P	= regret value when alternative K is selected and state of nature J is encountered
FMX	= this subprogram determines the largest value (YMAX) among a set of M values
PAYMTX(K, J)	= estimated profit when alternative K is selected and state of nature J is encountered
GMIN(K)	= minimum profit possible for alternative K
GMAX(K)	= maximum profit possible for alternative K
XC(J)	= maximum return possible for state of nature J
XMREG(K)	= $-C(K)$ where $C(K)$ is the maximum regret for alternative K
X(K)	= expected profit for action K
SUM	= PAYMTX(K, 1) + PAYMTX(K, 2) + $\cdots$ + PAYMTX(K, N)

The variables GMIN( ), GMAX( ), XMREG( ), and X( ) are one-dimensional arrays with M elements and XC( ) is a one-dimensional array with N elements. The subscripted variable PAYMTX(,) is a two-dimensional array of size M by N.

We might transmit the input data and print out the values by the following instructions:

```
      READ 100, M, N
      PRINT 1
      PRINT 2, M
      PRINT 3, N
      PRINT 4
      DO 110 I = 1, M
      READ 120, (PAYMTX(I, J), J = 1, N)
      PRINT 5, (PAYMTX(I, J), J = 1, N)
  110 CONTINUE
    1 FORMAT ( )
    2 FORMAT ( )
    3 FORMAT ( )
    4 FORMAT ( )
    5 FORMAT ( )
  120 FORMAT ( )
```

The following statements might be used to compute GMIN(K) for each alternative:

```
        DO 140 I = 1, M
        YMIN = 999999.99
        DO 150 J = 1, N
        IF (PAYMTX(I, J).GT.YMIN) GO TO 150
        YMIN = PAYMTX(I, J)
  150   CONTINUE
        GMIN(I) = YMIN
  140   CONTINUE
```

The maximum profit possible for each alternative, GMAX(K), might be determined by these instructions:

```
        DO 160 I = 1, M
        YMAX = -999999.99
        DO 170 J = 1, N
        IF (PAYMTX(I, J).LT.YMAX) GO TO 170
        YMAX = PAYMTX(I, J)
  170   CONTINUE
        GMAX(I) = YMAX
  160   CONTINUE
```

The optimum action as well as the expected profit for maximin criterion can be calculated and printed out by the following instructions:

```
        CALL FMX(GMIN, KPLAN, YMAX, M)
        PRINT 7, KPLAN
        PRINT 8, YMAX
    7   FORMAT ( )
    8   FORMAT ( )
```

The set of statements would cause the computer to calculate the optimum action and expected profit for maximax criterion and print out the computed values:

```
        CALL FMX(GMAX, KPLAN, YMAX, M)
        PRINT 7, KPLAN
        PRINT 8, YMAX
```

The index of optimism (α) of the decision maker is estimated, and it is utilized in determining the optimum action for the Hurwicz criterion. We could calculate the optimum action and the corresponding Hurwicz payoff for the 11 α values, starting from 0 through 1 in increments of 0.1, and print out the results thus:

```
      DO 400 K2 = 0, 10
      XK2 = K2
      ALPHA = XK2/10.0
      B = 1.0 − ALPHA
      DO 410 K3 = 1, M
      X(K3) = ALPHA * GMAX(K3) + B * GMIN(K3)
  410 CONTINUE
      CALL FMX(X, KPLAN, YMAX, M)
      PRINT 13, ALPHA, KPLAN, YMAX
  400 CONTINUE
   13 FORMAT ( )
```

The next step is to calculate the optimum action for the minimax regret criterion. The maximum return for each state of nature (MS_j) is calculated by the following instructions:

```
      DO 500 J = 1, N
      XMAX = −999999.99
      DO 510 I = 1, M
      IF (PAYMTX(I, J).LT.XMAX) GO TO 510
      XMAX = PAYMTX(I, J)
  510 CONTINUE
      XC(J) = XMAX
  500 CONTINUE
```

The following statements would cause the computer to select the optimum action and print out the chosen action as well as the corresponding expected outcome.

```
      DO 550 I = 1, M
      XMAX = −999999.99
      DO 560 J = 1, N
      P = XC(J) − PAYMTX(I, J)
      IF (P.LT.XMAX) GO TO 560
      XMAX = P
  560 CONTINUE
      XMREG(I) = −YMAX
  550 CONTINUE
      CALL FMX(XMREG, KPLAN, YMAX, M)
      YMAX = −YMAX
      PRINT 7, KPLAN
      PRINT 8, YMAX
```

To determine the optimum action for the Laplace criterion, the expected payoff for each alternative is calculated and the one that has the largest expected payoff is selected. We might utilize the following instructions to establish the optimum action and print out the calculated values:

```
        CN = N
        DO 600 I = 1, M
        SUM = 0
        DO 610 J = 1, N
        SUM = SUM + PAYMTX(I, J)
610     CONTINUE
        X(I) = SUM/CN
600     CONTINUE
        CALL FMX(X, KPLAN, YMAX, M)
        PRINT 7, KPLAN
        PRINT 8, YMAX
```

Subprogram FMX is designed to calculate the largest value (YMAX) for a set of M values.

```
        SUBROUTINE FMX (X, KPLAN, YMAX, M)
        DIMENSION X(10)
        YMAX = -999999.99
        DO 100 I = 1, M
        IF (X(I).LT.YMAX) GO TO 100
        YMAX = X(I)
        KPLAN = I
100     CONTINUE
        RETURN
        END
```

A complete computer program composed of our previously designed program segments is shown in Figure 16.3. The data given in Table 16.3 have been utilized to demonstrate the application of the computer model. There are four cards in the data deck; they have the following appearance:

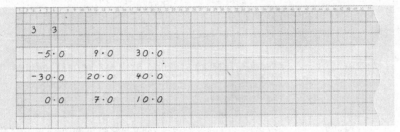

Figure 16.4 displays the resulting computer printout.

```
              DIMENSION    PAYMTX(10,10), XMREG(10)
              DIMENSION    GMAX(10), GMIN(10), X(10), XC(10)
C
C             READ AND PRINTOUT INPUT DATA
              READ 100,M,N
              PRINT 1
              PRINT 2,M
              PRINT 3,N
              PRINT 4
              DO 110 I=1,M
              READ 120, (PAYMTX(I,J),    J=1,N)
              PRINT 5, (PAYMTX(I,J), J=1,N)
          110 CONTINUE
C
C             CALCULATE AND PRINTOUT MAXIMIN CRITERION
              PRINT 6
              DO 140 I=1,M
              YMIN=999999.99
              DO 150 J=1,N
              IF (PAYMTX(I,J) .GT. YMIN)  GO TO 150
              YMIN=PAYMTX(I,J)
          150 CONTINUE
              GMIN(I) = YMIN
          140 CONTINUE
              DO 160 I=1,M
              YMAX=-999999.99
              DO 170 J=1,N
              IF (PAYMTX(I,J) .LT. YMAX)  GO TO 170
              YMAX=PAYMTX(I,J)
          170 CONTINUE
              GMAX(I)=YMAX
          160 CONTINUE
              CALL FMX(GMIN,KPLAN,YMAX,M)
              PRINT 7, KPLAN
              PRINT 8, YMAX
C
C             CALCULATE AND PRINTOUT MAXIMAX CRITERION
              PRINT 9
              CALL    FMX(GMAX,KPLAN,YMAX,M)
              PRINT 7, KPLAN
              PRINT 8, YMAX
C
C             CALCULATE AND PRINTOUT HURWICZ CRITERION
              PRINT 10
              PRINT 11
              PRINT 12
              DO 400 K2=0,10
              XK2=K2
              ALPHA=XK2/10.0
              B=1.0-ALPHA
              DO 410 K3=1,M
              X(K3)=ALPHA*GMAX(K3)+B*GMIN(K3)
          410 CONTINUE
              CALL FMX(X,KPLAN,YMAX,M)
              PRINT 13,ALPHA,KPLAN,YMAX
          400 CONTINUE
C
C             CALCULATE AND PRINTOUT MINIMAX REGRET CRITERION
              PRINT 14
              DO 500 J=1,N
              XMAX=-999999.99
              DO 510 I=1,M
```

FIGURE 16.3   Computer program for analyzing noncompetitive situations (illustrated for the construction company problem).

```
        IF(PAYMTX(I,J) .LT. XMAX)  GO TO 510
        XMAX=PAYMTX(I,J)
   510 CONTINUE
        XC(J) = XMAX
   500 CONTINUE
        DO 550 I=1,M
        XMAX=-999999.99
        DO 560 J=1,N
        P=XC(J)-PAYMTX(I,J)
        IF (P .LT. XMAX)    GO TO 560
        XMAX=P
   560 CONTINUE
        XMREG(I)=-XMAX
   550 CONTINUE
        CALL FMX(XMREG,KPLAN,YMAX,M)
        YMAX=-YMAX
        PRINT 7, KPLAN
        PRINT 8, YMAX
C
C       CALCULATE AND PRINTOUT LAPLACE CRITERION
        PRINT 15
        CN=N
        DO 600 I=1,M
        SUM =0
        DO 610 J=1,N
        SUM = SUM+PAYMTX(I,J)
   610 CONTINUE
        X(I)=SUM/CN
   600 CONTINUE
        CALL FMX(X,KPLAN,YMAX,M)
        PRINT 7, KPLAN
        PRINT 8, YMAX
C
     1 FORMAT(1H1,///,7X,37HANALYSIS OF NONCOMPETITIVE SITUATIONS)
     2 FORMAT(//,10X,19HNO. OF ALTERNATIVES,6X,1H=,I3)
     3 FORMAT(10X,23HNO. OF STATES OF NATURE,2X,1H=,I3)
     4 FORMAT(/,10X,10HINPUT DATA,/)
     5 FORMAT(18X,F5.1,2X,F5.1,2X,F5.1)
     6 FORMAT(/,10X,17HMAXIMIN CRITERION)
     7 FORMAT(/,15X,14HOPTIMUM ACTION,3X,1H=,I5)
     8 FORMAT(15X,16HEXPECTED OUTCOME,1X,1H=,F5.1)
     9 FORMAT(//,10X,17HMAXIMAX CRITERION)
    10 FORMAT(//,10X,17HHURWICZ CRITERION)
    11 FORMAT(/,15X,24HALPHA  OPTIMUM  EXPECTED)
    12 FORMAT(22X,14HACTION   VALUE,/)
    13 FORMAT(16X,F3.1,5X,I2,7X,F4.1)
    14 FORMAT(//,10X,24HMINIMAX REGRET CRITERION)
    15 FORMAT(//,10X,17HLAPLACE CRITERION)
   100 FORMAT(2I3)
   120 FORMAT(3F7.1)
        STOP
        END


        SUBROUTINE   FMX (X,KPLAN,YMAX,M)
C
C       DETERMINE THE LARGEST VALUE
        DIMENSION   X(10)
        YMAX=-999999.99
        DO 100 I=1,M
        IF(X(I) .LT. YMAX)    GO TO 100
        YMAX=X(I)
        KPLAN=I
   100 CONTINUE
        RETURN
        END
```

FIGURE 16.3 (Continued)

```
ANALYSIS OF NONCOMPETITIVE SITUATIONS


NO. OF ALTERNATIVES        =   3
NO. OF STATES OF NATURE    =   3

INPUT DATA

        -5.0      9.0     30.0
       -30.0     20.0     40.0
         0.0      7.0     10.0

MAXIMIN CRITERION

     OPTIMUM ACTION    =     3
     EXPECTED OUTCOME  =   0.0


MAXIMAX CRITERION

     OPTIMUM ACTION    =     2
     EXPECTED OUTCOME  =  40.0


HURWICZ CRITERION

     ALPHA   OPTIMUM   EXPECTED
             ACTION    VALUE

      0.0      3         0.0
      0.1      3         1.0
      0.2      3         2.0
      0.3      1         5.5
      0.4      1         9.0
      0.5      1        12.5
      0.6      1        16.0
      0.7      1        19.5
      0.8      2        26.0
      0.9      2        33.0
      1.0      2        40.0


MINIMAX REGRET CRITERION

     OPTIMUM ACTION    =     1
     EXPECTED OUTCOME  =  11.0


LAPLACE CRITERION

     OPTIMUM ACTION    =     1
     EXPECTED OUTCOME  =  11.3
```

FIGURE 16.4   Computer printout for the construction company problem.

ANALYSIS OF COMPETITIVE SITUATIONS

Competive situations involve adversaries pursuing conflicting interests such as military battles, political campaigns, contract bidding, and advertising campaigns by competing business firms. The decision maker in a competitive environment does not have complete control of the factors. The final outcome depends on his course of action as well as that of his opponents. For instance, the outcome of a military battle between army A and army B depends on the combination of strategies selected by A and B. Nature acts as a neutral opponent under noncompeting situations, but the opposing forces are out to get the decision maker in competing environments. The fates of the adversaries are interlocked in situations of conflict. Game theory is a mathematical approach for competitive situations, and it deals with the mathematical aspect of the decision-making process of the adversaries. The theory was developed by John von Neumann and Oskar Morgenstern during World War II.

If there are n adversaries or players, the game is called an n-person game. These may be individuals, teams, firms, or armies. If $n = 2$, it is referred to as a two-person game. A two-person zero-sum game is one in which two players are involved and the amount of gain of one player offsets the amount of loss of the other player. Thus, the qualification zero sum implies that the total payoff of the two players is zero. Extensive research has been conducted on two-person zero-sum games, and methods are available to determine solutions.

TWO-PERSON ZERO-SUM GAMES

In this section we shall present the characteristics of the two-person zero-sum game and the solution procedure for establishing the best course of action for each player. Each decision maker is the direct adversary of the other. One player's gain is exactly equal to the other player's loss. Their interests are strictly competitive. Each player has alternate courses of action (also called strategies or moves) available to him. Both players know these strategies and the rules that specify what moves can be selected. When each player selects a strategy, it leads to a final outcome. The final outcomes for all possible combinations of strategies are presented in numerical terms in a payoff matrix such as is shown in Table 16.8. When player A has m strategies and player B has n strategies, the outcomes are recorded in a payoff matrix of size $m \times n$. The value in each cell indicates the payoff to player A. Since this is a

TABLE 16.8   Payoff Matrix for the Advertising Campaign Problem

		B_1	B_2	B_3	B_4
	A_1	7	6	8	9
	A_2	−4	−3	9	10
A's strategy	A_3	3	0	4	2
	A_4	10	5	−2	0

B's Strategy spans the columns B_1, B_2, B_3, B_4.

zero-sum game, the payoff to player B for any combination of strategies is equal to "minus the corresponding payoff to player a." For example, note in Table 16.8 that if A selects A_1 and B selects B_1, the outcome to A is a gain of 7. Therefore, the outcome to B is a loss of 7 (payoff is −7). Likewise, the payoff to A for the combination of moves A_2 and B_1 is −4 and the payoff to B is 4.

It is assumed that the data presented in the payoff matrix are known to each player before starting the game. Each player wishes to select his move to maximize his expected payoff. Each player is to choose his strategy without the complete knowledge of his rival's move. Because of the pessimistic outlook, each decision maker selects his course of action to maximize his minimum gain or minimize his maximum loss. The two players are rational decision makers and equally intelligent. The competitive situations that occur frequently in business and industry do not have all of the characteristics of the two-person zero-sum game. Therefore, the application of game theory has been very limited in business and industrial environment. Game theory has been employed successfully in the military. Although game theory has not been used extensively for real life situations, it provides an improved framework for analyzing conflict situations. It also forms a basis for understanding the complex n-person games, nonzero sum games, as well as constant-sum games; these tend to have the characteristics of real life situations. We shall present various solutions methods for the two-person zero-sum games in the following section.

Example 16.2

Two firms, A and B, each have two million dollars allotted for advertising their sporting goods in a certain market region. The managements of the two firms wish to determine how much must be allotted to different advertising media. Each firm is considering four strategies. A strategy may be stated as "spend 80% of cost for television, 10% for newspapers and magazines, and the remaining 10% for billboards." Each element in the payoff matrix portrayed in Table 16.8 represents the amount of additional revenue estimated in millions of dollars above cost. Recall that the outcomes are the payoffs to firm A.

Dominance On examining the payoffs to player A from strategy A_1 and A_3, we note that the payoff of A_1 is preferred to A_3 regardless of what player B selects. Therefore, A will not select A_3 under any circumstance. A_1 is the dominant strategy and A_3 is the dominated strategy. Since the dominated course of action will never be selected, it is discarded from consideration. Table 16.9(a) gives the reduced payoff matrix. A strategy A_i is said to dominate another strategy A_k, if the payoff of A_i is preferable to or equal to the payoff of A_k for each course of action of his adversary. Likewise, the strategies of player B in the reduced matrix are examined for dominance, and the dominated strategies, if any, are eliminated. Because the payoff to B is "minus the payoff to A," we note that B_3 dominates B_4. Therefore, the dominated strategy B_4 is removed from consideration. The strategies of A and B are examined for dominance successively until all dominated strategies are discarded. The first step in the computational procedure eliminates the dominated moves because this reduces the amount of calculations in the following steps.

Pure Strategies and Saddle Points

Game theory tells that the following line of reasoning is used in establishing the strategies for the two players. If player A selects A_2, he may gain as much as 9. Since B is a rational person, he would select his move so that he receives the largest gain from A or the smallest loss from A if a gain is not possible. Therefore, B might choose B_1 to gain 4 from A. This results in a loss of 4 to A. If we select A_4, A may receive a payoff of 10. However, B would select B_3, yielding a loss of 2 to A. But, if A selects A_1, he is guaranteed that he will win at least 6. If B selects any one of the three strategies, he loses in the game. Hence, he would choose the strategy that will minimize his loss. Therefore, A will play A_1 and B will play B_2 in each trial.

We can use a similar line of reasoning for B and examine its effects. If B selects B_1, he may gain 4 from A. His rational opponent might choose A_4, and it results in a maximum loss of 10 to B. Likewise, the maximum loses to B by employing B_2 or B_3 are 6 and 9, respectively. Among the three alternatives, strategy B_2 gurantees that the loss in each game cannot be greater than 6. Therefore, B would be better off by selecting B_2 in each

TABLE 16.9   Reduced Payoff Matrix for the Advertising Campaign Problem

(a) A_1 dominates A_3

	B_1	B_2	B_3	B_4	
A_1	7	6	8	9	←Dominates A_3
A_2	−4	−3	9	10	
A_3	3	0	4	2	←Dominated by A_1
A_4	10	5	−2	0	

(b) B_3 dominates B_4

	B_1	B_2	B_3	B_4
A_1	7	6	8	9
A_2	−4	−3	9	10
A_4	10	5	−2	0

↑ Dominates B_4 ↑ Dominated by B_3

(c) Reduced Payoff Matrix

B's Strategy

		B_1	B_2	B_3
	A_1	7	6	8
A's strategy	A_2	−4	−3	9
	A_4	10	5	−2

play. Given that B selects B_2, A would employ A_1 since it yields the maximum gain to A. Each player employs the same strategy in each play, A_1 for player A and B_2 for player B; these are called pure strategies. The element that lies at the intersection of the pure strategies is called the *equilibrium value* or *the saddle point*. It indicates the average payoff to player A over a large number of trials. The value of the advertising game, therefore, is 6 and it indicates that A wins 6 and B loses 6 in each game. The value of a game is zero in a fair game. In an unfair game such as the advertising problem, the average payoff to a player in each play is a gain and the average payoff to the other player is an equivalent loss.

The pure strategies and saddle points can be established by performing the following operations:

STEP 1. Determine the row minimum for each row.

STEP 2. Establish the maximum of the row minimum and the corresponding row (A_r).

STEP 3. Determine the column maximum for each column.

STEP 4. Establish the minimum of the column maximum and the corresponding column (B_c).

STEP 5. If the maximum of the row minimum equals the minimum of the column maximum, it indicates that the game has a saddle point and pure strategies are available. If they are unequal, it denotes that a saddle point does not exist and pure strategies are not available for the game.
The pure strategies are A_r and B_c. The saddle point is the common element in row A_r and column B_c.

Table 16.10 presents the calculated values for our advertising campaign problem. In the payoff matrix note that the saddle point is both the minimum in its row and the maximum in its column. Neither player can improve his position by choosing any other strategy. When B plays his pure strategy and A selects any alternative other than his pure strategy A_1, his gain is smaller than 6. On the other hand, when A plays his pure strategy A_1 and B employs any alternative other than his pure strategy B_2, his loss

TABLE 16.10   Payoff Matrix with the Pure Strategies and the Saddle Point

		B's Strategy			Row minimum
		B_1	B_2	B_3	
	A_1	7	⑥	8	6 ← Maximin value
A's strategy	A_2	-4	-3	9	-4
	A_4	10	5	-2	-2
Column maximum		10	6	9	

Minimax value

Strategy A_r $= A_1$
Strategy B_c $= B_2$
Value of the game $= 6$

is more than 6. Therefore, no benefit can be derived by either player from changing their pure strategies. The two players, hence, would not change their alternatives. Players A and B would therefore select A_1 and B_2, respectively, in each play.

When a game does not have an equilibrium point, it implies that the pure strategy approach is not applicable to the competing situation. In such situations, the players can increase their payoffs by adopting a mixed strategy. Each player would play two or more strategies according to some probability distribution in a mixed strategy. In the next section, methods for determining the optimal strategies and game values for games with mixed strategies are discussed.

Mixed Strategies and Game Values

The payoff matrix of a two-person zero-sum game is examined to see if any dominated strategies exist. If any dominated strategies are found, they are eliminated from consideration. Next, the resulting matrix is tested to see if it has a saddle point. In case the game has a saddle point, the pure strategies for the players as well as the game value are established by applying the procedure presented in the previous section. If no saddle point exists, the mixed strategies and their probabilities can be determined by various methods. The sum of probabilities for the set of strategies of a player must equal one. A course of action having zero probability indicates that it will not be played in the game. We shall discuss the following methods for the three kinds of payoff tables:

1. (2×2) games by arithmetic method.
2. $(2 \times n)$ and $(m \times 2)$ games by graphical method.
3. $(m \times n)$ game by linear programming method.

Arithmetic Method to (2×2) Games　　In 2×2 game each player has two strategies. Now consider a (2×2) game whose payoff matrix is shown in Table 16.11.

TABLE 16.11　A Payoff Matrix for a (2×2) Game

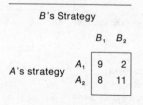

		B's Strategy	
		B_1	B_2
A's strategy	A_1	9	2
	A_2	8	11

Notice that this game does not have dominated strategies and saddle point. For (2×2) games such as this, we can perform the following calculations to establish the solution:

STEP 1. (a) For each row, determine the absolute difference between the elements of the row (ADR_i).

(b) For each column, determine the absolute difference between the elements of the column (ADC_j).

STEP 2. Calculate the sum of the row difference values $(\sum ADR_i)$ and the sum of the column difference values $(\sum ADC_j)$. $\sum ADR_i$ must be equal to $\sum ADC_j$.

STEP 3. (a) For each row, obtain a fraction (FR_i) by forming its ADR_i as the numerator and $\sum ADR_i$ as the denominator

$$FR_i = ADR_i/\sum ADR_i$$

(b) For each column, obtain a fraction (FC_j) by forming its ADC_j as the numerator and $\sum ADC_j$ as the denominator

$$FC_j = ADC_j/\sum ADC_j$$

STEP 4. (a) Interchange the fractions $(FR_i s)$ of the first row and the second row. The new fraction associated with each row refers to the probability for that course of action.

(b) Interchange the fractions $(FC_j s)$ of the first column and the second column. The new fraction associated with each column denotes the probability for that alternative.

The calculations for the (2×2) game problem are illustrated in Table 16.12. Note that the optimal mixed strategy for player A is using A_1 for $3/10$ of the time and A_2 for the remaining $7/10$ of the time. Similarly, the optimal mixed strategy for B is selecting B_1 and B_2 for $9/10$ and $1/10$ of the time, respectively. Each player can employ a chance device such as a random number table for selecting a strategy in each play.

The game value can be determined by using the joint probability concept. The probability that A will play A_1 is 0.3 and the probability that B will play B_1 is 0.9. Therefore, the joint probability of the two players selecting A_1 and B_1 simultaneously is 0.27 $(= 0.3 \times 0.9)$. Likewise, the joint probability for all combinations of strategies is calculated as follows:

$$
\begin{array}{cc}
B_1 & B_2 \\
0.9 & 0.1
\end{array}
$$

$$
\begin{array}{c}
A_1 \ 0.3 \\
A_2 \ 0.7
\end{array}
\left[
\begin{array}{cc}
0.3 \times 0.9 = 0.27 & 0.3 \times 0.1 = 0.03 \\
0.7 \times 0.9 = 0.63 & 0.7 \times 0.1 = 0.07
\end{array}
\right]
$$

The sum of joint probabilities must be equal to one.

$$0.27 + 0.03 + 0.63 + 0.07 = 1.00$$

TABLE 16.12   Application of Arithmetic Method to a (2 × 2) Game

(a) ADR_i and ADC_j values

	B_1	B_2	ADR_i
A_1	9	2	$9 - 2 = 7$
A_2	8	11	$11 - 8 = 3$

ADC_j $9 - 8$ $11 - 2$ $\boxed{10} \leftarrow \sum ADR_i = \sum ADC_j$
          $= 1$ $= 9$

(b) FR_i and FC_j values

	B_1	B_2	FR_i
A_1	9	2	7/10
A_2	8	11	3/10

FC_j 1/10   9/10

(c) Probabilities for the strategies

	B_1 (9/10)	B_2 (1/10)
A_1 (3/10)	9	2
A_2 (7/10)	8	11

Next, the joint probabilities and the payoff values are employed to determine the game value.

$$\begin{array}{cc} & B_1 \qquad\qquad\qquad B_2 \\ \begin{array}{c} A_1 \\ A_2 \end{array} & \left[\begin{array}{cc} 0.27 \times 9 = 2.43 & 0.03 \times 2 = 0.06 \\ 0.63 \times 8 = 5.04 & 0.07 \times 11 = 0.77 \end{array} \right] \end{array}$$

Value of the game $= 2.43 + 0.06 + 5.04 + 0.77$
$= 8.30$

Player A gains 8.3 in each play. Because this is a zero-sum game, player B loses an equivalent amount in each play.

Graphical Method to (2 × n) and (m × 2) Games   When the payoff matrix for a two-person zero-sum game has two strategies for one player and three or more strategies for the other player, the graphical approach can be used to establish the optimal strategies and the game value. Consider that player A has two strategies and player B has three strategies. Table 16.13 shows a payoff matrix for the game. Observe that dominating

TABLE 16.13   A Payoff Matrix for a (2 × 3) Game

	B_1	B_2	B_3
A_1	8	4	−2
A_2	−2	−1	3

strategies and saddle point are not present for the illustration. Figure 16.5 portrays the graphical solution for player A in the game of Table 16.13. The probability of A's playing strategy A_1, namely $P(A_1)$, is shown by the horizontal axis. Because $P(A_1)$ must be nonnegative and cannot exceed 1, the axis ranges from 0 to 1. The vertical axis represents the payoff to A. The vertical scale on the left end refers to the payoff to A when $P(A_1) = 0$. The vertical scale on the right end refers to the payoff to A when $P(A_1) = 1$. Player A selects either A_1 or A_2. Therefore, we have

$$P(A_1) + A_2 = 1$$

Thus, when A plays A_1, $P(A_1) = 1$; when A plays A_2, $P(A_1) = 0$. If A selects A_2 and B employs B_1, the payoff to A is -2. This corresponds to $P(A_1) = 0$. On the other hand, if A selects A_1 and B selects B_1, the payoff

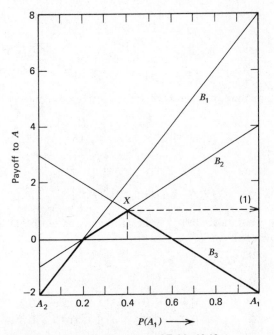

FIGURE 16.5   Graphical solution to the game of Table 16.13.

to A is 8. This corresponds to $P(A_1) = 1$. These two payoff values (-2 and 8) are plotted on the vertical scales at the left end and the right end of the horizontal axis, joined by a straight line and labeled B_1. Likewise, the two payoffs associated with each B_j are plotted on the vertical scales, and the respective B_j lines are drawn on the graph as shown in Figure 16.5. Player A would select the strategies that maximizes his minimum gain. The expected payoff to A is shown by a heavy line on the graph. The optimum point is shown by point X. A vertical line is drawn through X to determine the optimum $P(A_1)$ value. The optimum $P(A_1)$ equals 0.4. Therefore, $P(A_2)$ at optimum is equal to 0.6. The payoff to A is established by drawing a horizontal line through X and reading the corresponding payoff on the vertical scale. Note that the payoff to A or the game value is 1.

The graph indicates that the optimum point X is formed by the intersection of the two lines that correspond to B_2 and B_3. It discloses the fact that the optimal strategies for player B are B_2 and B_3. Now, we can solve the 2 × 2 game, in which A has optimal strategies A_1 and A_2 and B has optimal strategies B_2 and B_3, to determine the probabilities for these strategies and the game value. The FR_i and FC_j values are:

	B_2	B_3	ADR_i	FR_i
A_1	4	-2	6	6/10
A_2	-1	3	4	4/10
ADC_j	5	5	$\boxed{10} \leftarrow \sum ADR_i$	
FC_j	5/10	5/10		

The probability weights for the strategies are:

		B_2	B_3
		0.5	0.5
A_1	0.4	4	-2
A_2	0.6	-1	3

The joint probability values for various combinations of strategies are shown below:

		B_2	B_3
		0.5	0.5
A_1	0.4	0.4 × 0.5 = 0.2	0.4 × 0.5 = 0.2
A_2	0.6	0.6 × 0.5 = 0.3	0.6 × 0.5 = 0.3

Value of the game $= (0.2 \times 4) + 0.2 \times (-2) + 0.3 \times (-1)$
$$+ (0.3 \times 3) = 1.0$$

Recall that the graphical method has yielded the same optimal probabilities for A's strategies and game value.

The graphical method can be employed to solve $m \times 2$ games by following a line of reasoning similar to the one presented above. To illustrate the procedure, consider the 3×3 payoff matrix shown in Table 16.14a. Note that B_3 dominates B_2. Hence, B_2 is eliminated from consideration. Table 16.14b displays the reduced matrix with three strategies for A and two strategies for B. Note that this revised (3×2) game does not have a saddle point. Figure 16.6 portrays the graphical solution for player B in the (3×2) game. The horizontal axis represents the $P(B_1)$ value. The vertical axis denotes the payoff made by B to A. The vertical scale at the left end represents the payoff by B from selecting B_2 and it corresponds to $P(B_1) = 0$. Likewise, the vertical scale at the right end represents the payoff to B from selecting B_1 and it corresponds to $P(B_1) = 1$. The two payoff values for each A_i strategy are plotted on the vertical scales, and lines are drawn connecting these points. Player B would employ the strategies that minimize his maximum loss. The heavy line on the graph shows the expected payoff by B to A. Point Y on the graph specifies the optimum point for player B. Note that the optimum $P(B_1)$ is 0.9 and the game value of 2.3. Therefore, $P(B_2) = 0.1$.

TABLE 16.14   A Payoff Matrix for a (3 × 3) Game

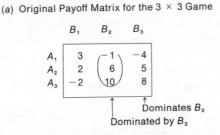

(a) Original Payoff Matrix for the 3 × 3 Game

	B_1	B_2	B_3
A_1	3	−1	−4
A_2	2	6	5
A_3	−2	10	8

Dominates B_2
Dominated by B_3

(b) Reduced Payoff Matrix

	B_1	B_3
A_1	3	−4
A_2	2	5
A_3	−2	8

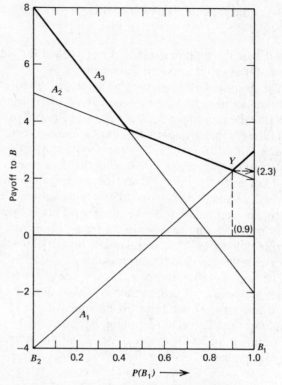

FIGURE 16.6 Graphical solution to the game of Table 16.14*b*.

The lines that intersect at Y reveal the optimal strategies for player A. Therefore, A_1 and A_2 are the optimal strategies for A. By solving the 2×2 game, which is formed by the strategies A_1, A_2, B_1, and B_3, the optimal probability weights for A_1 and A_2 can be determined.

Games with (2×2) payoff tables can also be solved by the graphical method; nevertheless, it would be advantageous to solve them by using the arithmetic method as it requires less calculations and effort.

Linear Programming for ($m \times n$) Games When a two-person zero-sum game with ($m \times n$) payoff table is to be solved, the methods discussed in the preceding sections are applied to eliminate the dominated courses of actions and to establish the saddle point. If a saddle point is found, the game is solved and the pure strategies and game value are established. If a saddle point does not exist, the suitability of using the arithmetic method and the graphical method are explored. If the payoff matrix of the revised game is of the size (3×3) or more, the game can be structured as a linear programming model. We can establish the optimal strategies and the game value solving the linear programming model by using the Computer program developed in Chapter 5.

To illustrate the transformation of a game into a linear programming model, consider the game shown in Table 16.15. Each element in the payoff matrix is nonnegative, and it implies that the game value will be some nonnegative quantity. Two linear programming models are formulated, one for each player, to determine the optimal strategies and the related probability values for both players. Note that the game does not have dominated strategies and saddle point.

Player A is interested in determining his optimal strategies and their probability weights.

TABLE 16.15   A Payoff Matrix for a (2 × 3) Game

	B_1	B_2	B_3
A_1	10	7	1
A_2	4	5	9

Let x_1 and x_2 denote the probabilities of strategies A_1 and A_2, respectively. Therefore,

$$x_1 + x_2 = 1$$

and

$$x_1, x_2 \geq 0$$

The expected gain to A when B selects B_1 is

$$10x_1 + 4x_2$$

Similarly, the gains to A when B selects B_2 or B_3 are

$$7x_1 + 5x_2$$

$$x_1 + 9x_2$$

Player A wishes to assign x_i values so that they maximize his minimum expected gain. When v is the expected gain to A, the value of the game is v. Then, we have

$$10x_1 + 4x_2 \geq v \qquad (16.1)$$

$$7x_1 + 5x_2 \geq v \qquad (16.2)$$

$$x_1 + 9x_2 \geq v \qquad (16.3)$$

$$x_1 + x_2 = 1 \qquad (16.4)$$

and
$$x_1, x_2, v \geqq 0$$

Since v is positive, we can divide both sides of relationships (16.1) to (16.4) by v and obtain the following:

$$10(x_1/v) + 4(x_2/v) \geqq 1 \tag{16.5}$$

$$7(x_1/v) + 5(x_2/v) \geqq 1 \tag{16.6}$$

$$(x_1/v) + 9(x_2/v) \geqq 1 \tag{16.7}$$

$$(x_1/v) + \quad (x_2/v) = (1/v) \tag{16.8}$$

We now define x'_1, x'_2, and z such that

$$x'_1 = (x_1/v)$$

$$x'_2 = (x_2/v)$$

$$z = (1/v)$$

Substitution of x'_1, x'_2, and z into the relationships (16.5) to (16.8) leads to

$$10x'_1 + 4x'_2 \geqq 1 \tag{16.9}$$

$$7x'_1 + 5x'_2 \geqq 1 \tag{16.10}$$

$$x'_1 + 9x'_2 \geqq 1 \tag{16.11}$$

$$z = x'_1 + x'_2 \tag{16.12}$$

Player A wishes to maximize v; this is same as minimizing z. Now, we can state the linear programming model to determine the optimal strategies of A and the game value as follows:

$$\text{Minimize } z = x'_1 + \quad x'_2$$

$$\text{Subject to} \quad 10x'_1 + 4x'_2 \geqq 1$$

$$7x'_1 + 5x'_2 \geqq 1$$

$$x'_1 + 9x'_2 \geqq 1$$

and

$$x_1, \quad x'_2 \geqq 0$$

We can solve the linear programming problem and obtain the following solution:

$$z = 0.1724$$

$$x'_1 = 0.0689$$

$$x'_2 = 0.1034$$

Therefore, the values of v, x_1, and x_2 are calculated as follows:

$$v = (1/z) = 1/0.1724 = 5.8$$

$$x_1 = x_1' \cdot v = (0.0689)(5.8) = 0.4$$

$$x_2 = x_2' \cdot v = (0.1034)(5.8) = 0.6$$

The optimal strategies of A are A_1 and A_2 and their probabilities are 0.4 and 0.6, respectively. The value of the game is 5.8.

We shall formulate another linear programming model to establish the optimal strategies of player B. Let y_1, y_2, and y_3 denote the probabilities of strategies B_1, B_2, and B_3, respectively. Hence,

$$y_1 + y_2 = 1$$

and

$$y_1, y_2, y_3 \geqq 0$$

The expected loss to B when A selects A_1 is

$$10y_1 + 7y_2 + y_3$$

The expected loss to B when A selects A_2 is

$$4y_1 + 5y_2 + 9y_3$$

The objective of player B is to choose the value of y_j so that he minimizes his maximum expected loss. If v is the expected loss to B, the value of the game is v. Therefore, we have

$$10y_1 + 7y_2 + \ y_3 \leqq v \qquad (16.13)$$

$$4y_1 + 5y_2 + 9y_3 \leqq v \qquad (16.14)$$

$$y_1 + \ y_2 + \ y_3 = v \qquad (16.15)$$

and

$$y_1, y_2, y_3, v \geqq 0$$

Since v is positive, we can divide each of the inequations 16.13 to 16.15 by v without changing the inequality signs of the solutions.

$$10(y_1/v) + 7(y_2/v) + \ (y_3/v) \leqq 1 \qquad (16.16)$$

$$4(y_1/v) + 5(y_2/v) + 9(y_3/v) \leqq 1 \qquad (16.17)$$

$$(y_1/v) + \ (y_2/v) + \ (y_3/v) = (1/v) \qquad (16.18)$$

If we let $y_1' = y_1/v$, $y_2' = y_2/v$, $y_3' = y_3/v$, and $z = 1/v$ and substitute them into the inequations 16.16 to 16.18, we obtain a linear programming problem of the form:

$$\text{Maximize } z = y_1' + y_2' + y_3'$$
$$\text{Subject to } 10y_1' + 7y_2' + y_3' \leq 1$$
$$4y_1' + 5y_2' + 9y_3' \leq 1$$

and

$$y_1', y_2', y_3' \geq 0$$

Using the simplex method of the linear programming for the problem, we can obtain the following solution:

$$z = 0.1724$$
$$y_1' = 0$$
$$y_2' = 0.1379$$
$$y_3' = 0.0345$$

Now, the game value and the optimal strategies for B are established by calculating v, y_1, y_2, and y_3.

$$v = (1/z) = 1/0.1724 = 5.8$$
$$y_1 = y_1' \cdot v = (0)(5.8) = 0$$
$$y_2 = y_2' \cdot v = (0.1379)(5.8) = 0.8$$
$$y_3 = y_3' \cdot v = (0.0345)(5.8) = 0.2$$

It is assumed that the game value is positive in structuring the linear programming model. When each element in a payoff matrix is positive, it indicates that the game value would be a positive quantity. However, if any element is negative, then we cannot say that the game value would be positive. In such situations, a sufficiently large constant (K) is added to each element so that the resulting table does not contain any negative element. This insures that the game value would be positive; hence, the linear programming approach can be employed to solve the game. This modification does not affect the optimal strategies of the players. By transforming the modified matrix to the corresponding linear programming models and solving the problems, the optimal strategies for the original game are established. The game value of the original game is obtained by subtracting K from the game value of the modified game.

We have discussed different methods to solve the two-person zero-sum games. Figure 16.7 shows the computational procedure for determining the optimal strategies of the players. Some assumptions of the game theory

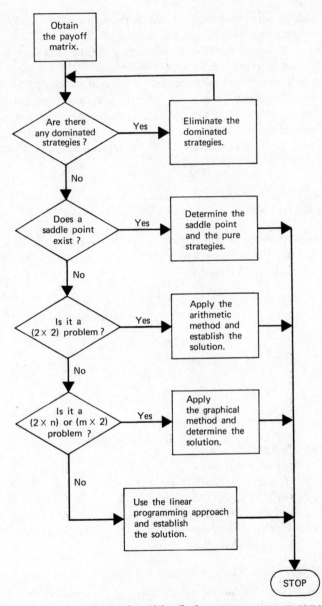

FIGURE 16.7   Summary of steps for solving the two-person zero-sum games.

such as only two persons or firms participating in a conflicting situation, the zero-sum outcome and each player acting or reacting rationally do not often represent the real life situations. Hence, two-person zero-sum game has not been more extensively applied. Nevertheless, game theory provides a well-conceived and systematic approach for analyzing conflicting situations. In recent years, much research has been done for different kinds of games such as n-person games and nonzero-sum games. Some real world competing situations can be described by these games. However, some analysis of these games is more complex, we shall not pursue the subject further.

PROBLEMS

16.1 The following payoff matrix indicates the profit values for an uncertain situation. Determine which alternative will be selected for the decision criteria stated below:

	States of Nature			
Alternative	S_1	S_2	S_3	S_4
A_1	3	-10	20	30
A_2	-20	40	35	-5
A_3	0	10	-5	8
A_4	5	2	3	0

(a) Maximin criterion
(b) Maximax criterion
(c) Hurwicz criterion for $\alpha = 0.7$
(d) Minimax regret criterion
(e) Laplace criterion
(f) Establish the optimum action for all possible optimism index values by graphical method.

16.2 A corporation has developed four products. In manufacturing each of these products, an ingredient must be procured from a foreign supplier. The price of the ingredient fluctuates significantly. The amount of profit by producing the items depends upon the price of the ingredient. The states of nature are low price, moderate price, and high price of the ingredient. The corporation wishes to produce and market one of these products. The following payoff matrix specifies the estimated profit values for the uncertain situation:

	States of Nature		
Product	Low Price (S_1)	Moderate Price (S_2)	High Price (S_3)
A	12	3	−4
B	7	5	4
C	6	4	2
D	10	6	2

(a) Apply the following criteria to find an optimum action:
  1. Maximin criterion
  2. Maximax criterion
  3. Hurwicz criterion for $\alpha = 0.9$
  4. Minimax regret criterion
  5. Laplace criterion
(b) Determine the optimum action for all possible optimism index values by graphical method.

16.3 Dreyfus Enterprises is planning to import electronic calculators and sell them in California. Management is considering to sell them by wholesale (A_1), to retail stores (A_2), or directly to users (A_3). The company wishes to select one of the three alternatives. The estimated gain in $100,000's for three states of nature is given in the following payoff matrix.

	States of Nature		
Alternative	Low Demand (S_1)	Moderate Demand (S_2)	High Demand (S_3)
A_1	1	3	4
A_2	−3	5	9
A_3	2	0	6

(a) Apply the following criteria to determine an optimum action:
  1. Maximin criterion
  2. Maximax criterion
  3. Hurwicz criterion for $\alpha = 0.8$
  4. Minimax regret criterion
  5. Laplace criterion

(b) Establish the optimum action for all possible optimism index values by graphical method.

16.4 Find the optimum strategies for players A and B as well as game value for the following game:

(a)

	B_1	B_2	B_3	B_4
A_1	5	3	8	5
A_2	-4	-3	12	9
A_3	8	2	-1	-5
A_4	3	-1	2	3

(b)

	B_1	B_2	B_3	B_4	B_5
A_1	7	6	5	8	9
A_2	8	-5	-4	9	6
A_3	-3	9	4	-1	7
A_4	3	2	-1	1	0

(c)

	B_1	B_2	B_3	B_4
A_1	2	6	5	4
A_2	8	10	9	11
A_3	-1	11	-2	12
A_4	7	0	12	2

16.5 Find the optimal strategy mixture for players A and B as well as game value for the following games:

(a)

	B_1	B_2
A_1	4	2
A_2	0	5

(b)

	B_1	B_2	B_3
A_1	7	8	3
A_2	6	4	5

(c)

	B_1	B_2	B_3
A_1	6	5	2
A_2	1	7	4
A_3	3	2	-3

16.6 Determine the optimum strategies for players A and B and the value of the game by the graphical method for the following games:

(a)

	B_1	B_2	B_3
A_1	1	8	4
A_2	5	1	3

(b)

	B_1	B_2
A_1	2	1
A_2	-2	8
A_3	6	-4

(c)

	B_1	B_2	B_3	B_4
A_1	-1	1	5	-4
A_2	2	3	0	5
A_3	1	-1	-2	4

(d)

	B_1	B_2	B_3
A_1	6	3	-3
A_2	0	1	5

16.7 Consider Problem 16.6. Find the optimum strategies for players A and B and game value by formulating linear programming models.

16.8 The ABC corporation and XYZ corporation are vigorous competitors in newspaper publishing business in a large city. Each of them is trying to increase their market share. Three strategies are available to each firm. There are nine combinations of strategies and the resulting increases or decreases in weekly sales for firm ABC is represented in the following payoff matrix:

<table>
<tr><td></td><td></td><td colspan="3">XYZs strategies</td></tr>
<tr><td></td><td></td><td>X_1</td><td>X_2</td><td>X_3</td></tr>
<tr><td>ABCs</td><td>A_1</td><td>20,000</td><td>20,000</td><td>10,000</td></tr>
<tr><td rowspan="2">strategies</td><td>A_2</td><td>10,000</td><td>60,000</td><td>$-60,000$</td></tr>
<tr><td>A_3</td><td>50,000</td><td>$-30,000$</td><td>40,000</td></tr>
</table>

Find the optimal strategy mixture for ABC and XYZ and the game value.

————————— SELECTED REFERENCES —————————

Chernoff, H., and L. Moses, *Elementary Decision Theory*, Wiley, New York, 1959.

Halter, A. N., and G. W. Dean, *Decisions Under Uncertainty*, South-Western Publishing, Cincinnati, 1971.

Kwak, N. K., *Mathematical Programming with Business Applications*, McGraw-Hill, New York, 1973.

Levin, R. I., and R. B. Desjardins, *Theory of Games and Strategies*, International Textbook Co., Scranton, 1970.

Luce, R. D. and H. Raiffa, *Games and Strategies*, Wiley, New York, 1957.

McKinsey, J. C. C., *Introduction to the Theory of Games*, McGraw-Hill, New York, 1952.

Shubick, M., The Uses of Game Theory of Management Science, *Management Science*, Vol. 2, October 1955, pp. 40–54.

Von Neumann, J., and O. Morgenstern, *The Theory of Games and Economic Behavior*, Third Edition, Wiley, New York, 1964.

Wagner, H., "Advances in Game Theory: A Review Article," *American Economic Review*, Vol. 48, 1958, pp. 368–387.

Williams, J. D., *The Complete Strategyst*, Second Edition, McGraw-Hill, New York, 1966.

CHAPTER 17

systems analysis-
an overview

WE HAVE DISCUSSED the nature of systems analysis and various quantitative tools that are useful in studying systems. Systems analysis provides a systematic approach to the solution of real world problems. Each step of the analysis is made explicit wherever possible. It is problem-oriented and considers every side of the problem. It is a comprehensive vehicle for complex problem solving and is concerned with the effective utilization of resources—men, money, machines, and materials—to achieve the desired goal. Systems approach enables the decision makers to make better decisions considering the dynamic environment, future uncertainties, and complex interrelationships. De Neufville and Stafford (1971, p. 3) suggest the use of systems analysis as follows:

1. It sharpens the designer's awareness of his objectives by forcing him to make explicit statements about what they are and how they are to be measured.
2. It seeks mechanisms for predicting the future demands of a system, which often are not observable in advance, but must be determined from an interaction of social and economic factors.
3. It establishes procedures for generating a large number of possible solutions and for determining efficient methods to search through them.
4. It assembles optimization techniques which can pick out favorable alternatives.
5. It suggests strategies of decision-making which can be used to select among possible alternatives.

Systems analysis methodology can be employed in the management of complex organizations—international situations, national, state, and local governments—apart from the business and industry problems. Duncanson (1970) discusses a systems approach to the computer control in manufacturing a wide range of chemical raw materials for the plastics and synthetic fiber industries. Johnson (1969) describes the development and operation of a systems analysis model using computer systems for operating a large industrial power station effectively.

The Ford Foundation (1973) reports on the use of systems analysis for global problems. Emerson (1971) has conducted a systems study of Caribbean transport. Most islands situated in the Commonwealth Caribbean Free Trade Region that extend from Guyana on the mainland of South America, along the island chain and across to Jamaica and the Central American coast are considered in the study. Worthington (1970) discusses the application of systems analysis for strategic planning to the U.K. wool textile industry that consists of public and private companies, partnerships, and individual proprietorships. The study demonstrates the potential use of systems concepts in formulating a framework within which an information system has been developed to the achievement of objectives.

The National Aeronautics and Space Administration (1968) reports the potential uses of systems analysis for urban and regional planning, transportation planning, city administration, public health programs, and educational programs. Montgomery (1971) presents several studies conducted by state and local governments for effective treatment of waterways carrying industrial wastes. De Neufville (1971) has guided a systems analysis project for designing the billion-dollar additions of New York City's water supply system. The study has yielded a net savings of over $100 million. Clark (1972) reports on the application of systems analysis for a regional wastewater collection and treatment problem. Sommers (1973) has applied systems analysis to municipal government to suggest system changes that will result in a better, more responsive local government. Shaw (1970) discusses a model to study the overall performance of a college of education. Allen (1972) presents the application of systems analysis techniques in the management of industrial libraries. Singh (1971) utilizes systems approach for improved management and better construction of facilities for post offices.

The development of real-time applications in the field of computer systems is best suited for numerous real-life situations. Real-time computer systems respond immediately to a transaction when it occurs by producing the requisite output. Some real-time examples are the entry and recording of airline reservations, and industrial applications in the field of continuous process control. Moreover, in the last two decades there has been a drastic reduction in cost per calculation—by a factor of 100,000—resulting from use of the computer, and the development of more powerful computers with large computer memory have significantly increased the usefulness of computers in the solution of complex problems through systems approach. The operations of modern organizations are influenced by such factors as the social responsibility, competition, interdependence, technological development, and the rapid development of multinational firms. Management can use computer-based systems analysis for making better decisions in planning, organizing, and controlling operations of complex situations.

SELECTED REFERENCES

ABT Associates Inc., *Applications of Systems Analysis Models—A Survey*, National Aeronautics and Space Administration, Washington, D.C., 1968.

Allen, A. H., "Systems to Manage Industrial Library," *Journal of Systems Management*, Vol. 23, pp. 24–27, June 1972.

Clark, III, M. S., "Systems Analysis: What Can It Do?" *Water and Wastes Engineering*, Vol. 9, pp. 31–33, October 1972.

De Neufville, R., "Systems Analysis of Large Scale Public Facilities: New York City's Water Supply Networks as a Case Study," *Journal of Systems Engineering*, Vol. 2, pp. 49–64, Summer 1971.

De Neufville, R., and J. H. Stafford, *Systems Analysis for Engineers and Managers*, McGraw-Hill, New York, 1971.

Duncanson, A., "A Systems Engineering Approach to the Computer Control of a Manufacturing Process," *Journal of Systems Engineering*, Vol. 1, pp. 55–66, Summer 1970.

Emerson, E. C., "A Systems Study of Caribbean Transport," *Journal of Systems Engineering*, Vol. 2, pp. 81–94, Summer 1971.

Ford Foundation, "Global Systems Analysis," *Ford Foundation Letter*, Vol. 4, November 1, 1973.

Forrester, J. W., *Advanced Industrial Dynamics Course*, Massachusetts Institute of Technology, Summer 1965.

Johnson, C. A., "The Use of Computers in the Operations Planning of a Large Industrial Power Station," *Journal of Systems Engineering*, Vol. 1, pp. 62–70, Autumn 1969.

Montgomery, A. H., "Systems Approach to Water Pollution Abatement," *Journal of System Management*, Vol. 22, pp. 8–11, March 1971.

Shaw, K. E., "A Systems Study of a College of Education," *Journal of Systems Engineering*, Vol. 1, pp. 43–54, Summer 1970.

Singh, M., "Systems Approach to Post Office Improvement," *Journal of Systems Management*, Vol. 22, pp. 31–34, May 1971.

Sommers, A. N., "Operations Research Applied to Municipal Government," *Industrial Engineering*, Vol. 5, pp. 30–32, March 1973.

Worthington, P. M., "Strategic Planning of the U.K. Wool Textile Industry," *Journal of Systems Engineering*, Vol. 1, pp. 19–28, Summer 1970.

APPENDIX I compound interest tables

1%

	To Find F Given P $(F/P, i, n)$	To Find P Given F $(P/F, i, n)$	To Find A Given P $(A/P, i, n)$	To Find A Given F $(A/F, i, n)$	To Find P Given A $(P/A, i, n)$	To Find F Given A $(F/A, i, n)$	
n							n
1	1.010	.99010	1.01000	1.00000	.990	1.000	1
2	1.020	.98030	.50751	.49751	1.970	2.010	2
3	1.030	.97059	.34002	.33002	2.941	3.030	3
4	1.041	.96098	.25628	.24628	3.902	4.060	4
5	1.051	.95147	.20604	.19604	4.853	5.101	5
6	1.062	.94205	.17255	.16255	5.795	6.152	6
7	1.072	.93272	.14863	.13863	6.728	7.214	7
8	1.083	.92348	.13069	.12069	7.652	8.286	8
9	1.094	.91434	.11674	.10674	8.566	9.369	9
10	1.105	.90529	.10558	.09558	9.471	10.462	10
11	1.116	.89632	.09645	.08645	10.368	11.567	11
12	1.127	.88745	.08885	.07885	11.255	12.683	12
13	1.138	.87866	.08241	.07241	12.134	13.809	13
14	1.149	.86996	.07690	.06690	13.004	14.947	14
15	1.161	.86135	.07212	.06212	13.865	16.097	15
16	1.173	.85282	.06794	.05794	14.718	17.258	16
17	1.184	.84438	.06426	.05426	15.562	18.430	17
18	1.196	.83602	.06098	.05098	16.398	19.615	18
19	1.208	.82774	.05805	.04805	17.226	20.811	19
20	1.220	.81954	.05542	.04542	18.046	22.019	20

1%

n	To Find F Given P (F/P, i, n)	To Find P Given F (P/F, i, n)	To Find A Given P (A/P, i, n)	To Find A Given F (A/F, i, n)	To Find P Given A (P/A, i, n)	To Find F Given A (F/A, i, n)	n
21	1.232	.81143	.05303	.04303	18.857	23.239	21
22	1.245	.80340	.05086	.04086	19.660	24.472	22
23	1.257	.79544	.04889	.03889	20.456	25.716	23
24	1.270	.78757	.04707	.03707	21.243	26.973	24
25	1.282	.77977	.04541	.03541	22.023	28.243	25
30	1.348	.74192	.03875	.02875	25.808	34.785	30
35	1.417	.70591	.03400	.02400	29.409	41.660	35
40	1.489	.67165	.03046	.02046	32.835	48.886	40
45	1.565	.63905	.02771	.01771	36.095	56.481	45
50	1.645	.60804	.02551	.01551	39.196	64.463	50

2%

n	To Find F Given P $(F/P, i, n)$	To Find P Given F $(P/F, i, n)$	To Find A Given P $(A/P, i, n)$	To Find A Given F $(A/F, i, n)$	To Find P Given A $(P/A, i, n)$	To Find F Given A $(F/A, i, n)$	n
1	1.020	.98039	1.02000	1.00000	.980	1.000	1
2	1.040	.96117	.51505	.49505	1.942	2.020	2
3	1.061	.94232	.34675	.32675	2.884	3.060	3
4	1.082	.92385	.26262	.24262	3.808	4.122	4
5	1.104	.90573	.21216	.19216	4.713	5.204	5
6	1.126	.88797	.17853	.15853	5.601	6.308	6
7	1.149	.87056	.15451	.13451	6.472	7.434	7
8	1.172	.85349	.13651	.11651	7.325	8.583	8
9	1.195	.83676	.12252	.10252	8.162	9.755	9
10	1.219	.82035	.11133	.09133	8.983	10.950	10
11	1.243	.80426	.10218	.08218	9.787	12.169	11
12	1.268	.78849	.09456	.07456	10.575	13.412	12
13	1.294	.77303	.08812	.06812	11.348	14.680	13
14	1.319	.75788	.08260	.06260	12.106	15.974	14
15	1.346	.74301	.07783	.05783	12.849	17.293	15
16	1.373	.72845	.07365	.05365	13.578	18.639	16
17	1.400	.71416	.06997	.04997	14.292	20.012	17
18	1.428	.70016	.06670	.04670	14.992	21.412	18
19	1.457	.68643	.06378	.04378	15.678	22.841	19
20	1.486	.67297	.06116	.04116	16.351	24.297	20

2%

n	To Find F Given P $(F/P, i, n)$	To Find P Given F $(P/F, i, n)$	To Find A Given P $(A/P, i, n)$	To Find A Given F $(A/F, i, n)$	To Find P Given A $(P/A, i, n)$	To Find F Given A $(F/A, i, n)$	n
21	1.516	.65978	.05878	.03878	17.011	25.783	21
22	1.546	.64684	.05663	.03663	17.658	27.299	22
23	1.577	.63416	.05467	.03467	18.292	28.845	23
24	1.608	.62172	.05287	.03287	18.914	30.422	24
25	1.641	.60953	.05122	.03122	19.523	32.030	25
30	1.811	.55207	.04465	.02465	22.396	40.568	30
35	2.000	.50003	.04000	.02000	24.999	49.994	35
40	2.208	.45289	.03656	.01656	27.355	60.402	40
45	2.438	.41020	.03391	.01391	29.490	71.893	45
50	2.692	.37153	.03182	.01182	31.424	84.579	50

3%

n	To Find F Given P $(F/P, i, n)$	To Find P Given F $(P/F, i, n)$	To Find A Given P $(A/P, i, n)$	To Find A Given F $(A/F, i, n)$	To Find P Given A $(P/A, i, n)$	To Find F Given A $(F/A, i, n)$	n
1	1.030	.97087	1.03000	1.00000	.971	1.000	1
2	1.061	.94260	.52261	.49261	1.913	2.030	2
3	1.093	.91514	.35353	.32353	2.829	3.091	3
4	1.126	.88849	.26903	.23903	3.717	4.184	4
5	1.159	.86261	.21835	.18835	4.580	5.309	5
6	1.194	.83748	.18460	.15460	5.417	6.468	6
7	1.230	.81309	.16051	.13051	6.230	7.662	7
8	1.267	.78941	.14246	.11246	7.020	8.892	8
9	1.305	.76642	.12843	.09843	7.786	10.159	9
10	1.344	.74409	.11723	.08723	8.530	11.464	10
11	1.384	.72242	.10808	.07808	9.253	12.808	11
12	1.426	.70138	.10046	.07046	9.954	14.192	12
13	1.469	.68095	.09403	.06403	10.635	15.618	13
14	1.513	.66112	.08853	.05853	11.296	17.086	14
15	1.558	.64186	.08377	.05377	11.938	18.599	15
16	1.605	.62317	.07961	.04961	12.561	20.157	16
17	1.653	.60502	.07595	.04595	13.166	21.762	17
18	1.702	.58739	.07271	.04271	13.754	23.414	18
19	1.754	.57029	.06981	.03981	14.324	25.117	19
20	1.806	.55368	.06722	.03722	14.877	26.870	20

3%

n	To Find F Given P $(F/P,i,n)$	To Find P Given F $(P/F,i,n)$	To Find A Given P $(A/P,i,n)$	To Find A Given F $(A/F,i,n)$	To Find P Given A $(P/A,i,n)$	To Find F Given A $(F/A,i,n)$	n
21	1.860	.53755	.06487	.03487	15.415	28.676	21
22	1.916	.52189	.06275	.03275	15.937	30.537	22
23	1.974	.50669	.06081	.03081	16.444	32.453	23
24	2.033	.49193	.05905	.02905	16.936	34.426	24
25	2.094	.47761	.05743	.02743	17.413	36.459	25
30	2.427	.41199	.05102	.02102	19.600	47.575	30
35	2.814	.35538	.04654	.01654	21.487	60.462	35
40	3.262	.30656	.04326	.01326	23.115	75.401	40
45	3.782	.26444	.04079	.01079	24.519	92.720	45
50	4.384	.22811	.03887	.00887	25.730	112.797	50

4%

n	To Find F Given P (F/P, i, n)	To Find P Given F (P/F, i, n)	To Find A Given P (A/P, i, n)	To Find A Given F (A/F, i, n)	To Find P Given A (P/A, i, n)	To Find F Given A (F/A, i, n)	n
1	1.040	.96154	1.04000	1.00000	.962	1.000	1
2	1.082	.92456	.53020	.49020	1.886	2.040	2
3	1.125	.88900	.36035	.32035	2.775	3.122	3
4	1.170	.85480	.27549	.23549	3.630	4.246	4
5	1.217	.82193	.22463	.18463	4.452	5.416	5
6	1.265	.79031	.19076	.15076	5.242	6.633	6
7	1.316	.75992	.16661	.12661	6.002	7.898	7
8	1.369	.73069	.14853	.10853	6.733	9.214	8
9	1.423	.70259	.13449	.09449	7.435	10.583	9
10	1.480	.67556	.12329	.08329	8.111	12.006	10
11	1.539	.64958	.11415	.07415	8.760	13.486	11
12	1.601	.62460	.10655	.06655	9.385	15.026	12
13	1.665	.60057	.10014	.06014	9.986	16.627	13
14	1.732	.57748	.09467	.05467	10.563	18.292	14
15	1.801	.55526	.08994	.04994	11.118	20.024	15
16	1.873	.53391	.08582	.04582	11.652	21.825	16
17	1.948	.51337	.08220	.04220	12.166	23.698	17
18	2.026	.49363	.07899	.03899	12.659	25.645	18
19	2.107	.47464	.07614	.03614	13.134	27.671	19
20	2.191	.45639	.07358	.03358	13.590	29.778	20

4%

n	To Find F Given P $(F/P, i, n)$	To Find P Given F $(P/F, i, n)$	To Find A Given P $(A/P, i, n)$	To Find A Given F $(A/F, i, n)$	To Find P Given A $(P/A, i, n)$	To Find F Given A $(F/A, i, n)$	n
21	2.279	.43883	.07128	.03128	14.029	31.969	21
22	2.370	.42196	.06920	.02920	14.451	34.248	22
23	2.465	.40573	.06731	.02731	14.857	36.618	23
24	2.563	.39012	.06559	.02559	15.247	39.083	24
25	2.666	.37512	.06401	.02401	15.622	41.646	25
30	3.243	.30832	.05783	.01783	17.292	56.085	30
35	3.946	.25342	.05358	.01358	18.665	73.652	35
40	4.801	.20829	.05052	.01052	19.793	95.026	40
45	5.841	.17120	.04826	.00826	20.720	121.029	45
50	7.107	.14071	.04655	.00655	21.482	152.667	50

5%

n	To Find F Given P $(F/P, i, n)$	To Find P Given F $(P/F, i, n)$	To Find A Given P $(A/P, i, n)$	To Find A Given F $(A/F, i, n)$	To Find P Given A $(P/A, i, n)$	To Find F Given A $(F/A, i, n)$	n
1	1.050	.95238	1.05000	1.00000	.952	1.000	1
2	1.103	.90703	.53780	.48780	1.859	2.050	2
3	1.158	.86384	.36721	.31721	2.723	3.153	3
4	1.216	.82270	.28201	.23201	3.546	4.310	4
5	1.276	.78353	.23097	.18097	4.329	5.526	5
6	1.340	.74622	.19702	.14702	5.076	6.802	6
7	1.407	.71068	.17282	.12282	5.786	8.142	7
8	1.477	.67684	.15472	.10472	6.463	9.549	8
9	1.551	.64461	.14069	.09069	7.108	11.027	9
10	1.629	.61391	.12950	.07950	7.722	12.578	10
11	1.710	.58468	.12039	.07039	8.306	14.207	11
12	1.796	.55684	.11283	.06283	8.863	15.917	12
13	1.886	.53032	.10646	.05646	9.394	17.713	13
14	1.980	.50507	.10102	.05102	9.899	19.599	14
15	2.079	.48102	.09634	.04634	10.380	21.579	15
16	2.183	.45811	.09227	.04227	10.838	23.657	16
17	2.292	.43630	.08870	.03870	11.274	25.840	17
18	2.407	.41552	.08555	.03555	11.690	28.132	18
19	2.527	.39573	.08275	.03275	12.085	30.539	19
20	2.653	.37689	.08024	.03024	12.462	33.066	20

5%

n	To Find F Given P $(F/P, i, n)$	To Find P Given F $(P/F, i, n)$	To Find A Given P $(A/P, i, n)$	To Find A Given F $(A/F, i, n)$	To Find P Given A $(P/A, i, n)$	To Find F Given A $(F/A, i, n)$	n
21	2.786	.35894	.07800	.02800	12.821	35.719	21
22	2.925	.34185	.07597	.02597	13.163	38.505	22
23	3.072	.32557	.07414	.02414	13.489	41.430	23
24	3.225	.31007	.07247	.02247	13.799	44.502	24
25	3.386	.29530	.07095	.02095	14.094	47.727	25
30	4.322	.23138	.06505	.01505	15.372	66.439	30
35	5.516	.18129	.06107	.01107	16.374	90.320	35
40	7.040	.14205	.05828	.00828	17.159	120.800	40
45	8.985	.11130	.05626	.00626	17.774	159.700	45
50	11.467	.08720	.05478	.00478	18.256	209.348	50

6%

n	To Find F Given P $(F/P, i, n)$	To Find P Given F $(P/F, i, n)$	To Find A Given P $(A/P, i, n)$	To Find A Given F $(A/F, i, n)$	To Find P Given A $(P/A, i, n)$	To Find F Given A $(F/A, i, n)$	n
1	1.060	.94340	1.06000	1.00000	.943	1.000	1
2	1.124	.89000	.54544	.48544	1.833	2.060	2
3	1.191	.83962	.37411	.31411	2.673	3.184	3
4	1.262	.79209	.28859	.22859	3.465	4.375	4
5	1.338	.74726	.23740	.17740	4.212	5.637	5
6	1.419	.70496	.20336	.14336	4.917	6.975	6
7	1.504	.66506	.17914	.11914	5.582	8.394	7
8	1.594	.62741	.16104	.10104	6.210	9.897	8
9	1.689	.59190	.14702	.08702	6.802	11.491	9
10	1.791	.55839	.13587	.07587	7.360	13.181	10
11	1.898	.52679	.12679	.06679	7.887	14.972	11
12	2.012	.49697	.11928	.05928	8.384	16.870	12
13	2.133	.46884	.11296	.05296	8.853	18.882	13
14	2.261	.44230	.10758	.04758	9.295	21.015	14
15	2.397	.41727	.10296	.04296	9.712	23.276	15
16	2.540	.39365	.09895	.03895	10.106	25.673	16
17	2.693	.37136	.09544	.03544	10.477	28.213	17
18	2.854	.35034	.09236	.03236	10.828	30.906	18
19	3.026	.33051	.08962	.02962	11.158	33.760	19
20	3.207	.31180	.08718	.02718	11.470	36.786	20

6%

n	To Find F Given P (F/P, i, n)	To Find P Given F (P/F, i, n)	To Find A Given P (A/P, i, n)	To Find A Given F (A/F, i, n)	To Find P Given A (P/A, i, n)	To Find F Given A (F/A, i, n)	n
21	3.400	.29416	.08500	.02500	11.764	39.993	21
22	3.604	.27751	.08305	.02305	12.042	43.392	22
23	3.820	.26180	.08128	.02128	12.303	46.996	23
24	4.049	.24698	.07968	.01968	12.550	50.816	24
25	4.292	.23300	.07823	.01823	12.783	54.865	25
30	5.743	.17411	.07265	.01265	13.765	79.058	30
35	7.686	.13011	.06897	.00897	14.498	111.435	35
40	10.286	.09722	.06646	.00646	15.046	154.762	40
45	13.765	.07265	.06470	.00470	15.456	212.744	45
50	18.420	.05429	.06344	.00344	15.762	290.336	50

7%

n	To Find F Given P $(F/P, i, n)$	To Find P Given F $(P/F, i, n)$	To Find A Given P $(A/P, i, n)$	To Find A Given F $(A/F, i, n)$	To Find P Given A $(P/A, i, n)$	To Find F Given A $(F/A, i, n)$	n
1	1.070	.93458	1.07000	1.00000	.935	1.000	1
2	1.145	.87344	.55309	.48309	1.808	2.070	2
3	1.225	.81630	.38105	.31105	2.624	3.215	3
4	1.311	.76290	.29523	.22523	3.387	4.440	4
5	1.403	.71299	.24389	.17389	4.100	5.751	5
6	1.501	.66634	.20980	.13980	4.767	7.153	6
7	1.606	.62275	.18555	.11555	5.389	8.654	7
8	1.718	.58201	.16747	.09747	5.971	10.260	8
9	1.838	.54393	.15349	.08349	6.515	11.978	9
10	1.967	.50835	.14238	.07238	7.024	13.816	10
11	2.105	.47509	.13336	.06336	7.499	15.784	11
12	2.252	.44401	.12590	.05590	7.943	17.888	12
13	2.410	.41496	.11965	.04965	8.358	20.141	13
14	2.579	.38782	.11434	.04434	8.745	22.550	14
15	2.759	.36245	.10979	.03979	9.108	25.129	15
16	2.952	.33873	.10586	.03586	9.447	27.888	16
17	3.159	.31657	.10243	.03243	9.763	30.840	17
18	3.380	.29586	.09941	.02941	10.059	33.999	18
19	3.617	.27651	.09675	.02675	10.336	37.379	19
20	3.870	.25842	.09439	.02439	10.594	40.995	20

7%

n	To Find F Given P $(F/P, i, n)$	To Find P Given F $(P/F, i, n)$	To Find A Given P $(A/P, i, n)$	To Find A Given F $(A/F, i, n)$	To Find P Given A $(P/A, i, n)$	To Find F Given A $(F/A, i, n)$	n
21	4.141	.24151	.09229	.02229	10.836	44.865	21
22	4.430	.22571	.09041	.02041	11.061	49.006	22
23	4.741	.21095	.08871	.01871	11.272	53.436	23
24	5.072	.19715	.08719	.01719	11.469	58.177	24
25	5.427	.18425	.08581	.01581	11.654	63.249	25
30	7.612	.13137	.08059	.01059	12.409	94.461	30
35	10.677	.09366	.07723	.00723	12.948	138.237	35
40	14.974	.06678	.07501	.00501	13.332	199.635	40
45	21.002	.04761	.07350	.00350	13.606	285.749	45
50	29.457	.03395	.07246	.00246	13.801	406.529	50

8%

n	To Find F Given P (F/P, i, n)	To Find P Given F (P/F, i, n)	To Find A Given P (A/P, i, n)	To Find A Given F (A/F, i, n)	To Find P Given A (P/A, i, n)	To Find F Given A (F/A, i, n)	n
1	1.080	.92593	1.08000	1.00000	.926	1.000	1
2	1.166	.85734	.56077	.48077	1.783	2.080	2
3	1.260	.79383	.38803	.30803	2.577	3.246	3
4	1.360	.73503	.30192	.22192	3.312	4.506	4
5	1.469	.68058	.25046	.17046	3.993	5.867	5
6	1.587	.63017	.21632	.13632	4.623	7.336	6
7	1.714	.58349	.19207	.11207	5.206	8.923	7
8	1.851	.54027	.17401	.09401	5.747	10.637	8
9	1.999	.50025	.16008	.08008	6.247	12.488	9
10	2.159	.46319	.14903	.06903	6.710	14.487	10
11	2.332	.42888	.14008	.06008	7.139	16.645	11
12	2.518	.39711	.13270	.05270	7.536	18.977	12
13	2.720	.36770	.12652	.04652	7.904	21.495	13
14	2.937	.34046	.12130	.04130	8.244	24.215	14
15	3.172	.31524	.11683	.03683	8.559	27.152	15
16	3.426	.29189	.11298	.03298	8.851	30.324	16
17	3.700	.27027	.10963	.02963	9.122	33.750	17
18	3.996	.25025	.10670	.02670	9.372	37.450	18
19	4.316	.23171	.10413	.02413	9.604	41.446	19
20	4.661	.21455	.10185	.02185	9.818	45.762	20

8%

n	To Find F Given P (F/P, i, n)	To Find P Given F (P/F, i, n)	To Find A Given P (A/P, i, n)	To Find A Given F (A/F, i, n)	To Find P Given A (P/A, i, n)	To Find F Given A (F/A, i, n)	n
21	5.034	.19866	.09983	.01983	10.017	50.423	21
22	5.437	.18394	.09803	.01803	10.201	55.457	22
23	5.871	.17032	.09642	.01642	10.371	60.893	23
24	6.341	.15770	.09498	.01498	10.529	66.765	24
25	6.848	.14602	.09368	.01368	10.675	73.106	25
30	10.063	.09938	.08883	.00883	11.258	113.283	30
35	14.785	.06763	.08580	.00580	11.655	172.317	35
40	21.725	.04603	.08386	.00386	11.925	259.057	40
45	31.920	.03133	.08259	.00259	12.108	386.506	45
50	46.902	.02132	.08174	.00174	12.233	573.770	50

9%

n	To Find F Given P $(F/P, i, n)$	To Find P Given F $(P/F, i, n)$	To Find A Given P $(A/P, i, n)$	To Find A Given F $(A/F, i, n)$	To Find P Given A $(P/A, i, n)$	To Find F Given A $(F/A, i, n)$	n
1	1.090	.91743	1.09000	1.00000	.917	1.000	1
2	1.188	.84168	.56847	.47847	1.759	2.090	2
3	1.295	.77218	.39505	.30505	2.531	3.278	3
4	1.412	.70843	.30867	.21867	3.240	4.573	4
5	1.539	.64993	.25709	.16709	3.890	5.985	5
6	1.677	.59627	.22292	.13292	4.486	7.523	6
7	1.828	.54703	.19869	.10869	5.033	9.200	7
8	1.993	.50187	.18067	.09067	5.535	11.028	8
9	2.172	.46043	.16680	.07680	5.995	13.021	9
10	2.367	.42241	.15582	.06582	6.418	15.193	10
11	2.580	.38753	.14695	.05695	6.805	17.560	11
12	2.813	.35553	.13965	.04965	7.161	20.141	12
13	3.066	.32618	.13357	.04357	7.487	22.953	13
14	3.342	.29925	.12843	.03843	7.786	26.019	14
15	3.642	.27454	.12406	.03406	8.061	29.361	15
16	3.970	.25187	.12030	.03030	8.313	33.003	16
17	4.328	.23107	.11705	.02705	8.544	36.974	17
18	4.717	.21199	.11421	.02421	8.756	41.301	18
19	5.142	.19449	.11173	.02173	8.950	46.018	19
20	5.604	.17843	.10955	.01955	9.129	51.160	20

9%

n	To Find F Given P $(F/P, i, n)$	To Find P Given F $(P/F, i, n)$	To Find A Given P $(A/P, i, n)$	To Find A Given F $(A/F, i, n)$	To Find P Given A $(P/A, i, n)$	To Find F Given A $(F/A, i, n)$	n
21	6.109	.16370	.10762	.01762	9.292	56.765	21
22	6.659	.15018	.10590	.01590	9.442	62.873	22
23	7.258	.13778	.10438	.01438	9.580	69.532	23
24	7.911	.12640	.10302	.01302	9.707	76.790	24
25	8.623	.11597	.10181	.01181	9.823	84.701	25
30	13.268	.07537	.09734	.00734	10.274	136.308	30
35	20.414	.04899	.09464	.00464	10.567	215.711	35
40	31.409	.03184	.09296	.00296	10.757	337.882	40
45	48.327	.02069	.09190	.00190	10.881	525.859	45
50	74.358	.01345	.09123	.00123	10.962	815.084	50

10%

n	To Find F Given P (F/P, i, n)	To Find P Given F (P/F, i, n)	To Find A Given P (A/P, i, n)	To Find A Given F (A/F, i, n)	To Find P Given A (P/A, i, n)	To Find F Given A (F/A, i, n)	n
1	1.100	.90909	1.10000	1.00000	.909	1.000	1
2	1.210	.82645	.57619	.47619	1.736	2.100	2
3	1.331	.75131	.40211	.30211	2.487	3.310	3
4	1.464	.68301	.31547	.21547	3.170	4.641	4
5	1.611	.62092	.26380	.16380	3.791	6.105	5
6	1.772	.56447	.22961	.12961	4.355	7.716	6
7	1.949	.51316	.20541	.10541	4.868	9.487	7
8	2.144	.46651	.18744	.08744	5.335	11.436	8
9	2.358	.42410	.17364	.07364	5.759	13.579	9
10	2.594	.38554	.16275	.06275	6.145	15.937	10
11	2.853	.35049	.15396	.05396	6.495	18.531	11
12	3.138	.31863	.14676	.04676	6.814	21.384	12
13	3.452	.28966	.14078	.04078	7.103	24.523	13
14	3.797	.26333	.13575	.03575	7.367	27.975	14
15	4.177	.23939	.13147	.03147	7.606	31.772	15
16	4.595	.21763	.12782	.02782	7.824	35.950	16
17	5.054	.19784	.12466	.02466	8.022	40.545	17
18	5.560	.17986	.12193	.02193	8.201	45.599	18
19	6.116	.16351	.11955	.01955	8.365	51.159	19
20	6.727	.14864	.11746	.01746	8.514	57.275	20

10%

n	To Find F Given P $(F/P, i, n)$	To Find P Given F $(P/F, i, n)$	To Find A Given P $(A/P, i, n)$	To Find A Given F $(A/F, i, n)$	To Find P Given A $(P/A, i, n)$	To Find F Given A $(F/A, i, n)$	n
21	7.400	.13513	.11562	.01562	8.649	64.002	21
22	8.140	.12285	.11401	.01401	8.772	71.403	22
23	8.954	.11168	.11257	.01257	8.883	79.543	23
24	9.850	.10153	.11130	.01130	8.985	88.497	24
25	10.835	.09230	.11017	.01017	9.077	98.347	25
30	17.449	.05731	.10608	.00608	9.427	164.494	30
35	28.102	.03558	.10369	.00369	9.644	271.024	35
40	45.259	.02209	.10226	.00226	9.779	442.593	40
45	72.890	.01372	.10139	.00139	9.863	718.905	45
50	117.391	.00852	.10086	.00086	9.915	1163.909	50

APPENDIX II   values of e^{-x}

z \ y	0	1	2	3	4	5	6	7	8	9
0	1.00000	.36788	.13534	.04979	.01832	.00674	.00248	.00091	.00034	.00012
.05	.95123	.34994	.12873	.04736	.01742	.00641	.00236	.00087	.00032	.00012
.10	.90484	.33287	.12246	.04505	.01657	.00610	.00224	.00083	.00030	.00011
.15	.86071	.31664	.11648	.04285	.01576	.00580	.00213	.00078	.00029	.00011
.20	.81873	.30119	.11080	.04076	.01500	.00552	.00203	.00075	.00027	.00010
.25	.77880	.28650	.10540	.03877	.01426	.00525	.00193	.00071	.00026	.00010
.30	.74082	.27253	.10026	.03688	.01357	.00499	.00184	.00068	.00025	.00009
.35	.70469	.25924	.09537	.03508	.01291	.00475	.00175	.00064	.00024	.00009
.40	.67032	.24660	.09072	.03337	.01228	.00452	.00166	.00061	.00022	.00008
.45	.63763	.23457	.08629	.03175	.01168	.00430	.00158	.00058	.00021	.00008
.50	.60653	.22313	.08208	.03020	.01111	.00409	.00150	.00055	.00020	.00007
.55	.57695	.21225	.07808	.02872	.01057	.00389	.00143	.00053	.00019	.00007
.60	.54881	.20190	.07427	.02732	.01005	.00370	.00136	.00050	.00018	.00007
.65	.52205	.19205	.07065	.02599	.00956	.00352	.00129	.00048	.00018	.00006
.70	.49659	.18268	.06721	.02472	.00910	.00335	.00123	.00045	.00017	.00006
.75	.47237	.17377	.06393	.02352	.00865	.00318	.00117	.00043	.00016	.00006
.80	.44933	.16530	.06081	.02237	.00823	.00303	.00111	.00041	.00015	.00006
.85	.42741	.15724	.05784	.02128	.00783	.00288	.00106	.00039	.00014	.00005
.90	.40657	.14957	.05502	.02024	.00745	.00274	.00101	.00037	.00014	.00005
.95	.38674	.14227	.05234	.01925	.00708	.00261	.00096	.00035	.00013	.00005

$x = y + z$

APPENDIX III   table of random digits

212	626	229	984	031	950	006	402	528	212
281	257	711	411	123	550	582	548	463	281
583	622	990	448	387	634	705	054	321	583
495	097	942	470	697	577	250	294	975	495
392	056	440	851	930	755	092	644	301	392
648	874	360	966	959	542	106	478	174	648
638	927	576	191	660	314	166	171	469	638
738	589	965	900	908	445	149	099	061	738
322	236	401	468	578	310	928	636	824	322
765	241	758	271	545	284	379	157	635	765
443	982	912	683	684	743	377	038	367	443
730	831	739	079	869	062	797	654	897	730
001	165	112	835	977	614	514	378	098	001
632	359	906	325	881	777	403	587	846	632
997	786	998	925	457	923	338	656	016	997
472	823	262	009	580	429	196	958	483	472
431	845	572	952	125	669	850	870	121	431
249	226	805	130	967	019	176	767	807	249
302	341	834	917	981	853	049	023	414	302
964	566	535	689	041	546	344	013	819	964
611	275	783	820	024	474	936	113	895	611
616	843	453	685	803	011	699	697	518	616
357	646	420	659	254	532	510	140	563	357
206	058	559	118	252	413	242	818	905	206
540	454	744	437	672	029	772	105	418	540
734	210	852	989	389	753	973	376	979	734
161	700	756	152	278	962	721	007	461	161
198	300	332	298	213	031	891	372	740	198
220	384	455	804	071	333	358	847	692	220
601	327	000	044	725	245	996	806	190	601
716	505	842	394	051	142	682	624	110	716
941	292	856	228	924	398	289	677	326	941
650	064	916	921	219	388	694	339	715	650
218	195	899	849	811	488	770	986	151	218
021	060	678	386	574	072	393	991	508	021
433	034	129	907	385	515	438	732	662	433
829	493	127	788	117	193	780	581	489	829
585	812	547	404	647	480	293	915	862	585
075	364	264	128	848	751	854	109	656	075
675	527	153	337	596	382	336	536	748	675
759	673	088	406	766	747	615	573	012	759
702	179	946	708	233	222	567	595	322	702
880	419	600	620	871	181	065	976	555	880
667	769	926	517	557	999	985	091	584	667
439	603	799	773	164	052	201	316	285	439
570	296	094	763	569	714	590	025	533	570
435	224	686	863	645	361	026	593	203	435
409	761	449	447	268	366	383	396	170	409
868	282	260	890	313	107	537	808	309	868
187	163	992	568	655	956	364	204	494	187
739	779	522	855	168	290	737	960	602	739
902	503	723	126	729	484	531	450	506	902
048	712	471	757	211	911	623	050	082	048

SUBJECT INDEX